4th International Conference on Ergonomics & 2nd International Conference on Industrial Engineering (ICE & ICIE 2019)

IOP Conference Series: Materials Science and Engineering
Volume 697

Kuala Terengganu, Malaysia
27 - 28 August 2019

ISBN: 978-1-7138-0989-0
ISSN: 1757-8981

Printed from e-media with permission by:

Curran Associates, Inc.
57 Morehouse Lane
Red Hook, NY 12571

Some format issues inherent in the e-media version may also appear in this print version.

Institute of Physics
Dirac House, Temple Back
Bristol BS1 6BE UK

Phone: 44 1 17 929 7481
Fax: 44 1 17 920 0979

techtracking@iop.org

Additional copies of this publication are available from:

Curran Associates, Inc.
57 Morehouse Lane
Red Hook, NY 12571 USA
Phone: 845-758-0400
Fax: 845-758-2633
Email: curran@proceedings.com
Web: www.proceedings.com

TABLE OF CONTENTS

INDUSTRIAL ENGINEERING

ERGONOMICS

AEROSPACE

OTHERS

IOP Conf. Series: Materials Science and Engineering **697** (2019) 011001 doi:10.1088/1757-899X/697/1/011001

4th International Conference on Ergonomics & 2nd International Conference on Industrial Engineering (ICE & ICIE 2019).

Preface

This conference was held on *26th - 28th August 2019* in Terengganu, Malaysia is a continuation of several series previously organised. The conference aims to provide a platform for the presentation of new advances and research work in the fields of ergonomics and industrial engineering. The theme represents universal consciousness in the design to meet the purpose of human needs as well as improving the human well-being within the engineering perspective. The conference has successfully brought together researchers, engineers and scientists in the area of interest from several countries including Malaysia, Indonesia, Taiwan and Saudi Arabia. Researchers among academia and Industries had presented their work and innovation. Full papers relating to broad areas of Ergonomics, Industrial Engineering and related discipline has been presented and published in several reputed journals and proceedings.

IOP Conf. Series: Materials Science and Engineering **697** (2019) 011001 doi:10.1088/1757-899X/697/1/011001

Organized by :

UNIVERSITY OF MALAYA

Co-organizers :

Sponsors :

IOP Conf. Series: Materials Science and Engineering **697** (2019) 011001 doi:10.1088/1757-899X/697/1/011001

Committees

Conference Chair

Assoc. Prof. Ir. Dr. Siti Zawiah Md Dawal

Universiti Malaya

Conference Co-Chair

Assoc. Prof. Dr. Wan Nor Liza Wan Mahadi

Universiti Malaya

Treasurer

Dr. Marizan Mubin

Universiti Malaya

Secretary

Assoc. Prof. Dr. Dian Darina Indah Darius

Unversiti Pertahanan Nasional Malaysia

Program Chair

Dr. Eida Nadirah Roslin

Universiti Kuala Lumpur

Technical Chair

Dr. Mirta Widia

Universiti Malaysia Pahang

Scientific Committee

Prof. Ir. Dr. Nukman Yusoff

Universiti Malaya

IOP Conf. Series: Materials Science and Engineering **697** (2019) 011001 doi:10.1088/1757-899X/697/1/011001

Dr. Ahmad Salahuddin Mohd Harituddin

Universiti Putra Malaysia

Dr. Karmegan Karuppiah

Universiti Putra Malaysia

Dr. Ezrin Hani Sukadarin

Universiti Malaysia Pahang

Secretariat

Dr. Sabariah Julai @ Julaihi

Universiti Malaya

Sponsorship

Dr. Ruhaizin Sulaiman

Universiti Putra Malaysia

Logistic and Industrial Visit

Dr. Nurul Shahida Ahmad Salahim

Universiti Malaysia Pahang

Promotion

Dr. Nurhayati Mohd Nur

Universiti Kuala Lumpur

Email: ice2019.my@gmail.com

Website: http://umconference.um.edu.my/ICEICIE2019

IOP Conf. Series: Materials Science and Engineering **697** (2019) 011001 doi:10.1088/1757-899X/697/1/011001

IOP Conf. Series: Materials Science and Engineering **697** (2019) 011002 doi:10.1088/1757-899X/697/1/011002

Peer review statement

All papers published in this volume of *IOP Conference Series: Materials Science and Engineering* have been peer reviewed through processes administered by the proceedings Editors. Reviews were conducted by expert referees to the professional and scientific standards expected of a proceedings journal published by IOP Publishing.

IOP Conf. Series: Materials Science and Engineering **697** (2019) 012001 doi:10.1088/1757-899X/697/1/012001

Eco-Process Innovation Performance: Production Waste Investigation through Discrete Event Simulation

S M Dahan[1] and S M Yusof[2]

[1] Faculty of Industrial Management, Universiti Malaysia Pahang, Malaysia.
[2] Razak Faculty of Technology and Informatics, Universiti Teknologi Malaysia, Malaysia.
Email: suziedahan@gmail.com

Abstract. Eco-process innovation has been recognised as an effective green strategy for altering manufacturing processes so as to improve a firm's environmental performance. As many recent eco-innovation studies focusing on perceptual approach of investigating the performance indicators, new study opportunity knocked the door for new knowledge disclosure of operational approach, since it has been largely ignored. This paper aims to assess production waste improvement brought by eco-process innovation implementation in a manufacturing facility. For this purpose, the discrete event simulation approach has been applied to model and simulate the waste rate of previous state and current state of an eco-innovatively improved production line, using the real operational data. Comparison between both states of production system demonstrated a reduction of 0.14% and 0.02% in waste rate at visual mechanical inspection and final test inspection points consecutively, hence confirmed that eco-process innovation practice enabled the improvement of the environmental performance of manufacturing firm. The study is a small part of a larger research work of which the authors are developing indicators for measuring eco-process innovation performance at the firm level.

1. Introduction

While manufacturing operations are essential for elevating the humankind quality of life, they are proven to also contribute to the harmful impacts on the environment. Thus, in the recent decade, manufacturers are proactively seeking a secret recipe for improving their operational performances without jeopardising the ecosystem's sustainability. Studies have revealed that there is a positive association of eco-process innovation (also recognised as environmentally friendly changes taking place in production processes) with firms' competitive advantage [1–5]. Cheng et al. [6] referred eco-process innovation as the introduction of new production processes or modification of existing processes so as to minimise the negative impacts of environment. It is a type of technical improvement which utilising environmental friendly technologies to avoid or reduce environmental degradation [7] and promoted internally at the level of the individual firm [8]. This paper covers eco-process innovation that is defined by [9] as eco-innovative improvement done on the production processes which resulted in the enhanced firm's performance of economic, environmental and social.

In view of eco-process innovation performance, [10–13] revealed that its implementation could lead to better wealth generation, and environmentally and socially responsible firm. Adoption of eco-process innovation practice could not only results in better environmental quality but also enhanced the economic achievement and social through the responsible use of resources and the reduced hazardous waste, pollution and emission during the production process. According to [6], the performance measure of eco-process innovation covers the assessment of production process which minimises the generation of harmful waste and substances, recycles and reuses waste and emission and improves resources usage such as raw material, water, electricity, coil or oil.

In the context of the assessment approach of eco-process innovation performance, a comprehensive system to track and monitor its effectiveness needs to be established. Such an assessment system should consist of indicators which are measurable and quantifiable [14,15], relevant [14,16], reliable and accessible [15,16]. However, various versions of indicators introduced and employed to measure eco-process innovation, in which they are still broad in nature and vary with contexts where they have been applied [17–20]. This study addresses the production waste, an indicator to represent the environmental performance of the production process. This study aims to assess the improvement in the waste rate following the eco-innovative changes done on the production processes. For the purpose, the actual production data of a manufacturing plant was used to quantify the improvement. Changes in the waste produced by a production line of interest have been modelled and simulated using a popular simulation-based analysis, namely discrete event simulation (DES) method.

2. Literature Review

Firms' commitment to undertaking eco-innovation initiatives is causing additional costs to them. However, such investments are worthwhile as studies demonstrated that eco-innovation practices brought positive effects to the firms [3,21]. Eco-process innovation has been proven to cause the firm to achieve a better performance in the aspect of economic, environmental and social [10–13,22–25]. Prior study by [24] on manufacturing firms by taking the triple bottom line (TBL) into consideration, discovered a significant connection between eco-innovative changes done on processes and firms' sustainability performance, and they recommended firms to deliberate in such initiatives to obtain better performance of economic, environmental and social. Similarly, a recent review of eco-innovation concept by [12], who equalised eco-innovation with sustainable innovation, viewed eco-innovation as promoting sustainable development concept by enhancing consumers' quality of life, thus improving not only the environmental and economic aspect but also the norms and values of society. [13] added that eco-innovation adoption contributed to protecting the social well-being by using the natural resources responsibly and refraining the harmful impacts on the environment.

In the context of environmental impacts, ecological innovation introduced in the manufacturing process has been proven to minimise the harmful impacts on the environment. It was noted that productive resource usage, less pollution and waste and more efficient production processes could be achieved through the conduct of eco-process innovation [11]. Most manufacturers are investing in eco-process innovation technologies to gain waste reduction and energy efficiency [26]. The technology and process upgrades may cause more efficient manufacturing processes and logistics, and greater resource productivity, thus leading to less waste and much cost savings [27,28]. Reduction of production costs may also be possible through the practice of reusing, recycling and remanufacturing. These practices done on resources such as materials causes a reduction in the consumption of new materials, provide a greater capacity of acquiring other resources and less waste production by the process.

3. Methodology

A single case study method has been applied to assess the production waste using the actual operational and production data. A Malaysian electronic components manufacturer has been chosen as the case study company. The company produces a wide variety of customised and standard common mode chokes, transformers and power inductors for medical, automotive, alternative energy and industrial sectors applications.

The waste assessment was carried out on a common mode choke production line. The line is the longest and long run production line due to the continuous and large order quantity of the product model. Previously, most processes were done manually by the operators. Later, as the company adopting the continuous process improvement to enhance their performance, most critical processes were gradually upgraded and transformed. The production operation starts with the arrival of insulated wires at the

IOP Conf. Series: Materials Science and Engineering **697** (2019) 012001 doi:10.1088/1757-899X/697/1/012001

stripping and cutting process and come to an end when the finished common mode chokes are placed into the protecting packages and labelled for transfer to the warehouse department. All defective items found at the visual mechanical inspection (VMI) and final test inspection is rejected, recorded as scrap, discarded and transferred to the scrap yard. The rejected parts due to the mechanical or electrical defects are not reworked or repaired, hence considered as production waste. This study refers to production waste as the defective items found and rejected at VMI and final test inspection points, thus the rate is computed based on the rejected quantity per the inspected quantity.

Simulation models were developed using Arena Version 15 simulation software. According to Thiede et al. [29], Arena software is a recommended simulation software due to its capability to evaluate the diverse production performances and simultaneously coping with the assessment of environmental aspects such as waste and energy usage.

3.1. Data collection

Various types of data, gathered from numerous sources in the case study company were used so as to ensure the real world production line was accurately visualised by the Arena modules:

i. Structural data was used to design the structure of the model such as the production line layout and process control plan.
ii. Process parameter which include production planning and control and processing time.
iii. Process parameter relationships such as process improvement projects details to visualise the interdependencies of the parameters.

The processes cycle time were directly measured through time study conducted on all processes. The computation of minimum sample size of observations was performed based on the following formula which suggested by [30], to make sure that the measured cycle time was representative.

$$n = \left(\frac{40\sqrt{n'\sum x^2 - \left(\sum x\right)^2}}{\sum x} \right)^2 \qquad (1)$$

where, n is the sample size, n' is the number of readings taken in the preliminary measure, $\sum$ is the sum of values and x is the value of the readings. Preliminary measure of cycle time was performed on each processes, whereby the readings then were used to compute the appropriate sample size for the actual measure of cycle time.

3.2. Analysis of input data

Before the observed data could be applied as input to the simulation models, they were evaluated and categorised as either deterministic (non-random data) or stochastic (random data). Deterministic data for instance the resource capacity was used directly. Stochastic data like the time between arrivals, on the other hand, was fitted to a probability distribution by performing the goodness-of-fit test in Input Analyzer (an Arena software distinct application). Three numerical measures were used, such as suggested by [31], to analyse the results of goodness-of-fit test and decide if the data fits the proposed probability distribution:

i. Square error - good fit is described by smaller value.
ii. Chi-Square and Kolmogorov-Smirnov (K-S) – good fit is corresponded by greater than 0.05 p-value.

3.3. Development of Simulation Models

Two models were constructed to visualise the production line of interest, referred as previous state to represent the production line prior to the implementation of the eco-innovative improvement, and current state to illustrate the post-eco-process innovations production line.

3.3.1 Previous State Modelling.

Figure 1 shows the previous state model which consists of 20 processes.

3.3.2 Current State Modelling.

The structural and quantitative change was used to represent the eco-improvements in the current state model (illustrated in Figure 2). Structural change refers to the model structure change, such as the change of process order. Whereas quantitative change involves the input value change for example the change of process cycle time [32]. The eco-innovative changes conducted on the production line and the corresponding changes made on the model are summarised in Table 1.

Table 1. Eco-process innovation and the respective modelling change

Eco-Process Innovation	Type of Model Change	Model Change
Efficient Equipment Manual to auto winding jig	Quantitative	Shorter process cycle time
Process Automation: Manual to auto tinning machine	Quantitative	Shorter process cycle time
Process Integration: Curing 1 & Curing 2	Structural	Reduced process from 2 to 1
Process Integration: Cooling 1 & Cooling 2	Structural	Reduced process from 2 to 1

3.4. Model assumptions

The following assumptions were made when running the simulation models:

i. Breaks period were excluded from the daily production operation of 8 hours and 50 minutes.

ii. Process cycle time has considered the entity transfer time.

Both models were simulated in the steady-state simulation setup with 100 replications; warm up period 10,000 seconds; 1-day production operation replication length and 11.6112 hour per day (31,800 seconds plus 10,000 seconds). Such as recommended by [31], the initialisation bias of steady-state simulation needs to be eliminated by setting up the warm-up period. Preliminary runs with multiple replications of both models were performed and Output Analyzer was applied to produce the statistical plot of the saved data in order to examine the appropriate warm-up period.

3.5. Verification and validation of simulation models

Concerning the models' verification, processes sequence of the real production line process sequence was compared and verified by the line leader (of the case study company) to ensure that the models' structural logic was accurately representing the real production line under investigation. Apart from that, the detail modules setting was also examined through animation running and the generated SIMAN coded files.

Both models were validated by performing the goodness-of-fit tests on the stochastic cycle time. Moreover, the collected actual data was compared with the simulation results, whereby a minimal difference or error justifying the models' validity.

IOP Conf. Series: Materials Science and Engineering **697** (2019) 012001 doi:10.1088/1757-899X/697/1/012001

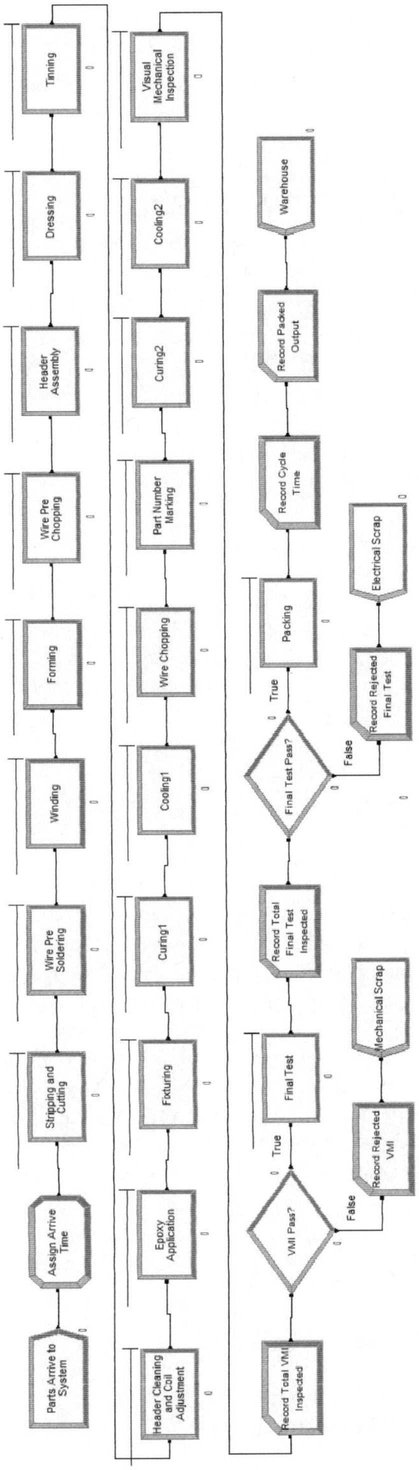

Figure 1. Simulation model of previous state production line

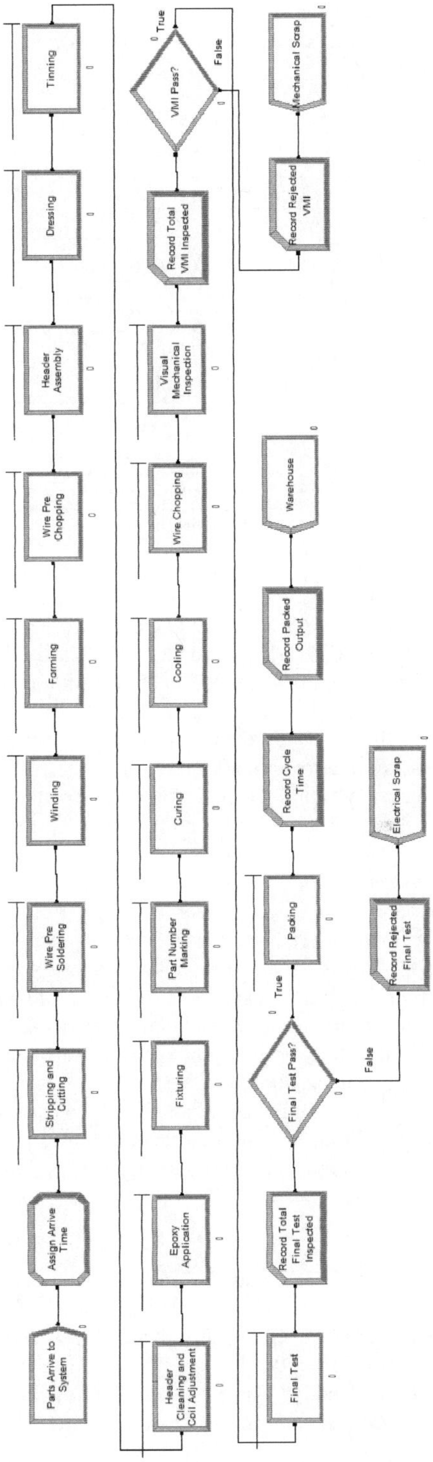

Figure 2. Simulation model of current state production line

3.6. The improvement results

Table 2 shows the simulation results of production waste for previous and current state models of production line.

Table 2. Simulation results of production waste

Production Waste		Previous State	Current State
Rejected VMI	Rate	2.46%	2.32%
	Average	172.3	163.01
	Half-width	2.60	2.62
	Minimum Average	141.00	128.00
	Maximum Average	197.00	196.00
Rejected Final Test	Rate	0.08%	0.06%
	Average	5.29	4.23
	Half-width	0.47	0.37
	Minimum Average	1.00	1.00
	Maximum Average	12.00	9.00

It is apparent from the above table that the total rejected entities at VMI counter has reduced from 172.3 to 163.01 entities (with 7,012.09 inspected entities) during the simulation period. The result recorded the improvement of VMI rejection by 0.14% after the implementation of eco-process innovations in the production line. Besides that, at final test counter, a reduction of 0.02% has been demonstrated with the reduction of rejected entities from 5.29 (with 6,839.83 inspected entities) to 4.23 (with 6,849.10 inspected entities). Overall, this study discovered differences in the production waste rate before the eco-process innovation and after the improvement of a few production processes. These improvements suggest that the continuous eco-process innovation implementation can help to reduce the production waste rate, hence improves the environmental performance of manufacturing firm.

4. Conclusion

This paper concerns the assessment of eco-process innovation environmental performance, denoted by the improvement of production waste rate in particular. DES simulation results demonstrated reductions in the waste rate after the eco-process innovation than those generated by production operation prior to the eco-process improvements. The reduction of waste rate after the adoption of eco-process innovation is suggestive of better environmental performance following the continuous eco-improvement of processes. This study enriches the operational method of measuring the improvement in the production line resulted from the implementation of eco-process innovation. Moreover, this study also highlights the potential application of simulative study to assess the performance of eco-process innovation in E&E manufacturing firm. In a future investigation, it might be possible to conduct a similar study approach to measure other environmental indicators of eco-process innovation such as energy usage.

References

[1] Y.S. Chen, S.B. Lai and C.T. Wen, *The Influence of Green Innovation Performance on Corporate Advantage in Taiwan*, J. Bus. Ethics 67 (2006), pp. 331–339.

[2] S.B. Karaarslan, *Eco-Innovation Competitiveness Inland shipping*, Erasmus University Rotterdam, (2015).

[3] J. Hojnik and M. Ruzzier, *The Driving Forces of Process Eco-Innovation and Its Impact on Performance : Insights from Slovenia*, J. Clean. Prod. 133 (2016), pp. 812–825.

[4] J. Horbach, *Impacts of Regulation on Eco-Innovation and Job Creation*, IZA World Labor (2016), pp. 1–10.
[5] J. Hojnik, M. Ruzzier and T.S. Manolova, *Internationalization and Economic Performance : The Mediating Role of Eco-Innovation*, J. Clean. Prod. 171 (2018), pp. 1312–1323.
[6] C.C.J. Cheng, C. Yang and C. Sheu, *The Link Between Eco-Innovation and Business Performance: A Taiwanese Industry Context*, J. Clean. Prod. 64 (2014), pp. 81–90.
[7] T. Kong, T. Feng and C. Ye, *Advanced Manufacturing Technologies and Green Innovation: The Role of Internal Environmental Collaboration*, Sustainability 8 (2016), pp. 1056.
[8] M. Sanni, *Drivers of Eco-Innovation in the Manufacturing Sector of Nigeria*, Technol. Forecast. Soc. Change 131 (2018), pp. 303–314.
[9] S.M. Dahan, S.M. Yusof and M.Y. Taib, Performance Measure of Innovation :Insights from a Literature Review, in UTP-UMP SES 2017, 131 (2017).
[10] S. Scarpellini, J. Aranda-Usón, M.M. Fondevila, A. Aranda-Usón and E. Llera-Sastresa, *Eco-Innovation Indicators for Sustainable Development: The Role of the Technology Institutes*, Int. J. Innov. Sustain. Dev. 10 (2016), pp. 40.
[11] E. Calik and F. Bardudeen, *A Measurement Scale to Evaluate Sustainable Innovation Performance in Manufacturing Organizations*, Procedia CIRP 40 (2016), pp. 449–454.
[12] A. Kowalska, *Implementing Eco-Innovations. Determinants and Effects*, Ann. Polish Assoc. Agric. Agribus. Econ. XVI (2017), pp. 153–158.
[13] A. de Jesus, P. Antunes, R. Santos and S. Mendonça, *Eco-Innovation in the Transition to a Circular Economy: An Analytical Literature Review*, J. Clean. Prod. 172 (2018), pp. 2999–3018.
[14] A. Davidescu, A. Paul, R.-M. Gogonea and M. Zaharia, *Evaluating Romanian Eco-Innovation Performances in European Context*, Sustainability 7 (2015), pp. 12723–12757.
[15] M.S. Park, R. Bleischwitz, K.J. Han, E.K. Jang and J.H. Joo, *Eco-Innovation Indices as Tools for Measuring Eco-innovation*, (2017), pp. 1–29.
[16] C.B. Joung, J. Carrell, P. Sarkar and S.C. Feng, *Categorization of Indicators for Sustainable Manufacturing*, Ecol. Indic. 24 (2013), pp. 148–157.
[17] V. Oltra, R. Kemp and F.P. Vries, *Environmental Innovation* France: Dynamics of Institutions and Markets in Europe (DIME), (2009).
[18] A. Reid and M. Miedzinski, *Eco-innovation: Final Report for Sectoral Innovation Watch Project*, Technopolis, (2008), Available at: www.technopolis-group.com.
[19] A. Arundel and R. Kemp, *Measuring Eco-Innovation*, ISSN 1871-9872, Maastrich, Netherlands: United Nations University, (2009).
[20] L.F.C. Basso, A.C.S. Braga, D.F.L. Santos and H. Kimura, *Eco-Innovation in Brazil. The Creation of an Index*, Bus. Manag. Rev. 4 (2013), pp. 1–17.
[21] J. Conding, N.F. Habidin, A.F. Mohd Zubir, S. Hashim and N.A. Seri Lanang Jaya, *The Structural Analysis of Green Innovation (GI) and Green Performance (GP) in Malaysian Automotive Industry*, Res. J. Financ. Account. 3 (2012), pp. 172–179.
[22] B. Bossink, *Eco-Innovation and Sustainability Management*, Taylor & Francis Group, New York, (2012).
[23] D. Gadenne, L. Mia, J. Sands, L. Winata and G. Hooi, *The Influence of Sustainability Performance Management Pactices on Organisational Sustainability Performance*, J. Account. Organ. Chang. 8 (2012), pp. 210–235.
[24] B. Sezen and S.Y. Çankaya, *Effects of Green Manufacturing and Eco-Innovation on Sustainability Performance*, Procedia - Soc. Behav. Sci. 99 (2013), pp. 154–163.
[25] S.P. Ganapathy, J. Natarajan, A. Gunasekaran and N. Subramanian, *Influence of Eco-Innovation on Indian Manufacturing Sector Sustainable Performance*, Int. J. Sustain. Dev. World Ecol. 21 (2014), pp. 198–209.
[26] J.S. Keshminder and V. Chandran, *Eco-Innovation in the Chemical Manufacturing Firms : Insights for Policy Response*, Institutions Econ. 9 (2017), pp. 21–42.

IOP Conf. Series: Materials Science and Engineering **697** (2019) 012001 doi:10.1088/1757-899X/697/1/012001

[27] A.N. Sarkar, *Promotion of Eco-Innovation to Leverage Sustainable Development of Eco-Industry and Green Growth*, Eur. J. Sustain. Dev. 2 (2013), pp. 171–224.

[28] Y. Dong, X. Wang, J. Jin, Y. Qiao and L. Shi, *Effects of Eco-Innovation Typology on Its Performance: Empirical Evidence from Chinese Enterprises*, J. Eng. Technol. Manag. 34 (2014), pp. 78–98.

[29] C.M. Luminea, J. Donovan and D. Tormey, *Application of Systems Simulation for Predicting and Optimising Energy Requirements for HDF Production*, J. Ind. Eng. 16 (2015).

[30] B. Haleh. *Work Measurement Methods* Georgia, United States: Institute of Industrial & Systems Engineer, (2014).

[31] D.W. Kelton, R.P. Sadowski and D.T. Sturrock, *Simulation with Arena*, Third Edn. McGraw-Hill Higher Education, New York, USA, (2004).

[32] N. Diaz-Elsayed, A. Jondral, S. Greinacher, D. Dornfeld and G. Lanza, *Assessment of Lean and Green Strategies by Simulation of Manufacturing Systems in Discrete Production Environments*, CIRP Ann. - Manuf. Technol. 62 (2013), pp. 475–478.

Key performance indicator of sustainability in the Malaysian food supply chain

Anis Anizah Mohammad Baba[1], Azanizawati Ma'aram[1*], Fatin 'Izzati Ishak[1], Rozlina Md. Sirat[1], and Aini Zuhra Abdul Kadir[1]

[1]School of Mechanical Engineering, Faculty of Engineering, Universiti Teknologi Malaysia, 81310 UTM Johor Bahru, Johor, Malaysia.

Abstract. The food supply chain in Malaysia plays a significant function in the nation's economy. Since the industry is dominated by small and medium sized enterprises (SME), the local food supply chain must satisfy consumer demand. In order to help Malaysia to become more self-sufficient, improvements in the supply chain in terms of key performance indicators (KPIs) are important. However, it is common that strategic and operational KPIs are different and a disparity often exists between them. To be more successful for measuring the entire supply chain, the linkage of strategic objectives to operational objectives must be aligned to ensure the appropriate KPIs measure a proper connection between what they do and what the business achieves. The objective of this article is to propose a preliminary key indicator to measure sustainable performance within the perspective of Malaysian food supply chain. Key indicators have been summarized based on the previous researches focusing on the issue between 2007 until 2018. An overall of nine perspectives on sustainable performance measures in the food supply chain involved employee, supplier, health and safety, community, customer, cost effectiveness, quality, compliance and efficiency. These perspectives can be classified into three hierarchies; namely, strategic, tactical and operational. This paper proposes integration of key indicators as a starting point to evaluate the generic performance of Malaysia food supply chain.

1. Introduction

The food industry portrays a crucial part in supplying both essentials and necessities for sustaining various individual activities and behaviours [1,2]. Once harvested and manufactured, the food ought to be stored, distributed, then put on the market to reach the final customers by the due date [2]. According to the most current statistics announced by the Food and Agriculture Organization of the United Nations [3], each year, all around the world, 1.3 billion tonnes of food are *"lost"* or wasted. This number amounts to a third of total food produced for human ingestion. Two-thirds of the wasted food occurs in the supply chain, such as harvesting, shipping and storage [2,4]. Over the coming decades, the world population is expected to increase from 7.4 billion at the end of 2016 to 10.5 billion in 2050 [3]. To ensure a food supply for the increasing number of people, there is a need for expansion and intensification of agriculture and food production, which will put pressure on the environment and on societies [5,6].

According to previous researches [7,8], sustainability practices have turn out to be one of the highly critical factors affecting customer's judgments and decision-making activity. Besides, customer involvement refers to their feedback in terms of effort, acquaintance, time, or other types of sources associated to service production as well as distribution [9]. Research on customer engagement in sustainable supply chain management (SSCM) recently has stretching beyond the customers' role of purchasing sustainable product [10]. The level of customer involvement in food supply chain sustainability has expanded to social paired with functional role where customer involvement have being driven by the customer's satisfaction or business interest in the past. Customers attitudes and behaviour involving satisfaction, loyalty and commitment to pay for premium or sustainable food are tending to be motivated by sustainable food supply chain. Recently, the concern with ethical, environmental and health issues amongst customers around the world are becoming increasing dramatically [11,12,13]. A heightened demand for green manufactured goods has also been stated with exclusive reference to food industries [14], and organic products in the Malaysia have also testified a growth in sales due to the same reasons [15].

Today, the food supply chain in Malaysia must remain strong enough to be competitive in the market with the changes of customer needs and trend. One of the alternatives is to implement and monitor a performance indicator of the supply chain, concentrating on sustainability and its dimensions of economy, environment and society. To ensure efficient and sustainable development, there is also a need for scientifically validated and accepted indicators that take into account all three aspects of sustainability – economics, environment and society – to reduce costs, protect the environment and help societies to improve their standard of living [16] besides the alignment of customer demand and behaviour to the performance indicators of the supply chain.

Recently the evaluation approaches are primarily dedicated on the strategic level and offer slight insight into tactical and operational levels, which are acknowledged as significant areas for accomplishing strategic alignment. Additionally, majority of the current approaches are experimented in large organisations and little research is undertaken to assess the effectiveness of these methods in small and medium enterprises (SMEs) [17]. The Malaysian food processing and production sector has not yet received due attention, even though it is a significant contributor to issues relating to global sustainability [18]. In addition, there is no method is available to enable the Malaysian food sector to evaluate its sustainability performance [19].

In order to evaluate the accomplishment of supply chains, a sufficient sustainable performance measurement system (SPMS) that consider customer needs and behaviour and align with strategic, tactical and operational levels to be developed [20]. Performance indicators occur to be as benchmarks where the performance in terms of services, products as well as correspondingly production practices are able to be assessed [20]. However, sparse literature is available focusing on alignment between strategic, tactical and operational levels for performance measures in organizations. This alignment is important for improving performance, enhancing efficiency and allowing organizations to be more competitive in their respective industries [17] as well as driven by the customer demand and behaviour [12, 13, 14]. The alignment of the three levels would contribute to a company's competitive strategies, but there is also no available research to assist in determining a set of performance measures to align these three levels in the Malaysian food supply chain together with the customer demand and behaviour. Therefore, this article aims to develop a key performance indicator, aligning strategic and operational levels and customer demand and behaviour in Malaysia in order to enable the whole supply chain to remain competitive in the market, from farmer/grower to end customer in terms of sustainability perspectives.

2. Existing food supply chain performance measure

A literature study was conducted to investigate the perspectives and indicators that are commonly used for evaluating performance around the context of food supply chain. Studies focusing in behaviour of customers regarding sustainability issues in purchasing food amongst customers are in dramatically increase trend whether in Malaysia [15] or developed countries [12, 13, 14]. The gap between the customer behaviour and performance indicators is there is no clear link to align them to make the performance indicators develop based on the customer needs. A summation of the general studies associated to food performance measures is shown in Table 1. It can be deduced that of the thirteen (13) articles, six (6) have categorised the employee and quality perspective as the most important aspect of a sustainable performance measure for the food supply chain.

Table 1. Summary of research perspectives on food sustainability performance measures.

Authors	Perspectives on food sustainability performance measures								
	Employee	Supplier	Health and Safety	Community	Customer	Cost Effectiveness	Quality	Compliance	Efficiency
N. Yakovleva (2007)	√								
Aramyan et al. (2007)							√		√
Carter and Rogers (2008)						√	√		
Baldwin (2009)	√	√	√				√		√
Hassini et al. (2012)		√							
Bloemhof et al. (2013)	√		√			√	√		√
Van der Vorst (2013)	√						√		
M. Bourlakis et al. (2014)					√	√	√		
Xu and Gursoy (2015)					√				
Ju and Chang (2016)					√				
Maestrini et al. (2017)	√					√			
Ahmad et al. (2017)	√	√		√	√	√	√		√
Rashid et al. (2017)			√	√				√	√

Therefore, the linkage with the customer behaviour with the sustainable food supply chain and its performance indicators is crucial especially on the Malaysian perspective.

2.1. Sustainability performance measures for the Malaysian food supply chain
Referring on the research review from former studies as shown in Table 1, this study has developed a conceptual framework of key performance indicators (KPI) for the Malaysian food supply chain. The conceptual framework of the KPIs consists of three main components; external factors, attitude and consumer behaviour and sustainability performance measurement with three main aspects which are social, economic and environmental. The external factors that may influence the attitude and consumer behaviour such as government policy [21], environmental degradation issues [21, 22], and social value [23]. While the attitude and consumer behaviour can categorize into five perspectives [15, 24] including performance, convenience and price sensitivity, reducing packaging waste, reusing/recycling, reducing carbon footprint and beneficial, safe and health to individual and community. Then, the nine perspectives of preliminary performance measurement of sustainable food supply chain is introduced as presented in Figure 1.

IOP Conf. Series: Materials Science and Engineering **697** (2019) 012002 doi:10.1088/1757-899X/697/1/012002

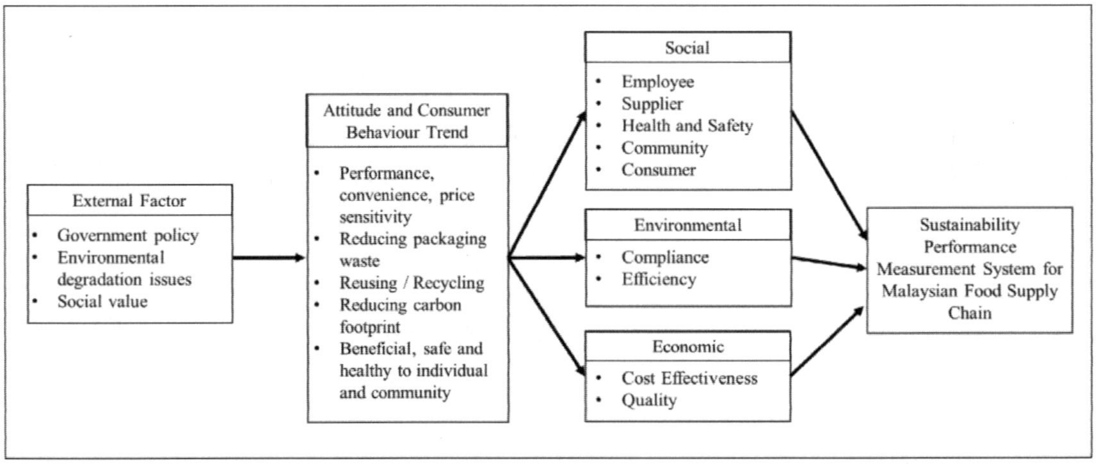

Figure 1. Preliminary key performance indicators for sustainable performance measures in the Malaysian food supply chain.

The KPI adopts a decision-making approach by dividing the indicators into three hierarchies; strategic, tactical and operational. It is believed that a balanced and integrated KPI benefits organizations by creating competitive advantage, essential for a sustainable organization. Moreover, a hierarchical manner will ensure that performance measures have direct linkage and a clear impact on business strategy and profit generation.

Finally, the preliminary KPI of the Sustainable Performance Measure for Malaysian Food Supply Chain is proposed with 113 indicators in total, which are categorized in the social, economic and environmental factors. These indicators were adopted and modified to fit the KPI based on relevant literature in Table 1 and attitude and customer behaviour that influenced by external factors.

Table 2 presents the social dimension of the preliminary key performance indicators of the sustainable performance measurements for the Malaysian Food Supply Chain. In this dimension, five perspectives namely; employee, supplier, health and safety, community and customer are introduced. These perspectives consist of 52 indicators and are categorized into strategic, tactical and operational aspects.

Table 2. Social dimension of preliminary key performance indicators for a sustainable performance measure for the Malaysian food supply chain.

	SOCIAL FACTOR	
STRATEGIC	**TACTICAL**	**OPERATIONAL**
Employee Perspective		
Learning and Growth	Skill development	Improved employee education and skills
		Training per employee
		Sustainability training
Recruitment program	Human rights observance	Fair pay systems
		Male-female ratio
Employee satisfaction	Employee well-being	Turn-over rate
		Absence

IOP Conf. Series: Materials Science and Engineering **697** (2019) 012002 doi:10.1088/1757-899X/697/1/012002

	SOCIAL FACTOR	
STRATEGIC	**TACTICAL**	**OPERATIONAL**
Supplier Perspective		
Supplier relation	Improved supplier commitment	Percentage of suppliers with up-to-date sustainable development policies Increase certified suppliers with ISO certification
	Percentage of contracts with provincial suppliers	Utilize local supplier
	Percentage of purchase orders placed with Aboriginal companies	Strengthen relationship with aboriginal people.
Health and Safety Perspective		
Increased public health and safety practices	Follow guidelines for Good Manufacturing Practice (GMP)	
Increased occupational health and safety	Improved work environment	
	Improved work safety	Number of accidents
Community Perspective		
Community well-being	Community Projects	Number of community projects
	Environmental fines	Number of fines received
Customer Perspective		
Customer well-being/Customer Satisfaction	Customer satisfaction rate	Customer complaints
	Health and safety assessment of products	Product recall
	Customer Loyalty	Positive Recommendation of product to others Repurchasing product
	Customer willingness to pay	Willing to pay premium/ pay more to buy sustainable product
Sustainable food preparation	Sustainable food ingredients	Use certified organic vegetables Use environment friendly fed meats and dairy products Use on-site food produce
Food service organization	Green Packaging	No disposable food service item Disposable food service items for reusable or recycled materials Disposable food service items for bio-based materials

Conversely, Table 3 presents the economic dimension of the preliminary key performance indicators of sustainable performance measures for the Malaysian Food Supply. This dimension shows another 2 perspectives; specifically, cost effectiveness and quality. These perspectives consist of 48 indicators, also classified into strategic, tactical and operational aspects.

IOP Conf. Series: Materials Science and Engineering **697** (2019) 012002 doi:10.1088/1757-899X/697/1/012002

Table 3. Economic Factor of Preliminary Key Performance Indicators of Sustainable Performance Measures for the Malaysian Food Supply Chain.

ECONOMIC FACTOR		
STRATEGIC	**TACTICAL**	**OPERATIONAL**
Cost Effectiveness Perspective		
Manufacturing budget	Operational and capital cost	
	Raw Material cost	Total purchase of raw material
	Labor cost	Average wage
		Overtime
Distribution budget	Transportation cost	Average travel cost per shipment
		Average fuel consumption cost per shipment
	Handling cost	Average handling cost in warehouse
	Storage cost	Average holding cost in warehouse
	Transaction cost	
	Delivery cost	Average delivery cost of shipment
Quality Perspective		
Food quality	Appearance	Color
		Size and form
		Firmness
		Lack of blemishes and damage
		Taste
	Shelf life	
	Product safety and health	Salubrity
		Product safety
	Product reliability	Ratio of actual product composition and product description
Process quality	Production system characteristic	Traceability
		Storage condition
		Transport condition
		Working condition
	Environmental aspect	Energy usage
		Water usage
		Pesticide usage
		Recycling/reuse
Packaging quality	Labeling	Clearer product information
	Safety	Material used
		Types of food to be pack

Finally, 13 indicators for the environmental aspect are presented in Table 4. The measures demonstrate two perspectives namely; compliance and efficiency. Indicators in the environmental aspect are also alienated into three hierarchies; strategic, tactical and operational.

Table 4. Environmental factors of preliminary key performance indicators of sustainable performance measures for the Malaysian food supply chain.

ENVIRONMENTAL FACTORS		
STRATEGIC	**TACTICAL**	**OPERATIONAL**
Compliance Perspective		
Compliance with regulations	Improved compliance with environmental standards	Qualified by professional body
Efficiency perspective		
Eco-efficiency	Resource consumption	Reduced water usage
		Reduced waste water emission
		Reduced energy consumption
		Reduced non-renewable resources usage
		Reduced hazardous inputs usage
		Reduced solid waste
		Land use
		Reduced emission of polluting gases

3. Discussion and conclusions

This study has explored available sustainability perspectives and indicators, focusing on the Malaysian food supply chain. These indicators are divided, in a balanced manner, into three factors of sustainability namely; social, economic and environmental, and simultaneously cascaded into three hierarchies; specifically, strategic, tactical and operational which generated from customer demand attitude and behaviour as well as external factors. This is a generic classification, which could be used together with an in-depth organization analysis as a basis for improving business strategies and profit generation.

On the other hand, this study proposes nine (9) perspectives of sustainable performance measures, which can contribute to improved performance of the Malaysian food supply chain. A questionnaire survey are planned to be carried out to collect data from the customer to derive their attitude and behaviour that will be aligned with the perspective of supply chain players in food industries for the conclusive arrangement of key performance measures specially tailored for the Malaysian food supply chain.

4. Acknowledgements

This work was supported by the Fundamental Research Grant Scheme (FRGS) under the Ministry of Education (MOE) Malaysia with project code R.J130000.7824.4F975.

5. References

[1] Cooper M C and Ellram L M 1993 Characteristics of Supply Chain Management and the Implications for Purchasing and Logistics Strategy *Int. J. Logist. Manag.*, **4** 13–24

[2] Zhong R Y, Xu X and Wang L 2017 Food supply chain management: systems, implementations, and future research *Ind. Manag. Data Syst.*, **117** 2085–2114
[3] Alexandratos N and Bruinsma J 2012 World Agriculture towards 2030/2050: the 2012 revision (Rome) 12–03
[4] Fritz M and Schiefer G 2008 Food chain management for sustainable food system development: A European research Agenda *Agribusiness* **4** 440–452
[5] Montero M, Schmalenberg A C, Quirós O, and Doluschitz R 2018 Identification of Supply Chain Performance Indicators: Case Study of Costa Rican Coffee Production *Univers. J. Ind. Bus. Manag.* **6** 1–10
[6] Yakovleva N 2007 Editorial introduction: Measuring the sustainability of the food system," *J. Environ. Policy Plan.* **9** 1–3
[7] Berezan O, Raab C, Yoo M, and Love C 2013 Sustainable hotel practices and nationality: The impact on guest satisfaction and guest intention to return *Int. J. Hosp. Manag.* **34** 227–233
[8] Xu X and Gursoy D 2015 Influence of sustainable hospitality supply chain management on customers' attitudes and behaviors *Int. J. Hosp. Manag.* **49** 105–116
[9] Dong B, Sivakumar K, Evans K R, and Zou S, 2015 Effect of Customer Participation on Service Outcomes: The Moderating Role of Participation Readiness *J. Serv. Res.* **18** 160–176
[10] Kim H and Lee C W 2018 The effects of customer perception and participation in sustainable supply chain management: A smartphone industry study *Sustainability* **10** 2271–90
[11] Lyons K, Burch D, Lawrence G A, and Lockie S 2004 Contrasting paths of corporate greening in Antipodean agriculture: organics and green production *Agribusiness and Society. Corporate Responses to Environmentalism, Market Opportunities and Public Regulation* ed K. Jansen and S. Vellema (London: Zed Books) pp. 91–113.
[12] Mollá-Bauzá M M B, Martínez-Carrasco L, Martínez-Poveda A, and Pérez M R 2005 Determination of the surplus that consumers are willing to pay for an organic wine *Spanish J. Agric. Res.* **3** 43–51
[13] Vermeir I and Verbeke W 2006 Sustainable food consumption: exploring the consumer 'attitude – behavioral intention' gap *J. Agric. Environ. Ethics* **19**, 169–194
[14] Sharples L 2009 Organic Wines — The UK Market: A Shift from 'niche market' to 'mainstream' Position? *Int. J. Wine Mark.* **12**, 30–41
[15] Shaharudin M R, Pani J, Mansor S W, and Elias S 2010 Factors Affecting Purchase Intention of Organic Food in Malaysia's Kedah State *Cross-Cultural Commun.* **6** 105–116
[16] Gupta V, Abidi N, and Bandyopadhayay A 2013 Supply Chain Management - A Three-Dimensional Framework *J. Manag. Res.* **5** 76–97
[17] Gutierrez A, Orozco J, Serrano A, and Serrano A 2006 Using tactical and operational factors to assess strategic alignment: An SME study *European and Mediterranean Conference on Information Systems* pp. 1–10.
[18] Ahmad S, Wong K Y, and Elahi H 2017 Sustainability assessment and analysis of Malaysian food manufacturing sector—A move towards sustainable development *Adv. Sci. Lett.* **23** 8942–46
[19] Maestrini V, Luzzini D, Maccarrone P, and Caniato F 2017 Supply chain performance measurement systems: A systematic review and research agenda *Int. J. Prod. Econ.* **183** 299–315
[20] Aramyan L H, Lansink A, van der Vorst J, and van Kooten O 2007 Performance measurement in agri-food supply chains: a case study *Supply Chain Manag. Int. J.* **12** 304–315
[21] Coad A, de Haan P, and Woersdorfer J S 2009 Consumer support for environmental policies: An application to purchases of green cars *Ecol. Econ.* **68** 2078–86
[22] Shove E 2010 Beyond the ABC: Climate change policy and theories of social change *Environ. Plan. A* **42** 1273–85
[23] Biswas A and Roy M 2015 Green products: An exploratory study on the consumer behaviour in emerging economies of the East *J. Clean. Prod.* **87** 463–8

[24] Nordin N and Selke S 2010 Social aspect of sustainable packaging *Packag. Technol. Sci.* **23** 317–326
[25] Carter C R and Rogers D S 2008 A framework of sustainable supply chain management: Moving toward new theory *Int. J. Phys. Distrib. Logist. Manag.* **38** 360–387
[26] Baldwin C 2009 *Sustainability in the Food Industry*. Wiley-Blackwell and IFT Press
[27] Hassini E, Surti C, and Searcy C 2012 A literature review and a case study of sustainable supply chains with a focus on metrics *Int. J. Prod. Econ.* **140** 69–82
[28] van der Vorst J G A J, Peeters L, and Bloemhof J M 2013 Sustainability Assessment Framework for Food Supply Chain Logistics: Empirical Findings from Dutch Food Industry *Int. J. Food Syst. Dyn.* **4** 130–9
[29] Bourlakis M, Maglaras G, Aktas E, Gallear D, and Fotopoulos C 2014 Firm size and sustainable performance in food supply chains: Insights from Greek SMEs *Int. J. Prod. Econ.* **152** 112–130
[30] Ju S and Chang H 2016 Consumer perceptions on sustainable practices implemented in foodservice organizations in Korea *Nutr. Res. Pract.* **10** 108–114
[31] Ahmad A, Hami A N, Shafie S, and Mat Yamin F 2018 A proposed framework of sustainable manufacturing practice and sustainability performance among Malaysian SMEs in the manufacturing sector *Proceedings of the 2nd Conference on Technology & Operations Management* pp. 159–168.
[32] Abdul-Rashid S H, Sakundarini N, Raja Ghazilla R A, and Thurasamy R 2017 The impact of sustainable manufacturing practices on sustainability performance: Empirical evidence from Malaysia *Int. J. Oper. Prod. Manag.* **37** 182–204
[33] Abdullah R, Hassan M G, and Johari N A 2014 Exploring the Linkage of Supply Chain Integration between Green Supply Chain Practices and Sustainable Performance: a Conceptual Link *Int. Proc. Chem. Biol. Environ. Eng.* **61**
[34] Handfield R, Sroufe R, and Walton S 2005 Integrating environmental management and supply chain strategies *Bus. Strateg. Environ.*, **14** 1-19
[35] Chopra S, Laux C, Schmidt E, and Rajan P 2017 Perception of Performance Indicators in an Agri-Food Supply Chain: A Case Study of India's Public Distribution System *Int. J. Food Syst. Dyn.* **8** 130–145
[36] Sari E, Shaharoun A M, and Ma'aram A 2014 Preliminary Framework of Sustainable Maintenance Performance Measurement Systems for Automotive Companies," *Adv. Mater. Res.* **845** 590–595
[37] Dong B, Sivakumar K, Evans K R, and Zou S 2015 Effect of Customer Participation on Service Outcomes: The Moderating Role of Participation Readiness," *J. Serv. Res.* **18** 160–176

IOP Conf. Series: Materials Science and Engineering **697** (2019) 012003 doi:10.1088/1757-899X/697/1/012003

Analysis of Port Fare Increases On Container Yard Services Using Logistic Regression Model

S Zuhri[1*], P D Sentia[1], N A Lubis[1] S D Permai[2]

[1]Industrial Engineering Department, Universitas Syiah Kuala, Banda Aceh, 23111, Indonesia

[2]Statistics Department, School of Computer Science, Bina Nusantara University, Indonesia

[*]Email: sarika.zuhri@unsyiah.ac.id

Abstract. The port company used in this research is the only port where container ship contained in Aceh and has been operating for more than 1 year. The problem with this was the large buildup of containers with long periods of time in the cultivation field that would cause losses. Losses due to the duration of the buildup are influenced by several factors, such as the yard of ratio and the length of time the container is reviewed based on the large number of containers in the container yard. For payment of the first five days of service, the container accumulation fare shall be borne by container owner company as stakeholder of port company. After more than five days, the customer will be charged the container accumulation rate calculated per box without any time consideration. This causes the customers don't take the goods immediately, therefore it was necessary to do research on the analysis of port fare increase in port company in Aceh using logistic regression model as the basic of decision making. Based on the prediction of logistic regression model using software R3.4.3, there was a loss cost in the container yard, therefore it was necessary to take decision of the application of the port fare increase on container yard service. Based on the results of these predictions, the resulting model was a good model with an accuracy of 74.8%. The factor that affect of the loss cost was the length of time the container in the container yard period II and III.

1. Introduction

Indonesia is an archipelago with a strategic location so that, Indonesia needs a good logistics system. Ports have an important role as one of the distribution chain of goods, which is a logistics process. Based on the logistics performance index of Indonesia in 2016 decreased by 10 ratings ie the order of 63 of the 160 countries [1]. To handled this, presidential of Indonesia plans decreasing in logistic costs by 20 to 25%.

A way the government is lowering logistics costs by simplifying the examination by pressing the number of time loading and unloading and set fines of up to 900 percent of port rates at Tanjong Priok Port as the reference to the three ports such as Tanjong Perak, Belawan and Soekarno Hatta port. The presence of the port fare increase plan make some pros and cons.

Port company used in this research is one of 24 ports that includes in the program sea motorway. With the marine highway program is expected to improve the performance of logistics and lowering the loading and unloading time so, no more excessive accumulation of goods for a long time inport.The accumulation of goods is too long portis one of the things that should be minimized. In addition to bringing the ineffectiveness also cause of loss cost for itself. Losses incurred due to the length of the buildup allegedly influenced by several factors: the yard of the ratio and the length of time the container that is period I for 1 to 5 days, the period II for 6 to 9 days, and the period III for the great time equal to 10 days that reviewed by the large number of containers in the container yard. The load capacity can be stacked in the container yard of Port company is 500 (Twenty-Foot Equivalent Units).

Based on current conditions, for payment of container yard service usage after more than five days, it is not covered by container owner company as stakeholder of port company. So, the customer will be charged a container stacking which are calculated per box without any consideration of the time. Because there is no consideration of time in the stacking fare, it makes the customer does not take the goods immediately. To minimize the inefficiencies and losses because of stacking container are too long, the necessary basis for decision making in determining the rate increase on services which consider the accumulation of time in port company in Aceh.

2. Methodology and Conceptual Framework

Port company in Aceh has problem that can be seen based on survey data. Survey data is the data from port company that is used in this research. Survey data consist of two part such as dependent and independent variable. Each data of survey data consists of 12 month that started from May 2017 to April 2018.

Making the decision in a firm is very important thing. Logistic regression is one of tools as part of regression that can be used for decision making. The difference between logistic regression and linear regression is a dependent variable of the linear regression is continuous and dependent variable in logistic regression is dichotomous,It can be seen in a regression analysis when variable dependent is the dichotomy so a regression equation must be formulated to be limited between 0 and 1. The distribution is binomial distribution, not normal. It describe the distribution of the error and will be a statistical distribution is the basis of its analysis, and principles used in the linear regression will also be a guide in logistic regression [2].

Based on previous research logistic regression model is a model that can be used in solving the problem of decision making [3]. Logistic regression model is also an effective model to detect and predict an issue [4,5,6,7] and a logistic regression model can also be used in evaluating the risk that logistic regression is a model that is attractive and effective techniques [8,9,10].

The variables used in this study to build a logistic regression model that is

Table 1. Defining Variables

Notation	Name	Definition	Information
Y	There is or there is no loss cost	Variable presence or absence of costs or loss arising from the container stacking in the container yard.	0= No loss 1= loss
X_1	*Yard of Ratio*	Comparison of stacking yard capacity consumed with overall capacity container yard	Continuous
X_2	Number of Containers	Number of existing container Stacking in the container yard	Continuous
X_3	The length of time the container	The length of time of the container that's located in the container yard	0= Period I 1= Period II 2= Period III

The logistical function can ensure that whatever is needed, there will always be between 0 and 1 which is the reason that the logistic model is available to estimate the opportunity [10]. The specific form of the logistic regression model or regression model is [2]

$$\pi(x) = \frac{e^{\beta_0+\beta_1 x}}{1+e^{\beta_0+\beta_1 x}} \qquad (1)$$

Making more easier to estimate regression parameters, π(x) is transformed using logit transformation. Logit transformation in the logistic regression model is

$$g(x) = \beta_0 + \beta_1 x \tag{2}$$

Where,
$\pi(x)$: Probability of ocuurrence
$g(x)$: Logit estimation value
β_0, β_1 : Coefficient values for the constant variables obtained using the estimation method
x : Independent variable

what can be seen in a regression analysis is a regression equation must be formulated to be limited between 0 and 1, binomial distribution not normal, describes the distribution of errors and will be a statistical distribution which is the basis of the analysis, the principle used in regression linear will also be a guide in logistic regression. Here is the value of the logistic regression model if the independent variable is dhycotomous

Table 2. Value of Logistic Regression Model If Independent Variable is Dhycotomous

Dependent variable (Y)	Independent variable (X)	
	X=1	X=0
Y=1	$\pi(1) = \frac{e^{\beta_0+\beta_1}}{1+e^{\beta_0+\beta_1}}$	$\pi(0) = \frac{e^{\beta_0}}{1+e^{\beta_0}}$
Y=0	$1-\pi(1) = \frac{1}{1+e^{\beta_0+\beta_1}}$	$1-\pi(0) = \frac{1}{1+e^{\beta_0}}$
Total	1,0	1,0

Logistic regression significance test consists of two stages: test the significance of the model parameters together and test the significance of the model parameters separately. A likelihood ratio goodness of fit (G) test statistic is a test of the significance of the model parameters together which is used to examine the role of predictor variables in the model together. The formula for the G likelihood ratio test is

$$G = -2ln\left[\frac{L_0}{L_p}\right] \tag{3}$$

Where,
L_0 : *Likelihood* with no independent variable
L_p : *Likelihood* with p independent variable

Another test phase is Wald test which is obtained by comparing the estimation of maximum likelihood from the slope parameter to the estimated standard of errors. Wald test is an individual test on the significance of the model parameters defined by

$$W = \left[\frac{\widehat{\beta_\iota}}{SE(\widehat{\beta_\iota})}\right]^2 \tag{4}$$

Where,
$\widehat{\beta_\iota}$: Estimator of $\widehat{\beta_\iota}$
$SE(\widehat{\beta_\iota})$: Default error estimator of $\widehat{\beta_\iota}$

IOP Conf. Series: Materials Science and Engineering **697** (2019) 012003 doi:10.1088/1757-899X/697/1/012003

One of the classification accuracy tests used for categorical data is apparent error rate (APER). A good model is a model that produces a slight misclassification. APER is defined as a comparison of observations in a sample of training data that is incorrect by the sample classification function where, APER is a value that is used to see opportunities for errors in classifying objects. Apparent error rate (APER) can be easily calculated from the confusion matrix, which shows the actual group membership versus the predicted one. The form of the confusion matrix is

Table 3. Confusion Matrix

		Prediction member	
		π_1	π_1
Actual member	π_1	n_1C	$n_1M = n_1 - n_1C$
	π_2	$n_2M = n_2 - n_2C$	n_2C

Where,
n_1C = Number of data π_1 correctly classified as data π_1
n_1M = Number of data π_1 incorrectly classified as data π_2
n_2C = Number of item π_2 correctly classified
n_2M = Number of item π_2 incorrectly classified

The formula to calculate APER is

$$APER = \frac{n_1M + n_2M}{n_1 + n_2} \qquad (5)$$

3. Result and Discussion

The following are the results of modeling and parameter estimation using logistic regression models. Modeling is done using one dependent variable and three independent variables.

Table 4. Output of logistic regression model

β_0	X_1	X_2	X_{32}	X_{33}
-3,491	0,004	-0,003	4,386	3,230

Based on Table 2 model created is:

$g(x) = -3{,}491 + 0{,}004X_1 - 0{,}003X_2 + 4{,}386X_{32} + 3{,}230X_{33}$

Based on the logistic regression model, the output summary conclusions model which can be seen in value of p-value of each variable.

Table 5. Result of logistic regression model test

Variable	*p-value*	Information
Yard of Ratio (X_1)	0,659	Not significant
Number of Container (X_2)	0,538	Not significant
The length of time the container period II (X_3)	4,22e-12	Significant
The length of time the container period III (X_3)	1,40e-07	Significant

Based on Table 5 it can be seen that the independent variables that influence whether or not the loss cost is variable X3, namely the length of the time period II and III. It can be seen from the p-value $\leq \alpha$ that is 4,22e $\times 10^{-12}$ dan 1,40e $\times 10^{-07}$. Variables that do not affect whether or not the loss is a X_1 variable that is yard of ratio and X_2 that is the number of containers in the container yard. It can be seen from the vaue of p-value that is greater than $\boldsymbol{\alpha}$. Here's a likelihood ratio test output which is a significance test parameters simultaneously

Table 6. Output of *ratio likelihood test* in model

LogLik	Chisq	Pr(>Chisq)
-174,26		
-233,24	117,96	<2,2e$\times 10^{-16}$

The hypothesis are:

H_0 = No influence of variables - independent variables to whether or not the loss cost
H_1 = At least there is one independent variable which affects whether or not loss cost

To calculate the value of G that is by comparing the value of G obtained in the Table

G = -2 (-233,24-(-174,26))
G = 117,96

Rated G obtained in the amount of 117.96 that will be compared with the value of chi-square Table with df = $4(X^2{}_{0,05;4})$ is 9.488. If G $\geq$ Chi square Table then H_0 rejected and H_1 accepted. Based on the results of the likelihood ratio test G $\geq$ Chi square Table value is equal to 117.96 $\geq$ 16.92. This means that, at least there is one independent variable which affects the loss cost in the container yard.

Wald test is used to determine the independent variables that affect the model. H_0 is rejected if P(Wald's *Test*) $\leq \alpha$. Hyphotesis in Wald test are :

H_0 = Independent variable k on g(x) is not significant effected on the model.
H_1 = Independent variable k on g(x) is significant effected on the model.

Table 7. Output of Wald test in model

Variabel	P(Wald's Test)
Yard of Ratio (X_1)	0,659
Number of Container (X_2)	0,538
The length of time the container period II (X_3)	<0,001
The length of time the container period III (X_3)	<0,001

Based on the Wald test output in Table 7, namely yard of X_1 and X_2 ratio is the number of containers is an independent variable that does not affect the model because, the value of P (Wald's Test) $\geq \alpha$ are 0.659 and 0.538. X_3 is the length of time the buildup of future category II and III period are independent variables that affect the model, because nillai P (Wald's Test) $\leq \alpha$ is 0.001.

Fitting model test is performed to evaluate the model fit with data. Model fit test that used was Hosmer and Lemeshow test. H_0 is rejected if the p-value $\leq \alpha$. Hypotheses used are:

H_0 = Logistic regression model fit to the data
H_1 = Logistic regression model did not fit to the data

Here is the output from the model fit test using Horsmer and Lameshow test.

Table 8. Output of Horsmer and Lameshow test in model

	P-value
Horsmer and Lameshow	$<2{,}2e\times10^{-16}$

Based on Table 8, the value of p-*value* $\leq \propto$, that is $2{,}2e\times 10^{-16} \leq 0{,}05$. It means that the logistic regression model formed is not fit to the data. This happens because of from the five independent variables were used only two variables that affect whether or not the cost of loss.

The means used to interpret the parameter is to look at the value of the odds ratio. Odds ratio results of the data are shown in Table 9

Table 9. Result of odds ratio test in model

Variabel	Odds Ratio
Yard of Ratio (X_1)	1,004
Number of Containers (X_2)	0,997
The length of time the container Period II (X_3)	80,35
The length of time the container Period III (X_3)	25,27

According to the Table 9, can be interpreted that each increase of one unit of a variable length of time the container category period II, the opportunity cost of the loss due to accumulation of containers in port company is 80.35 times compared to category period III.

Evaluation of the regression model described by pseudo R^2, but because of the logistic regression using R3.4.3 there are not pseudo R^2 value, then the index McFadde can be used to assess the percentage of independent variables able to explain the dependent variable of the logistic regression model that has been formed. McFadden output results can be seen in Table 10.

Table 10. Result of evaluation of model

G^2	McFadden	R^2ML
117,961	0,253	0,276

R^2 McFadden value = 0.253 it means the independent variable of the existing logistic regression model can explain the variability of the dependent variables of 25.3%, while the remaining 74.7% is explained by other variables not included in the model.

How to read the results predicted in Figure 1 is to compare the predicted results using a logistic regression model with dependent variable survey data.

1	2	3	4	5	6	7	8	9	10	11	12	13	14	15	16	17	18	19	20
0	0	0	0	0	0	0	0	0	0	0	0	0	0	0	0	0	0	0	0
21	22	23	24	25	26	27	28	29	30	31	32	33	34	35	36	37	38	39	40
0	1	1	1	1	0	0	0	0	0	0	0	0	0	0	0	0	0	0	0
41	42	43	44	45	46	47	48	49	50	51	52	53	54	55	56	57	58	59	60
1	1	1	1	0	0	0	0	0	1	1	1	1	0	0	0	0	0	0	0
61	62	63	64	65	66	67	68	69	70	71	72	73	74	75	76	77	78	79	80
0	0	0	0	0	0	1	1	1	1	0	0	0	0	0	0	0	0	0	0
81	82	83	84	85	86	87	88	89	90	91	92	93	94	95	96	97	98	99	100
0	0	0	1	1	1	1	0	0	0	0	0	0	0	0	0	0	0	0	0
101	102	103	104	105	106	107	108	109	110	111	112	113	114	115	116	117	118	119	120
1	1	1	1	0	0	0	0	0	0	0	0	0	0	0	0	0	0	0	0
121	122	123	124	125	126	127	128	129	130	131	132	133	134	135	136	137	138	139	140
1	1	1	1	0	0	0	0	0	0	0	0	0	0	0	0	0	0	1	1
141	142	143	144	145	146	147	148	149	150	151	152	153	154	155	156	157	158	159	160
1	1	0	0	0	0	0	0	0	0	0	0	0	0	0	0	0	0	1	1
161	162	163	164	165	166	167	168	169	170	171	172	173	174	175	176	177	178	179	180
1	1	0	0	0	0	0	0	0	0	0	0	0	0	0	0	0	0	0	0
181	182	183	184	185	186	187	188	189	190	191	192	193	194	195	196	197	198	199	200
0	0	0	0	0	0	0	0	0	1	1	1	1	0	0	0	0	0	0	0
201	202	203	204	205	206	207	208	209	210	211	212	213	214	215	216	217	218	219	220
0	0	0	0	0	0	1	1	1	1	0	0	0	0	0	0	0	0	0	0
221	222	223	224	225	226	227	228	229	230	231	232	233	234	235	236	237	238	239	240
0	0	0	0	0	0	0	0	0	0	0	1	1	1	1	0	0	0	0	0
241	242	243	244	245	246	247	248	249	250	251	252	253	254	255	256	257	258	259	260
0	0	0	0	0	1	1	1	1	1	0	0	0	0	0	0	0	0	0	0
261	262	263	264	265	266	267	268	269	270	271	272	273	274	275	276	277	278	279	280
0	0	0	1	1	1	1	0	0	0	0	0	0	0	0	0	0	1	1	1
281	282	283	284	285	286	287	288	289	290	291	292	293	294	295	296	297	298	299	300
1	0	0	0	0	0	0	0	0	0	1	1	1	1	0	0	0	0	0	0
301	302	303	304	305	306	307	308	309	310	311	312	313	314	315	316	317	318	319	320
0	0	0	1	1	1	1	0	0	0	0	0	0	0	0	0	1	1	1	1
321	322	323	324	325	326	327	328	329	330	331	332	333	334	335	336	337	338	339	340
0	0	0	0	0	0	0	0	0	1	1	1	1	0	0	0	0	0	0	0
341	342	343	344	345	346	347	348	349	350	351	352	353	354	355	356	357	358	359	360
0	0	0	0	1	1	1	1	0	0	0	0	0	0	0	0	0	1	1	1
361	362	363	364	365															
[illegible]	0	0	0	0															

Figure 1. Prediction result of dependent variable based on survey data

Based on the predicted results in Figure 1, there is the loss cost by the symbol 1. Cost loss arising on the predicted outcome means that the losses incurred due to the length of the stacking containers in container yard of port company so, the decision to port fare increase on container yard services is the best decision.

Calculate the classification accuracy of the model is done by calculating the value of the apparent error rate (APER). Confusion matrix for survey data can be seen in Table 11.

Table 11. Confusion matrix of surey data

Observation result	Prediction	
	Loss	No
Loss	58	27
No	65	215

The results of the model's accuracy survey data is:

$$
\begin{aligned}
1\text{-APER} &= 1-\left(\frac{27+65}{58+27+65+215}\right) \\
&= 1-0{,}252 \\
&= 0{,}748
\end{aligned}
$$

Based on the results of the accuracy survey data, the classification accuracy of determining the presence or absence of the loss cost in determining the need for a pors fare increase is 74, 8%. This shows that the model is already well in the presence or absence of classifying the loss cost in determining the need for a decision on the port fare increase in container yard services of port company.

4. Conclusions

The conclusions of this research are $g(x)= -3{,}491 + 0{,}004X_1\ \ 0{,}003X_2 + 4{,}386X_{32} + 3{,}230X_{33}$. The factors affecting the rise of the loss cost in the container yard of port company are the duration of the container period II and period III. Based on the prediction results using logistic regression there is the loss cost. Therefore it is necessary to take the decision of the port fare increase in container yard service. So the resulting model is a good model with an accuracy of 74.8%.

IOP Conf. Series: Materials Science and Engineering **697** (2019) 012003 doi:10.1088/1757-899X/697/1/012003

Acknowledgments
The authors are thankful to the industrial engineering department of syiah kuala university and bina nusantara university for having funded this study.

References
[1] Arvis, J. Mustra, M. Panzer, J. Ojala, L. Naula, T. (2016) 'Connecting To Compete: Trade Logistics In The Global Economy', *World Bank*, pp. 1–76.
[2] Hosmer, D. W. And Lemeshow, S. (2000) 'Applied Logistic Regression'.John *Wiley Series In Probability And Sattistics*, Inc, New York.
[3] Mehrjoo, S. and Bashiri, M. (2013) 'An Application Of Principal Component Analysis And Logistic Regression To Facilitate Production Scheduling Decision Support System: An Automotive Industry Case'. Journal of Industrial Engineering., pp. 2-12
[4] Chandrasekaran, R. (2012) 'Application Of Logistic Regression To Predict Over Target Baseline Of Software Projects', 44(April), pp. 1–6.
[5] Agarwal, S., Kachroo, P. And Regentova, E. (2016) 'A Hybrid Model Using Logistic Regression And Wavelet Transformation To Detect Traf Fi C Incidents', *IATSS Research.* International Association Of Traffic And Safety Sciences, 40(1), pp. 56–63.
[6] Christine, L. Aderson, A. Karoui, H. Podgorski, J. Yacouba, H. (2018) 'Catena Prediction Model For Cyanide Soil Pollution In Artisanal Gold Mining Area By Using Logistic Regression', *Catena.* Elsevier, 162 (November 2017), pp. 40–50.
[7] Salas-Eljatib, C. Fuentez-Ramirez, Andres. Gregoire, Timothy. Altamirano, Adison. Yaitul, Valeska. (2018) 'A Study On The Effects Of Unblanced Data When Fitting Logistic Regression Models In Ecology', 85(August 2017), pp. 502–508.
[8] Goh, A. T. C. Zhang, Yanmei. Zhang, Runhong. Zhang, Wengang. Xiao, Yang. (2017) 'Evaluating Stability Of Underground Entry-Type Excavations Using Multivariate Adaptive Regression Splines And Logistic Regression', *Tunnelling And Underground Space Technology.* Elsevier, 70(August), Pp. 148–154.
[9] Morotti, S. and Grandi, E. (2017) 'Methodsx Logistic Regression Analysis Of Populations Of Electrophysiological Models To Assess Proarrythmic Risk', *Methods X.* Elsevier B.V., 4, pp. 25–34.
[10] Zhu, Z. Lin, C. Zhang, X. Wang, K. Xie, J. Wei, S. (2018) 'Evaluation Of Geological Risk And Hydrocarbon Favorability Using Logistic Regression Model With Case Study', *Marine And Petroleum Geology*. Elsevier, 92(February), Pp. 65–77.

IOP Conf. Series: Materials Science and Engineering **697** (2019) 012004 doi:10.1088/1757-899X/697/1/012004

Palm Oil Marketing Model through Performance Analysis Approach in Simalungun Regency

Ira Apriyanti[1], , Juwita Ramadhani[1],
[1] Faculty of Agriculture, University of Muhammadiyah Sumatera Utara, Indonesia
*E-mail : iraapriyanti@umsu.ac.id

ABSTRACT. The one of indicator that can demonstrate market performance is profit gained on industry. Market performance analysis can be seen by calculating marketing margin and farmer share. The length of existing marketing channels may not necessarily indicate that the marketing performance is inefficient. If the marketing chain is long and able to increase consumer satisfaction then it can be said that the marketing system is efficient.

Some problems on marketing of palm oil commodity in Simalungun Regency are: producer's farmers are generally small farmers who have weak bargaining power, not much know the market situation and price certainty. The market in palm oil commodities are still dominated by large farmers and other market participants who have a network or links with the consumer makes the low of farmer share.

The purpose of this research is to determine the marketing strategy of palm oil using marketing performance approach to: (a) Analyze the margin and farmer share, to know the performance of palm oil industry in Simalungun Regency. (b) Analyze the relationship between structure, conduct and performance of the palm oil industry in Simalungun Regency.

The analytical method used in this research is marketing channel analysis and analysis of market structure that seen from market concentration and barrier in and out of market, analysis of market behavior seen from marketing function that is exchange function, physical function and facility function. Market performance analysis is seen from farmer share and marketing margin.

1. Introduction

Simalungun Regency is an area of palm oil plantations belonging to the people that have an area of 1025.91 ha of palm oil plantations (BPS, 2015) [1]. With this area the problems that are often present in palm oil marketing. The form of a monopoly or oligopoly market is considered to be detrimental to farmers. This loss can be seen from the low share of farmers. Evidence of the share is assessed from the difference between the price of farmers and factories reaching Rp. 400 / kg. The second problem lies in the limitations of capital. Farmers have low capital so that the marketing function carried out is an exchange function.

Fluctuations in the price of agricultural products are also a major problem in the marketing system. Farmers often suffer losses due to price fluctuations. Price fluctuations occur because palm oil production is only concentrated in certain regions, inappropriate production patterns, inadequate facilities, and the length of the marketing chain of agricultural products (Sinaga et al., 2014) [2].

Industrial structure, conduct and performance are variables that are interconnected with each other and these three variables will influence each other. The Structure approach, Conduct, Performance (SCP) is able to explain how the steps should be taken, because knowing the structure, behavior, and market performance can be known the most appropriate marketing policy strategy to do.

IOP Conf. Series: Materials Science and Engineering **697** (2019) 012004 doi:10.1088/1757-899X/697/1/012004

2. METHODS AND MATERIALS

Market Performance Analysis

Market performance (market performance) is a condition as a result of market structure and behavior in everyday reality as indicated by prices, costs, and production volumes that provide an assessment of whether or not a marketing system is good (Dahl and Hammond 1977) [3]. Market performance analysis, namely margin and farmer share (Sinaga 2014) [4]. This calculation is done to see the difference in prices paid by consumers with prices received by farmers. Marketing margins according to Robinson (1990) [5] provide two alternatives, namely the difference in prices paid by consumers with prices received by producers and the price of a collection of marketing services as a result of the demand and supply of marketing services. Asmarantaka (2012) [6] explains the marketing margin concept is the difference in prices at the producer farmer level with prices at the final consumer level or at the retail level. Margin is the overall approach of the marketing system of agricultural products, starting from the farmer level as the primary producer until the product is in the hands of the end consumer so that it is often said to be the Total Marketing Margin (MT).

Margin can provide an overview of the activities carried out by business actors in the marketing system, namely a collection of remuneration for productive activities (adding or creating use value) in the flow of agribusiness products from the farmer to the final consumer. Margin can also show value added in agribusiness products both at the farmer (producer) level to the final consumer. Not only used as an indicator of the efficiency of the marketing system, margins can also evaluate marketing functions that occur and customer satisfaction or end products must be equal.

Sitorus (2012) [7] measured the production function and structure, conduct and performance (profitability) of the truck industry that transported agricultural commodities and frozen food products in the United States in the period 1994-2003. To achieve these objectives, estimation of the stochastic frontier production function and and structure, behavior and performance (SCP) models is measured by measuring output and efficiency as endogenous variables with Battese and Coelli specifications to test the effects of several variables, including risk, market share concentration and material expenditure fuel to measure business profitability in the form of efficiency. The results of the analysis show that the average distance, load average and market concentration variables significantly affect the efficiency of companies with 2-8 years for each company. The production function variable positively affects output, with a few exceptions.

The structure, conduct and performance approach was also carried out by Sayaka (2008) [8] in the corn seed industry in East Java Province. The results showed that the market structure of corn seed producers in East Java Province was very oligopolistic. There are 3 (three) multinational companies that dominate the industry. The corn seed producer gets a high profit / profit rate even though the risk of unsold products is quite high. Wholesalers buy and sell corn seeds at a low price level and get higher profits compared to retail traders. In general, the market for corn seeds in East Java is less efficient.

Teka (2009) [9] conducted an analysis to measure the efficiency of the market chain of papaya, garlic and tomatoes in the Alamata District, Ethiopia with a structure, behavior and performance performance (SCP) approach. The analysis shows that potential profits are under imperfect marketing conditions. Market behavior is characterized by unethical practices of fraudulent collusion and information that cause uncompetitive market behavior even though the calculation of market concentration ratios does not indicate the behavior of the oligopsonist market 24.56%. For this reason, some measurement corrections are requested by the government and other institutions such as cooperatives.

One indicator that can show market performance is the profit gained in an industry. Market performance analysis can be seen by calculating marketing margins and farmer shares. The length of the existing marketing channels does not necessarily indicate that the marketing performance is inefficient. If the chain of advice is long and is able to increase customer satisfaction, it can be said that the marketing system is efficient.

a. Revenue Analysis

Total revenue is the sum of all company revenues from the sale of a number of products (goods produced). The way to calculate total revenue can be done by multiplying the number of products with the selling price of the product and the fresh fruit of Natalia's A.J. research (2016) [10]. Formulated as follows:

TR = Y x Py .... (1)

TR = Total company revenue (Rp)
Y = Number of products produced (Kg)
Py = Selling price of fresh fruit bunches (Rp).

b. Profit Analysis

The total cost component consists of variable costs (variable costs) and fixed costs. Variable costs are costs that totally change in proportion to changes in activity, in other words variable costs are costs whose magnitude is influenced by the amount of production produced, but variable costs per unit are constant. The profit formula is as follows;

π = TR - TC .... (2)

π = Farmer's profit (Rp.)
TR = Total Revenue (Rp)
TC = Total cost (total cost) (Rp).

Labor cost formula, namely;

(P X Q) / O .... (3)

P = Price of wages (Rp)
Q = Amount of production (kg)
O = Amount of labor wages (people)

The total cost formula is;

TC = TFC + TVC .... (4)

TC: Total Cost (Total Cost) (Rp)
TFC: Total Fixed Cost (Total Fixed Cost) (Rp)
TVC: Total Variable Cost (Total Variable Cost) (Rp)

Marketing Margin

Asmarantaka (2012) [11] explains the definition of margin as a reflection of business activities or marketing functions carried out in the marketing system. In addition, margin is a collection of remuneration due to productive activities (adding or creating use value) in the flow of agribusiness products from farmers to the end consumers.

Margin can show value added from farmers until it reaches consumers. This margin analysis can be used to analyze marketing systems from a macro perspective (product marketing from farmers to consumers). The formula for marketing margins can be seen as follows:

$$MT = Pr - Pf \quad (5)$$

MT = Total Margin
Pr = Palm oil Prices at the final consumer level (Rp. / Kg)
Pf = Palm oil prices at farm level (Rp. / Kg)

Farmer share
Farmer share is the percentage of prices received by farmers at prices paid by consumers. Some things that affect the farmer share, including the level of processing, transportation costs, number of products, and product durability. Farmer shares have a negative relationship with marketing margins. If the marketing margin is getting higher, then the portion received by farmers is getting lower. Farmer shares can be calculated using the formula:

$$FS = P_f / P_r \times 100 \text{ percent} \quad (6)$$

FS = Percentage received by Farmers
Pr = Price of palm oil at the consumer level (Rp. / Kg)
Pf = Price of palm oil at the farm level (Rp. / Kg)

Market Performance Analysis
Market performance is a condition as a result of the structure and behavior of markets in everyday reality as indicated by prices, costs, and volumes of production that provide an assessment of whether or not a marketing system is good (Dahl and Hammond 1977) [12]. Market performance analysis can be seen from the marketing margin and farmer share.

3. Results and discussions

Marketing Margin Analysis
Determining the level of efficiency of a marketing system can be done through the marketing margin analysis approach. Marketing margin is the sum of all marketing costs incurred by marketing institutions and the amount of profits taken in commodity distribution activities from one marketing agency to another. The marketing margin calculated in this study is based on the marketing channel pattern that is formed in the marketing activities of palm oil fresh fruit bunches (FFB) in Simalungun Regency. In this study, marketing margins can be seen in each marketing channel.
Marketing margin is the difference between the price received by the farmer and the price paid by the agent. The margin for each channel for fresh fruit bunches (FFB) can be seen in Tables 1,2,3 and 4.
In the marketing channel I, farmers directly sell fresh fruit marks (FFB) to the factory, meaning that they do not go through intermediaries in marketing the results. Therefore, the price received by farmers is Rp. 1,403, the total cost of Rp. 961, - with a margin of 68.39 percent, farmers' profits in the marketing channel I amounted to Rp. 433 with a margin of 31.61 percent.

IOP Conf. Series: Materials Science and Engineering **697** (2019) 012004 doi:10.1088/1757-899X/697/1/012004

Tabel 1. Marketing Margin Channel I

Description	Price Spread Rp/Kg	Share Margin %
Manufacture Selling Price	1.403	
Total Cost	961	68,39
Farmer Profit	433	31,61
Amount	2.797	100
Magin		0

Source : Primary Data Analysis

In marketing channel II, farmers directly sell to large agents, and the average price received by farmers is Rp. 1,172 per kg, and the average total cost spent by farmers is Rp. 821 per kg, with a margin of 70.05 percent, with a farmer‘s profit of Rp. 351 per kg with a margin of 29.95 percent. And the purchase of a sign of fresh fruit by a large agent is Rp. 1,172 per kg with a margin of 83.53 and an average cost of Rp.129 per kg with a margin of 9.19 percent and the income of large agents as much as Rp. 102 per kg with a margin of 7.27 percent. On channel II the margin is as much as Rp. 231 with a margin of 16.46.

Tabel 2. Marketing Margin Channel II

Description	Price Spread Rp/Kg	Share Margin %
Farmer's Selling Price	1.172	
Total Cost	821	70,05
Farmer Profit	351	29,95
Amount		100
Purchase Price	1.172	83,53
Total Cost	129	9,19
Agent Income	102	7,27
Factory Price	1.403	100
Amount		100
Margin	231	16,46

Source : Primary Data Analysis

Margin on Marketing Channel III can be seen in Table 3. It can be seen that the selling price of farmers is Rp. 925 per kg. The average cost incurred by farmers is Rp.681 with a margin of 73.79 percent. Farmer's income of Rp.236 per kg with a margin of 26.21 percent. The purchase price of a small agent is Rp. 925 with a margin of 78.78 percent and an overall cost of Rp.193 per kg with a margin of 16.11 percent. And income of Rp. 67 per kg with a margin of 5.11 percent. The price difference on this marketing channel III is Rp. 478 with a total margin of 34.06.

Tabel 3. Marketing Margin Channel III

Description	Price Spread Rp/Kg	Share Margin %
Farmers Selling Price	925	
Total Cost	681	73,79
Farmer Profit	236	26,21
Amount		100%
Purchase Price	925	78,78
Total Cost	193	16,11
Small Agent Income	67	5,11
Small Agent Selling Price	1.172	
Amount		100
Prices received by large agent	1.403	
Margin of small farmer agents	247	21,07
Marjin of small agents	231	16,46
Total Margin	478	34,06

Source : Primary Data Analysis

From the statement of primary data analysis above, the marketing margin of channel I has 0, channel II has a margin of 16.46 percent, while in channel III the margin is 36.04 percent, the margin is lowest because the channel of marketing institutions is short which is on channel I (farmers), and the highest margin is in channel III because the channel glue is long (farmers - small agents - large agents - factories) The higher the marketing margin, the lower the share farmer. And the lower the margin obtained, the channel will be efficient, the efficient channel is marketing channel I, Fitriani (2014) [13].

Farmer share

Farmer's share is a comparison of the level of prices received by farmers with prices paid by end consumers. Farmer's share is a concept of remuneration for farming activities. Based on the results of the study, the magnitude of the farmer's share of oil palm farmers in Tanah Jawa Subdistrict is different for each marketing channel. The higher the farmer's share, the lower the marketing margin, Sinaga (2014) [14].

Table 4 shows the highest farmer's share is in the marketing channel I (farmers - factories) which is equal to 100 percent with the selling price at the farmer level and the same factory, which is Rp. 1,403, - while the lowest farmer 's share is in channel III (farmers - small agents - large agents - factories) which is 65.94 percent. The difference in farmer share that occurs in each marketing channel is due to differences in prices received by each marketing agency, the number of marketing institutions involved and functions performed so as to increase prices at the consumer level, Rahmawati (2013) [15].

Tabel 4. Farmer Share Analysis

Marketing Channel	Price of the Farm Level (Rp/kg)	Price of the Consumer Level (Rp/kg)	*Farmers share* (%)
I	1.403	1.403	100
II	1.172	1.403	83,53
III	925	1.403	65,94

Source : Primary Data Analysis

4. Conclusions

In this study, analysis of market performance (palm oil market performance) in Simalungun Regency shows that there are 3 marketing channels with different total margins. The relationship between behavioral structure and performance in Simalungun Regency can be seen from the Structure approach, Conduct, Performance (SCP). Relations between structure and behavior, namely farmers selling fresh fruit bunches to large agents, tight oligopoly market structure, linked to market behavior, farmers who use marketing channels II and III do not have the power to control prices, the relationship between behavior and performance is a short marketing channel (channel I) has a high farmer share, the relationship between structure and performance is the market used is a tight oligopoly market then the marketing margin is high.

References

[1] [BPS] Badan Pusat Statistik. 2015. Statistik Kelapa Sawit Indonesia. Medan.

[2] Sinaga et al. 2014. Fluktuasi Harga pada Produk Pertanian. Yogyakarta.

[3] Dahl DC, Hammond JW. 1977. Market and Price Analysis. Mc. Graw Hill, New York.

[4] Sinaga et al. 2014. Fluktuasi Harga pada Produk Pertanian. Yogyakarta

[5] Robinson, 1990. Agricultural Product Prices. Cornell University Press, Ithaca.

[6] Asmarantaka RW. 2012. Pemasaran Agribisnis (Agrimarketing). Departemen Agribisnis, Fakultas Ekonomi dan Manajemen. Institut Pertanian Bogor.

[7] Sitorus, A.S. 2012. Analisis Struktur, Perilaku, Dan Kinerja Industri Kakao Di Indonesia. Departemen Manajemen. Fakultas Ekonomi dan Manajemen. Institut Pertanian Bogor.

[8] Sayaka, Bambang. 2008. Memperbaiki Kinerja Pengelolaan Pemasaran Produk Pangan. Litbang Pertanian.

[9] Teka.2009. https://pdfs.semanticscholar.org/7911/340853961adcaa422175b7b53469265513dd.pdf

[10] Natalia, A.J. 2016. Analisis Kelayakan Finansial Usahatani Kopi Arabika (Coffea arabica). Program Studi Agribisnis Fakultas Pertanian Universitas Sumatera Utara. Medan.

[11] Asmarantaka RW. 2012. Pemasaran Agribisnis (Agrimarketing). Departemen Agribisnis, Fakultas Ekonomi dan Manajemen. Institut Pertanian Bogor.

[12] Dahl DC, Hammond JW. 1977. Market and Price Analysis. Mc. Graw Hill, New York.

[13] Fitriani A, Daryanto HK, Nurmalina R, Susilowati SH. 2014. Struktur, perilaku dan kinerja industri broiler indonesia: pendekatan model simultan. Jurnal Agro Ekonomi 32(2): 167–186. https://doi.org/10.21082/jae.v32n2.2014.167-186.

[14] Sinaga et. al. 2014. Fluktuasi Harga pada Produk Pertanian. Yogyakarta.

[15] Rahmawati. 2013. Analisis efisiensi pemasaran nenas studi kasus di Desa Cipelang, Kecamatan Cijeruk, Kabupaten Bogor. Skripsi. Institut Pertanian Bogor.

IOP Conf. Series: Materials Science and Engineering **697** (2019) 012005 doi:10.1088/1757-899X/697/1/012005

Analysis of product quality control using six sigma method

B.A.D. Putri[1], Qurtubi[1*] and D. Handayani[1]
[1]Department of Industrial Engineering, Universitas Islam Indonesia, 55584 Yogyakarta, Indonesia
*E-mail: qurtubi@uii.ac.id

Abstract. Quality control is the activity of management to measure the quality characteristics of the product, compare it with specifications or requirements, and take appropriate action if there is a difference between the actual appearance of the standard. Every company needs an operational system that can indicate and overcome defective or damaged products, the company must control quality in each of its production processes. This study aims to determine the quality control of production and actions that should be done by the company to reduce product failure. The research subject used is a case study at the company, this manufacturing company produces bottled water products. The data used in the study include primary and secondary data, and data collection methods used are interviews, observation and documentation. Analysis of the data used in this study is the Six Sigma method which includes five stages of DMAIC analysis namely define, measure, analyze, improve and control. This study revealed that there are five types of defects namely lid defects, cup defects, sliding machine, volume defects, and dirty water. Based on the calculation of sigma value, the average sigma value of company is 4.39 with 2,047.96 Defect per Million Opportunities (DPMO). Fishbone diagrams are carried out to analyze the causal factors of defective products, namely machinery, humans, methods, and materials.

1. Introduction

The stronger competition, provision of products, and quality services become competitive advantages of a company [1]. Every exists company must be able to maintain and increase the efficiency and effectiveness with the aim of remain their competency and competitiveness [2]. In business competition, the company's ability to get customer needs is very important. The company's ability to get customer needs itself is greatly influenced by the level of quality provided by a company to its customers such as product quality, price and timeliness of product delivery. Quality is the most basic thing from customer satisfaction and success in competition. In fact, quality is the thing that should be for all sizes of companies and for the purpose of developing quality practices and showing consumers that they are able to find higher expectations of quality [3].

The production process is an activity to process raw materials into finished products that have a high selling value, but in the production process there are product defects or product damage is something that cannot be avoided.

In the production process to be able to produce good products and meet established quality standards, quality control is needed. Quality control is very useful for the production process so there is no high disability. In addition to the disability of the value of a product, the product will experience a decrease in selling value because the company lacks knowledge of quality control. In addition, the cause of the production process has decreased can be caused by three categories, namely caused by humans, machinery and the environment.

Quality control can be applied with one method, namely Six Sigma. Recent history of quality development said that the quality improvement Six Sigma has been proved effective [4]. Six Sigma represents a parameter that measures the variance of the statistical distribution, namely standard deviation [5], traditionally sigma variables are used to measure variance in a production process [6]. Six sigma terminology states that the sigma level is reflected as a company performance [7].

Six Sigma itself is a method used in business management strategies that can increase the business of the company periodically by design and observe business activities to reduce waste and resources, to increase customer satisfaction. To reduce the number of defects, you can use the DMAIC concept, otherwise known as the define, measure, improve and control cycle. Through optimal process parameters, this method can reduce company costs to save business [8].

This research was conducted at a company engaged in the manufacture of bottled water products, to determine the quality of the product and the type of defects produced. In addition, this study uses a causal diagram method to find out the causes of defects so that it can provide suggestions for improvement to improve the quality of the product.

2. Methods

2.1. Preliminary research
Preliminary research was conducted by conducting a survey to the company. Preliminary research is carried out in two ways, namely the first observation of the company's business processes and general study of the company through preliminary research, it can be seen also the business processes that are running at the company. In addition to conducting direct observation, discussions were also held with the company.

2.2. Defining the problem
After making direct observations and interviews, several problems were encountered by the company. The problems found will be appointed as an identification of problems that will be solved in this study.

2.3. Literature review
After defining the problem, a literature study will be carried out related to the problems taken as the research topic. Library studies carried out are using literature books, journals, or the internet. Literature studies carried out are expected to be a foundation for solving problems that exist in the company.

2.4. Data collection
At this stage, data collection is needed to solve the problem. Data collection is done both directly and indirectly. Direct data collection is done by interviewing the parties related to the existing problems and further direct observation of business process activities. Indirect data collection is done by retrieving data that has been stored by the company. The data collected included data and defects production of 220 ml bottled water.

2.5. Data processing
At this stage through the DMAIC process, the final result of the sigma value is obtained. DMAIC is a method of measuring sigma values consisting of define, measure, analysis, improve and control. Define is the first step in improving quality where problems are identified. Measure is a measurement activity of the previous process (basic measurement), which aims to evaluate based on existing goals. In this step the information or data is collected. Some of the tools used in this step include using control charts, data collection forms, flow diagrams, pareto diagrams, scatter diagrams, frequency plots. Analyze is the stage where identification of the root cause of the problem is carried out based on data analysis. The results of the analysis can be used to create solutions in developing and improving the observed processes. Improve is the stage where testing and implementation of solutions is done to eliminate the causes of problems that exist and improve existing processes. And control is the last step taken in

improving quality using DMAIC. This final step aims to exercise control in every activity, so that good results are obtained and can reduce time, problems, and costs that are not needed.

Table 1. Explanation of Achieving Sigma Levels

Sigma Level of Achievement	DPMO	Information
1-*sigma*	691.462	Very Uncompetitive
2-*sigma*	308.538	Uncompetitive
3-*sigma*	66.807	Indonesian Industry Average
4-*sigma*	6.210	USA Industry Average
5-*sigma*	233	Japan Industry Average
6-*sigma*	3,4	World Class Industry

2.6. Analyze and evaluation
This stage analyzes and evaluates the production process, product quality, and identifies the causes of product defects that occur.

2.7. Conclusion and recommendation
In this final stage, the author concludes the results of data processing and also analysis to answer various problem formulations that have been determined in the previous stage. In addition, a number of suggestions were expected to be useful for the company.

3. Results and discussion

3.1. Define
At the define stage, we defining the types of defects and determining CTQ. In the production of bottled water company there are several types of defects that occur in the production process. The following are types of defects: (1) Lid defects. The lid is the top cover of the product, the lid is not made by the company but ordered from another company, the lid defect is in the form of a lid on a cup that has been filled with water that is incorrect or inappropriate, namely: lid leak, tilt lid, lid inside and lid outside. (2) Cup defects. The defect is a dent cup and an empty cup without lid. (3) Sliding machine. Defective production process because the product that has been through the cutting process does not want to fall into the conveyor still attached to the holder so that the product will fall on the gutter. (4) Volume defects. The standard volume of filling a 220 ml bottled water product is exactly 220 ml. Volume defects in the form of less volume and more volume, the product is said to be less volume when the product volume is less than 220 ml with a lower limit not exceeding the line in the cup, the product is said to be more volume when fully charged without a cavity. (5) Dirty water. Defects that occur in finished products in which there is dirt due to the powder from the cup and powder from the lid.

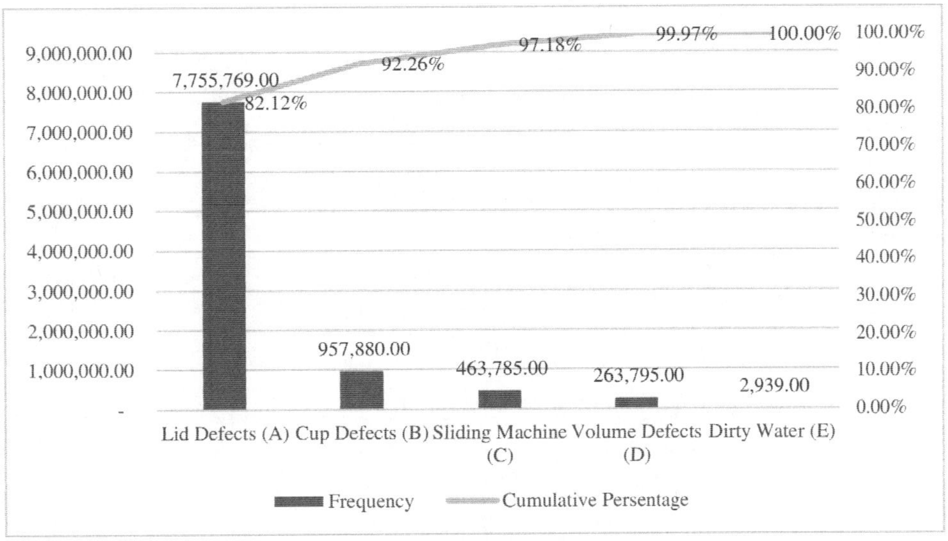

Figure 1. Pareto diagram of the type of defect

Based on data above it can be seen that the highest type of defect is lid defects (A) with a frequency of 7,755,769 units and the percentage of the whole is 82.12%. Followed by the type of cup defects (B) as much as 957,880 units with a percentage of 10.14%, then sliding machine (C) as many as 463,785 units with a percentage of 4.91%, volume defects (D) as many as 263,795 units with a percentage of 2.79%, and the dirty water (E) is 2,939 units with a percentage of 0.03%. So the number of CTQ obtained is 5 types of defects.

3.2. Measure

In the measurement process, a mapping of DPMO and sigma value per period was carried out, then a P control chart was mapping for all periods.

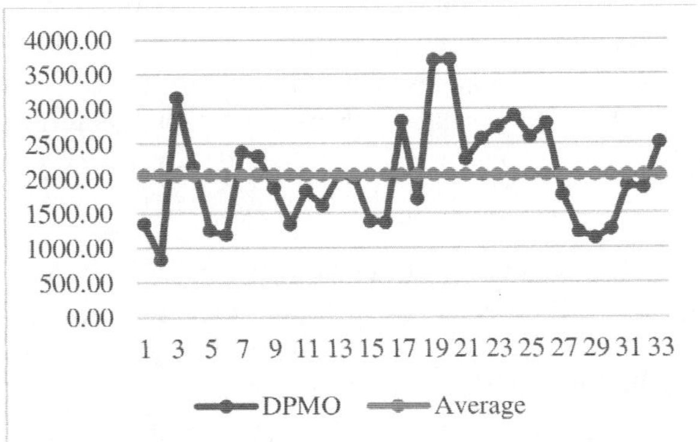

Figure 2. DPMO value pattern graph

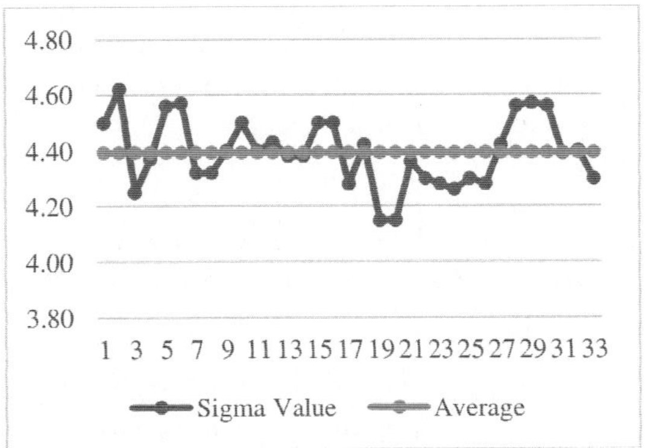

Figure 3. Sigma value pattern graph

Based on Fig. 3 and Fig. 4 show that there is a lack of process stability. For example, in the 19th and 20th periods there was the highest DPMO value which also caused a low sigma level value. The higher the sigma level, the better product manufacturing process carried out in the table is shown by the average DPMO = 2,047.96 and sigma value = 4.39.

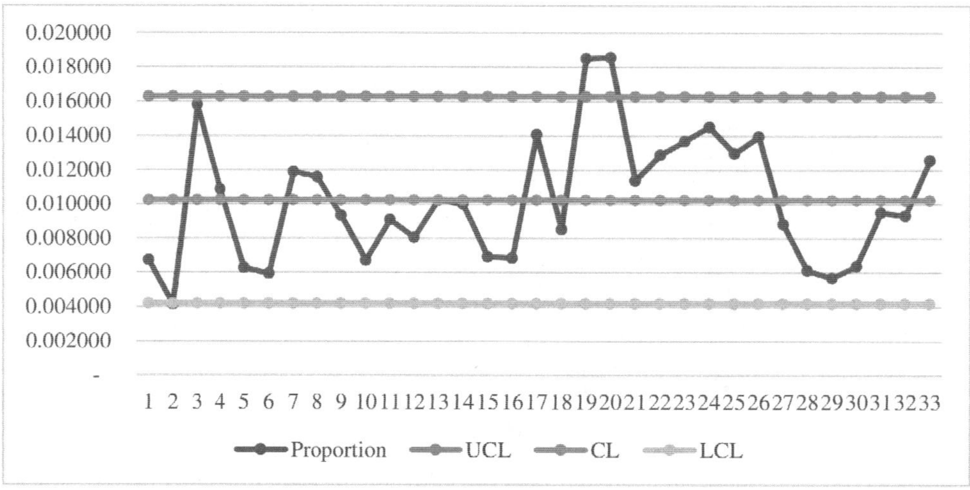

Figure 4. P control chart

On the p control chart above, it can be seen that the proportion of defects is included in the controlled category, but there is instability in the process. In the 19th and 20th periods, the proportion of defects appeared to exceed the upper limit. This is because the number of defects in the period is different from the others which tend to be at the control limit.

3.3. Analyze
In the analyze stage, DPMO analysis is carried out so that a sigma value will be obtained, then an analysis of production defects will be carried out.

3.3.1. Data analysis of DPMO and sigma value.
In this study there were five types of defects that characterize the potential CTQ namely lid defects (A), cup defects (B), sliding machine (C), volume defects (D), and dirty water (E). Based on the calculations performed, the average value of the production is 28,185,439 units with a sigma value of 4.39 and the average value of DPMO is 2,047.96, which means that the company production process of 220 ml bottled water is categorized at USA industry average level. In addition, based on the graph of the DPMO value and the sigma level it appears that there is instable. This shows that the production process has not been done high consistently.

3.3.2. Analysis of product defect causes.
In this stage the analysis carried out the product defect causes in bottled water products 220 ml. The type of product defect to be analyzed for the root cause is lid defects (A) with the highest number of defects, contributes 7,755,769 units from a total of 9,444,168 units, because this type of defect has the largest contribution of 82.12%. Causal analysis is carried out using a causal diagram which is often referred to as a fishbone diagram. The following are the results of a causal diagram of lid defects (A) in the process of producing bottled water 220 ml:

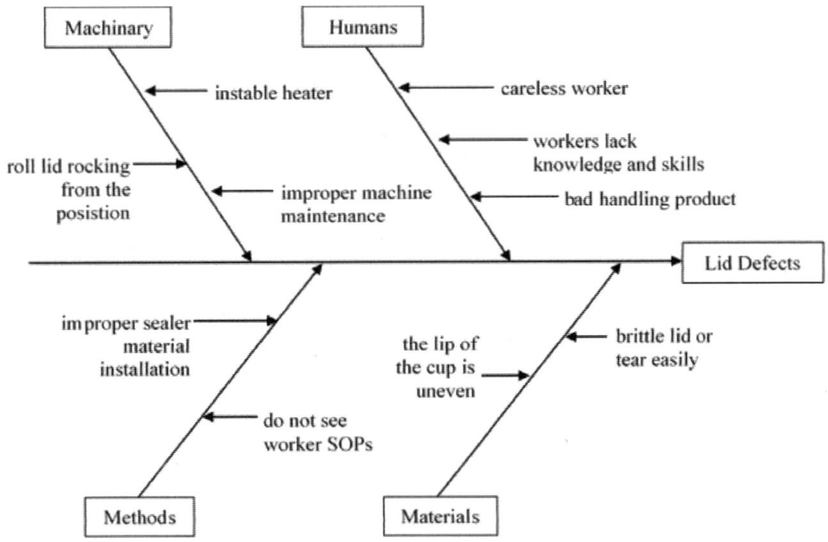

Figure 5. Fishbone diagram of lid defects

3.4. Improve

Based on the results of fishbone diagrams, proposed improvements are made of root causes of the 220 ml bottled water are machinery, humans, methods, and materials. The proposed improvements in accordance with each cause of the problem are as follows: (1) Machinery. Proposed improvements to the machine by conducting regular and scheduled inspections and maintenance. Engine components must be available so that when the engine experiences damage, it is replaced immediately without any delay in product production. (2) Humans. To increase focus and lack of careful work can be done research on the work environment. Proposed improvements for humans are by increasing workers' knowledge by conducting training (training for workers). (3) Methods. Proposed improvements for the method, namely by monitoring the workers who do not comply with existing SOPs. (4) Material. For material errors proposed improvements that can be applied is to standardize the material to be used and meet company requirements.

3.5. Control

Stage of control or take control measures for suggested improvements. As for several control measures that will be proposed, as follows: (1) Check before the production process. This check aims to prevent interference with the engine during the production process where this inspection is carried out on all engine components. (2) Proper guidance, strict supervision, and discipline. Due to the discovery of defective products, it is necessary to have supervision that aims to avoid high product defects caused by irresponsible workers. (3) Creating a work team. Aside from the products produced by the company benefiting from workers, this teamwork can add to the sense of belonging, responsibility and can strengthen openness, sharing and communication. (4) Monitor the course of production and analyze any problems that exist in the production floor by all workers involved in the problem. When it is found that a problem in the handling process of the problem can be dealt with as early as possible with the cooperation of all workers involved (work team) or interpreted as an act of supervision and prevention.

4. Conclusions

The results of product quality research conducted at the company produce several conclusions. The sigma value obtained is 4.39 which means that the company has a product quality equivalent to the USA industry average and is good enough. However, the production process has not run stably and still deviates from the upper and lower limits. The types of defects that occur are lid defects (A), cup defects

IOP Conf. Series: Materials Science and Engineering **697** (2019) 012005 doi:10.1088/1757-899X/697/1/012005

(B), sliding machine (C), volume defects (D), and dirty water (E). The causes of product defects caused by several factors such as machinery, humans, methods, and materials. Improvements that can be made are regular and scheduled machine checks and maintenance, conducting training for workers, and providing consistent supervision of workers who do not see or do their work in accordance with the SOPs.

5. Recommendations

Based on the results of the research that has been obtained, there are some suggestions for the company. The company can implement the proposal given in this study by exercising good control over the DPMO values and sigma levels regularly and continuously. The company can focus on companies with zero defects in order to compete with foreign companies by making continuous improvements, analyzing product quality properly, and conducting supervision and control in order to satisfy customers and in accordance with the targets expected.

References

[1] Jirasukprasert P, Garza-Reyes JA, Kumar V and Lim MK 2014 *Int. J. Lean Six Sigma.* **5(1)** 2-21
[2] Abbas 2009 *J. American Academy of Business.* **14(2)** 301-310
[3] Tata J, Motwani J and Prasad S *Multinational Business Review.* **8** 37-42
[4] Thakore R, Dave R, Parsana T and Solanki A 2014 *Int. J. Engineering Research and Applications.* **4(1)** 63-69
[5] Garrido-Vega P, Sacristan-Diaz M and Magana-Ramirez LM 2016 *A successful experience inaeronautics, Universia Business Review*
[6] Omachonu VK and Ross JE 2004 *CRC Press, Florida*
[7] Pyzdek T and Keller PA 2010 *McGraw-Hill Companies Inc., New York*
[8] Teplicka K, Culkova K and Zelenik O 2015 *Polish J. Management Studies.* **12(2)** 170-178

IOP Conf. Series: Materials Science and Engineering **697** (2019) 012006 doi:10.1088/1757-899X/697/1/012006

The reduction of waste on pile production process using value stream analysis tool (VALSAT) method

M.F. Septiani[1], Qurtubi[1*] and V.N. Helia[1]
[1]Department of Industrial Engineering, Universitas Islam Indonesia, 55584 Yogyakarta, Indonesia
[*]E-mail: qurtubi@uii.ac.id

Abstract. This paper discusses about the reduction of waste on the pile production process in PT. XYZ, a company that run its business in the field of engineering, production installation (EPI) of a concrete industry. The methods of data collection are observation and interview. The analysis that will be employed is Value Stream Analysis Tool (VALSAT). Based on the value stream mapping, it is identified that the cycle time data for product's manufacturing is 5.847 seconds with lead time of 1.380 seconds. Based on the result of observation and interview, it is identified that the biggest waste is resulted from the waste of excessive transportation. The movement process is executed on the same workstation and among workstations. While, the result of process activity mapping indicates that 27,6% of all activities are included in transportation activities. By conducting the mapping using fishbone diagram, later could be identified the factors that influence waste, in which, from human resource perspectives, it is caused by the less-evenly workers distribution at line 2 on pile production, unbalance production process that causes idle time or idled operators, the lack of understanding and workers' skill on machine utilization as well in handling the problems. In the perspective of method, the issue lies on layout factors and 5S system that runs inappropriately at line two. In terms of machine, it is stressed on the lack of numbers of tools or machines as well as the less upgraded machines. From the perspective of environment, rain definitely could affect the condensation maintaining process and evaporation.

1. Introduction

PT XYZ performs the concrete production process that facilitated with 4 plants/production line. This research is focused on line 2, the making of pile. At this line 2 production process, waste is found and could be minimized. The waste is emerged due to transportation process inside a workstation or among workstations. Hence, this research is carried out to optimize production process and to eliminate waste by employing lean system.

Lean manufacturing is one of the forms of line thinking to mitigate the waste, which is defined as an activity that absorbs resources but does not come up with value. Lean thinks about determining a value, arranging the actions to create best value and performing activities with less efforts, less machines, less time and minimum cost yet able to fulfil customers' preferences [1]. Waste can be defined as a work activity that unable to add value on production process. There are two types of waste, which are Type One Waste and Type Two Waste. Type One Waste is an activity that unable to add value on production process, but still cannot be eliminated for several reasons. While, Type Two Waste is an activity that unable to add value but could be eliminated [2].

Process activity mapping describes real situation on production process. The advantages are to clarify complicated process, highlight non-added value activities and to consider the improvement [3]. Value stream mapping is a technique to illustrate entire process in the company [4]. The process is noted with certain symbol. The objective of VSM is to identify production process, so it could run smoothly without significant issues, as well as to improve the productivity [5].

Based on above explanation, if lean manufacturing is implemented in a company, it will reduce non-added existing processes. Hence, this research is performed to identify waste at the company's production line and try to mitigate it by providing improvement recommendation that could be applied by company.

2. Data collection method

Data collection method that applied in this research is observation and interview. Observation is conducted by direct examination, while interview is performed by direct interview with relevant sources.

3. Result and discussions

Through the making process of current state value stream mapping, it could be identified the description of production process. It not only covers the production process but also information flow on the quantity of operators, time of process, total cycle time and lead time, or on other words, it involves existing entire business process. Based on the current state value stream mapping, it is identified that the cycle time of product is 5.847 seconds and the lead time process is 1.380 seconds.

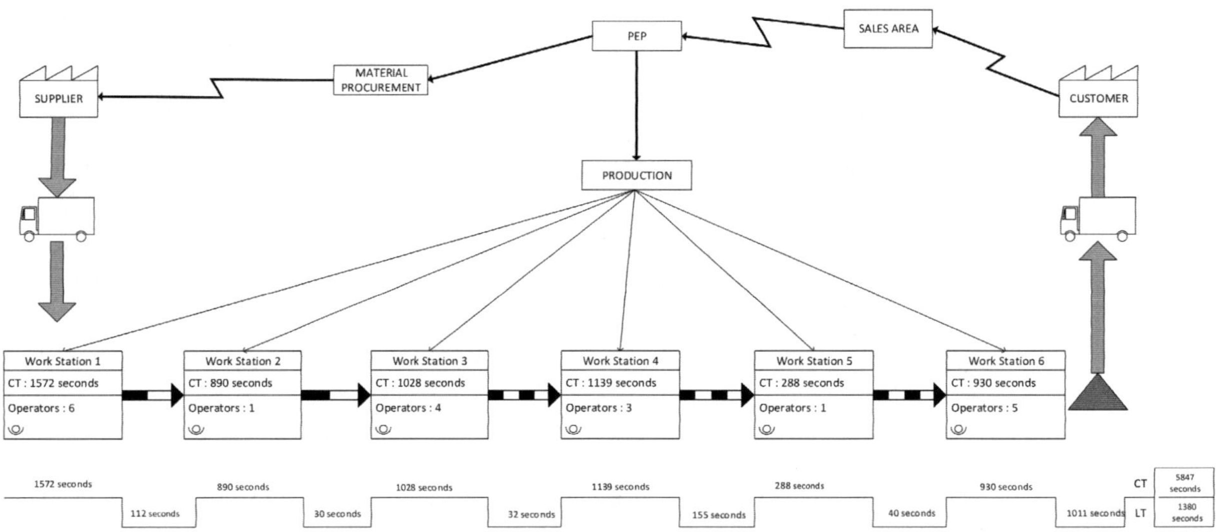

Figure 1. Current state value stream mapping of pile

3.1. The result of waste identification

The waste is classified into 7 (seven) types, which are, overproduction, defect, unnecessary inventory, inappropriate processing, excessive transportation, waiting and unnecessary motion. From data collecting or interview with experts, it is confirmed that the biggest waste of all pile's production processes is excessive transportation due to frequent movement of material or tools in the production process.

3.2. The result of identification on process activity mapping

It is found from the result of waste identification that waste of excessive transportation is the biggest waste produced. Correlation between waste of excessive transportation with the VALSAT tool can be depicted from the figure that has the biggest association, which is process activity mapping tool. Each activity in production process later be classified into group of operation, transportation, inspection, storage and delay. After being identified to specific group, further, they will be classified to activities

IOP Conf. Series: Materials Science and Engineering **697** (2019) 012006 doi:10.1088/1757-899X/697/1/012006

with value added (value added), activities with non-value added (non-value added) or activities that considered important with have no value added (necessary non-value added).

Based on the result of calculation, it is obtained the percentage for each activity; operation is resulted with 55,23%, transportation is 27,6% and delay is 17,17%. While for the classification of value-added existence, it is confirmed that value added activities resulted with 50,66%, 34,88% for non-value added activities and 14,46% for necessary non-added value. The recapitulation of process activity mapping could be seen in Table 1.

Table 1. Recapitulation of process activity mapping

Activity	Quantity	Total Time [seconds]	Percentage
Operation	17	3992	55,23%
Transportation	11	1994	27,6%
Inspection	0	0	0%
Storage	0	0	0%
Delay	3	1241	17,17%
VA	14	3661	50,66%
NVA	14	2521	34,88%
NNVA	3	1045	14,46%

3.3. The result of fishbone diagram identification

On the fishbone diagram, the causes of waste are clustered into 4 (four) categories; which are man, machine, method and environment. First cause of waste is man; the quantity of workers at the line 2 are considered as less-evenly in distribution. It means that there is a process that highly depended on human or machine. Frequent idle time or idle operators is emerged due to the unbalance production process. Besides, man factor is contributed to the waste due to the lack of understanding or skill in machine utilization and less expert in handling issues while working.

In the production process, issues happen frequently. For instance, the size of joint plate hole that matchless the mold. It is considered as significant obstacles since the workers will realize after joint plate have been installed at the iron reinforcement before being moved to concrete mold, and only several workers could handle this kind of problem. The second cause of waste is method; due to layout factor and inappropriate 5S system. Layout of line 2 in this production process is considered as narrow for handling the size of concrete products, at the same time 5S system run inappropriately by considering the simple things, such as unorganized tools.

The third cause of waste is machine; the lack of machine's quantity and the existence of less upgraded machines. It could easily be notified from the spinning machines or the compression of the machine, if it is improved with one type of single spinning machine, it could boost the production up to 10-15 additional products, batching plant that still integrated with line 1 causes longer waiting process for casting station, so it needs vacuum machine to pull out products from mold for effectiveness and efficiency purposes. The fourth cause is environment factor; the rain that could affect condensation maintaining process or evaporation. It is caused by low temperature and humidity that influence time of evaporation, besides, heavy rain could infiltrate the steam vessel.

IOP Conf. Series: Materials Science and Engineering **697** (2019) 012006 doi:10.1088/1757-899X/697/1/012006

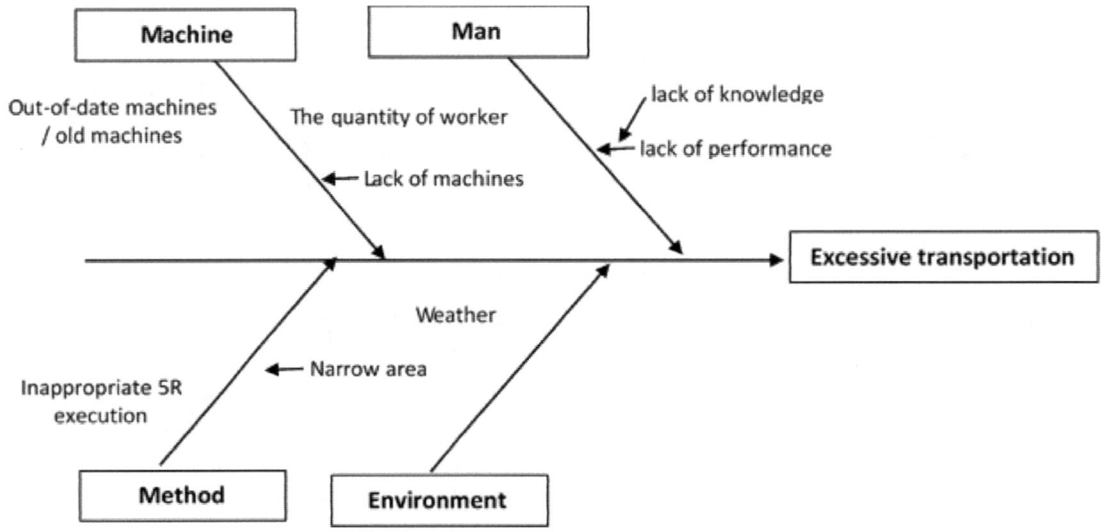

Figure 2. Fishbone diagram on waste of excessive transportation

3.4. Improvement recommendation
Based on the result of observation and interview with experts, it can be resumed that the waste in this production process could be identified as waste of excessive transportation. After the causes of waste are derived from fishbone diagram, hence improvement recommendations are suggested to reduce waste on the pile production process. Below is the improvement recommendation to mitigate the waste of excessive transportation.

3.5. The addition of tools/machines
The requirement of tools/machines is described as follows: the addition of crane and trolley to shorten the transportation time, hence the queueing for using tools could be removed from the process. The importance of disintegrating the batching plant with line 1 is to cut the waiting time on the casting process. The addition of spinning machine single type could improve the quantity of products up to 10-15 per day and reduce idle operators. The procurement of release and vacuum machine for work station 6 or mold opener could possibly reduce the process duration.

3.6. Training for workers
The training is designed to improve workers' ability. Training is divided into two types, which are soft skill training and technical training. Soft skill training delivers the knowledge on working in team and the way to communicate, while the technical training focuses on the ability to solve the problems and ability to master the tools or machines. Training is carried out gradually by the company. Not only executing the training, but the controlling towards the improvement of workers' ability and skills should be conducted as well by the company.

3.7. The implementation of optimal 5S
The implementation of 5S system at the production line, should be intensified. Such as the arrangement for placing tools/equipment to ease the searching whenever they needed. This could accelerate the workers in finding the right tools.

3.8. The arrangement of optimal layout
If the improvement of layout is possible, then the company needs the area expansion particularly for this production line. By considering bigger size of products require spacious room to store half-finished

IOP Conf. Series: Materials Science and Engineering **697** (2019) 012006 doi:10.1088/1757-899X/697/1/012006

products before being forwarded to the next process. It is designated to avoid excessive transportation in production process by utilizing space at the empty space.

4. Conclusions

Based on the results and discussion, it could be concluded that on the value stream mapping, it can be notified from the existing current state value stream mapping, cycle time for product is 5.847 seconds and the lead time of a process is 1.380 seconds. From the observation and interviews with the experts, indicated that the biggest waste is derived from the waste excessive transportation. It happens on the same workstation or among other workstations. The result of process activity mapping indicates that 27,6% of all activities are included in transportation activities.

The Mapping by employing fishbone diagram has resulted the causes of waste, which are: man; the quantity of workers at the line 2 are considered as less-evenly in distribution. It means that there is a process that highly depended on human or machine. Frequent idle time or idle operators is emerged due to the unbalance production process. Besides, man factor is contributed to the waste due to the lack of understanding or skill in machine utilization and less expert in handling issues while working. Method; due to layout factor and inappropriate 5S system. Layout of line 2 in this production process is considered as narrow for handling the size of concrete products, at the same time 5S system run inappropriately by considering the simple things, such as unorganized tools. Machine; the lack of machine's quantity and the existence of less upgraded machines. It could easily be notified from the spinning machines or the compression of the machine, if it is improved with one type of single spinning machine, it could boost the production up to 10-15 additional products, batching plant that still integrated with line 1 causes longer waiting process for casting station, so it needs vacuum machine to pull out products from mold for effectiveness and efficiency purposes. Environment; the rain that could affect condensation maintaining process or evaporation. It caused by low temperature and humidity that influence time of evaporation, besides, heavy rain could infiltrate the steam vessel.

The efforts or improvement are recommended as follows: The addition of tools/machine, the addition of specific tools/machines, such as vacuum machine, release machine to remove the screw, spinning machine to reduce duration of production. Besides, it needs additional lifting tools to anticipate limited numbers of crane. Training for workers; it is important to provide training to workers to enhance their skill and minimize human error while working. Optimal implementation of 5S (Seiri, Seiton, Seiso, Seiketsu, Shitsuke), as it need to be intensified for efficient and organized production process. Optimal layout; if the improvement of layout is possible, then the company needs the area expansion particularly for this production line. It is designated to avoid excessive transportation in production process by utilizing space at the empty space.

References

[1] Chauhan G 2016 *Int. J. Organizational Analysis.* **24** 107-122

[2] Gaspers V 2007 *Jakarta: Gramedia Pustaka Utama*

[3] Pude CM, Naik PR and Naik DG 2012 *Int. J. Modern Engineering Research*

[4] Rother M and Shook J 1998 *Massachusetts: The Lean Enterprise Institute*

[5] Womack J, Jones D and Roos D 1991 *New York: Harper Perennial*

IOP Conf. Series: Materials Science and Engineering **697** (2019) 012007 doi:10.1088/1757-899X/697/1/012007

Supplier selection and determination of quantity of raw material orders using analytical hierarchy process (AHP) and linear programming: A preliminary study

D. Handayani[1], Qurtubi[1*] and M. F. Alfaris[1]
[1]Department of Industrial Engineering, Universitas Islam Indonesia, 55584 Yogyakarta, Indonesia
[*]E-mail: qurtubi@uii.ac.id

Abstract. This article discusses the preliminary study about the selection and determination of order quantities of raw materials using the Analytical Hierarchy Process (AHP) and Linear Programming methods. AHP is used to carry out the weighting of each criterion and sub-criteria so that suppliers are found to be in accordance with the company, while linear programming is used to determine optimal raw material orders from each supplier that has been sorted. The purpose of writing this article is to provide a basis for the research that will be conducted. The method used in writing this article is a review literature. Based on inductive studies it can be concluded that the research that will be conducted has a difference with previous similar studies, namely the method used and the criteria set.

1. Introduction

There are three dominant industries that affect Indonesia's economic growth, namely manufacturing, mining, and agriculture. The manufacturing industry contributes to Indonesia's economic growth including employment, exports, and the National Gross Domestic Product. In 2013, the contribution of the manufacturing industry to the National Gross Domestic Product was 20.8% [11]. The number of companies in the manufacturing sector has led to the need for globalization and technology because it affects the manufacturing system and improves the company's business performance [3].

Higher business performance makes multi-product companies must be prepared for changes where there is a dynamic demand for agility, flexibility and product quality [6]. Now many processes, activities and services are outsourced to external providers to support the dynamic business environment in which the organization operates, thus increasing the complexity of inter-organizational and supply chain activities. Supply Chain Management (SCM) is intended to improve organizational sustainability and performance, and is supported by top management because it supports the execution and integration of suppliers, companies, customers and distribution networks [14].

Increasing competition and to continue to add value, places special emphasis on integration with suppliers, where suppliers assist with product and product innovation that is sustainable [16]. Then from the five main things in the Purchasing Supply Management (PSM), supplier management has the highest value of 0.72, then cross-functional integration 0.71, strategy development 0.68, human resource management 0.70, PSM control 0.71 [8].

The company must choose the most appropriate supplier, because the selection of suppliers that significantly reduces raw material costs and increases the competitiveness of the company, but inaccurate supplier selection can cause financial and operating problems. On the other hand, supplier selection can make the company more efficient and can produce higher quality products [17].

The problem of supplier selection and evaluation can be solved by several methods such as the total cost approach, mathematical programming methods, statistical methods, artificial intelligence methods, and linear weighting methods [12]. Based on the literature survey it can be explained that among the various methods and techniques for solving the problem of the supplier selection process, a model based on AHP and its combination is the most extensive method in the literature. The literature survey also explains that the AHP-based model can be combined with many other methods [1].

Other research on supplier selection and order allocation was carried out [9]. The purpose of the study was to use fuzzy analytic hierarchy process (AHP) to choose the best supplier company that gave the highest satisfaction for the specified criteria.

In a Turkish case study, purchasing managers from producers of goods were interviewed and given questionnaires to calculate the most important criteria managers use when choosing suppliers. To compare supplier companies, Fuzzy AHP is used. To create a good relationship between buyer's company policy and supplier selection, the Dynamic Total Value Purchase Model is recommended. [5]. This model indicates the number of orders to suppliers so that the total value of purchases (Total Value Purchasing) becomes maximum using integrated AHP and linear programming. This model also allows management to tradeoff between several real and unreal factors with different priorities.

Then the next study discusses the selection of green suppliers and the allocation of orders using discounts. Where this research is the availability of suppliers each period varies. The use of methods in this study using fuzzy TOPSIS for each weighting criterion then order allocation and discount are determined using integer linear programming [7].

The integrated approach of the analytical hierarchy process is enhanced by a multi-objective mixed linear integer program and the fuzzy set theory is proposed to determine the number of suppliers and the number of orders allocated to these suppliers simultaneously in several products, several sources, with supplier capacity constraints and several criteria. In this context, the supplier offers a discounted price for the total volume of the order [18].

Therefore, the research that will be carried out is about choosing suppliers using a method different from the others, namely the AHP method for decision making at suppliers, then for optimal order allocation, it is done using Linear Programming.

2. Research method

The method used in this study is the literature review. The review was conducted by exploring 18 papers related to the topic of the research to be conducted.

3. Result and discussion

This section discusses literature studies and the theoretical foundation used in the research that will be conducted. In addition, it presents a summary of the results of research that has been done before by other researchers who have links with this research.

3.1. Deductive study

3.1.1. Supply chain management (SCM)

Supply chain management requires traditionally separate material functions to be conveyed to an executive who has the responsibility to coordinate all process materials, and requires joint relationships with suppliers at various levels. Supply chain management is a concept that aims to manage and integrate material, flow, and material control with a total system perspective at various levels of suppliers and various functions. (Monczka et al., 1998).

According to Pujawan [13], the main activities included in the SCM classification are activities to design new products (product development), activities to obtain raw materials (procurement, purchasing, and or supply), activities for planning production and inventory (planning & control), conducting activities production (production), activities to carry out delivery / distribution (distribution), return management activities (return).

3.1.2. Purchasing
In the field of purchasing, purchase planning is done to determine the actions to be taken for future needs and actions taken to meet those needs. Purchasing planning is a cycle process of integrating overall purchases in the company's planning system.

Purchase planning is the formulation and implementation of a purchasing strategy. The purchasing strategy consists of the source, time, and life-cycle strategies. The source strategy is to pay more attention to the type and number of vendors they will use in the procurement of goods. The time strategy is more directed to the problem when the item will be received and the volume of goods to be accommodated in the company. Life-cycle strategy more adjusts the behavior when purchasing in the condition that the product will reach the consumer share [15].

3.1.3. Analitycal hierarcy process
The AHP method is a framework used in making decisions for complex problems by simplifying and speeding up the decision making process. The trick is to solve the problem into parts, then these parts are arranged in a hierarchical manner, then provide numerical values for subjective considerations about the importance of each variable and synthesize these subjective considerations to determine which variable has the highest priority and influences the results.

The AHP method helps solve complex problems by structuring a hierarchy of criteria, stakeholders, results and by taking considerations to develop priorities or weights. The AHP method combines the power of logic and feeling in a variety of problems, then synthesizes considerations into results that are intuitively predictable [4].

3.1.4. Linier programming
Linear programming is a mathematical technique used to find optimal solutions to problems that can be expressed using inequalities and linear equations. If the problem can be represented mathematically from a linear program accurately, this method will find the best solution. In a series of linear functions, there are only a few complicated problems that can be perfect. A linear program will provide a realistic representation of various problems, if little creativity is applied in the mathematical formulation of the problem [2].

3.2. Inductive study
Inductive studies contain affirmation of the state of the art or positioning of the research conducted compared to similar or previous research. The differences confirmed here can be reviewed based on the method used or supplier selection criteria. Differences from similar studies can be explained through Table 1.

Table 1. Similar research differences

No	Author	Method	Criteria
1	[9]	Fuzzy AHP	product performance, cost, quality
2	[5]	AHP Fuzzy & Linear Programming	Supplier flexibility, price, probability of sending supplier error
3	[7]	Fuzzy TOPSIS, Integer Linear Programming	Cost, flexibility, shipping, quality, payment method
4	[18]	Two-phase fuzzy & multi-objective linear programming (FMOLP).	Purchase fees, quality message fees, delivery time

4. Conclusions

The research that will be carried out is about choosing suppliers using a method different from the others, namely the AHP method for decision making at suppliers, then for optimal order allocation, it is done using Linear Programming. AHP is used to carry out the weighting of each criterion and sub-criteria so that suppliers are found to be in accordance with the company, while linear programming is used to determine optimal raw material orders from each supplier that has been sorted. Based on inductive studies it can be concluded that the research that will be conducted has a difference with previous similar studies, namely the method used and the criteria set.

References

[1] Bruno G, Esposito E, Genovese A and Passaro R 2012 *J. purchasing and supply management.* **18** 159-172

[2] Charnes A and Cooper WW 1962 *Naval Research logistics quarterly.* **9** 181-186

[3] Esmaeel RI, Zakuan N, Jamal NM and Taherdoost H 2018 *Procedia Manufacturing.* **22** 975-981

[4] Forman E and Peniwati K 1998 *Eur. J. operational research.* **108** 165-169

[5] Ghodsypour SH and O'Brien C 1998 *Int. J. production economics.* **56** 199-212

[6] Gungor ZE and Evans S 2015 *Procedia CIRP.* **26** 527-532

[7] Hamdan S and Cheaitou A 2017 *Computers & Operations Research.* **81** 282-304

[8] Hartmann E, Kerkfeld D and Henke M 2012 *J. purchasing and supply management.* **22** 22-34

[9] Kahraman C, Cebeci U and Ulukan Z 2003 *Logistics information management.* **16** 382-394

[10] Monczka RM, Petersen KJ, Handfield RB and Ragatz GL1998 *Decision sciences.* **29** 553-577

[11] Nurcahyo R and Wibowo AD 2015 *Procedia CIRP.* **26** 653-657

[12] Omurca SI 2013 *Applied Soft Computing.* **13** 690-697

[13] Pujawan IN 2017 *Edisi 3. Jakarta: Penerbit Andi*

[14] Reefke H and Sundaram D 2017 *Omega.* **66** 195-211

[15] Scheuing EE 1989 *Englewood Cliffs, NJ: Prentice-Hall*

[16] Schiele H 2012 *Research-Technology Management.* **55** 44-50

[17] Uhl C, Nabhani F, Kauf F, Shokri A and Hughes D 2017 *Procedia Manufacturing.* **11** 1366-1374

[18] Xia W and Wu Z 2007 *Omega.* **35** 494-504

IOP Conf. Series: Materials Science and Engineering **697** (2019) 012008 doi:10.1088/1757-899X/697/1/012008

Automatic Pick-and-Place Packaging System with Vacuum Lifter

Awais Farooqi[1,2*], Nukman Bin Yusoff[1,2] and Lim Siong Chung[3]

[1] Technocentric Solutions Sdn. Bhd., 50603 Kuala Lumpur, Malaysia

[2] Department of Mechanical Engineering, Faculty of Engineering, University of Malaya, 50603 Kuala Lumpur, Malaysia

[3] School of Engineering, Asia Pacific University of Technology & Innovation, TPM – Bukit Jalil, 57000 Kuala Lumpur, Malaysia
*Email: awaisfarooqi@hotmail.com

Abstract. Flexible packaging manufacturing is one of the most essential industry of the world nowadays. This research aims to develop an economical & ergonomically designed pick-and-place sorting system for bag-making machines in the packaging industry. It provides a standard universal setup that is feasible to adapt with any type of bag-making machine as a pick-and-place attachment. This automated system can be installed at pickup location of the bag-making machine for the operation of pick-and-place of packaging bags, which is able to distribute packaging bags into the accepted bags platform without the supervision of a physical operator. This pick-and-place sorting system is able differentiate between two coloured bags while accepting the correct dimension of it to place it at a location of 'accepted bag' else at a location of 'rejected bag'. The proposed machine system can sort-out three different dimensions of test samples by accepting bags with a fiducial mark and correct dimension only, while rejecting the bags without fiducial mark and of incorrect dimensions. A standalone mechanical structure based on stepper motors, electronics design and embedded control system was developed that includes an IR-proximity sensor, as a fiducial mark/colour detector and a loadcell based dimension analyser, to differentiate between different dimensions of the test samples.

1. Introduction

In every industry pick and place applications are commonly used for different type of products. Usually this pick and place process is mostly done by packaging operator at small industries therefore due to operator errors most of the times those products are being packed into the boxes for delivery which are faulty and do not match to the standards of the desired client or the required designs. This research document focuses on the development of an ergonomically designed application of a pick-and-place system for flexible packaging plastic film bags that is able to save time and wastage of material during bag making process, by the setup of bags dimension and other parameters into the control system of the machine. The miss-alignments errors are folded bags, wrongly or irregular bags packed by the operator without being noticed for delivery. An invention of a system, that is able to pack bags automatically while reducing these errors is needed. Automatic Pick & Place Packaging System with Vacuum Lifter is able to pick and place the plastic bags; at a same time rejects or separates these bags in accordance with their difference of colours and correct dimension (i.e. the bag length and width comparison is focused in this research paper only).

The vacuum suction end-effector(lifter) picks the plastic bag and move towards the location for placing the bag by following two degree of freedom (DOF) in a linear motion i.e. left and right & forward and backward (X-axis and Y-axis Motion). The vacuum suction pickup action is at the Z-Direction (axis), which is additional ½ degree of freedom to the system. The control system algorithm is able to categorize the bags by differentiating the two colours of bags and correct dimensions at pickup location and place the correct colour of bag with selected dimension to the area of accepted bags whereas it is also able to place various colored bags or of wrong dimensioned bags to the area of rejected bags.

The flexible packaging film reel which is placed in the bag making machine has already passed through the printing process and it is testified by printing machine automatically in contrast to its colour depth and quality. But still there are chances of errors in the bag design pattern and alignments most probably during the bag making process of the running machine. In the bag making machine (see Figure 1) process, the bags might be folded wrongly or irregularly and manually packed by the operator without noticing errors. Therefore, a novel application is needed to beneficial the company by incorporating this machine invention with the current setup of bag making machine. It adds-up a further investigation stage to minimize the errors for customer satisfaction with the improvement in serviced quality of the manufactured material and process sustainability by reducing the raw material wastage. These bags are of plastic poly-ethane/PVC material produced for the following products such as packaging material for tea, cooking oil, ice creams, biscuits, toffees, noodles etc.

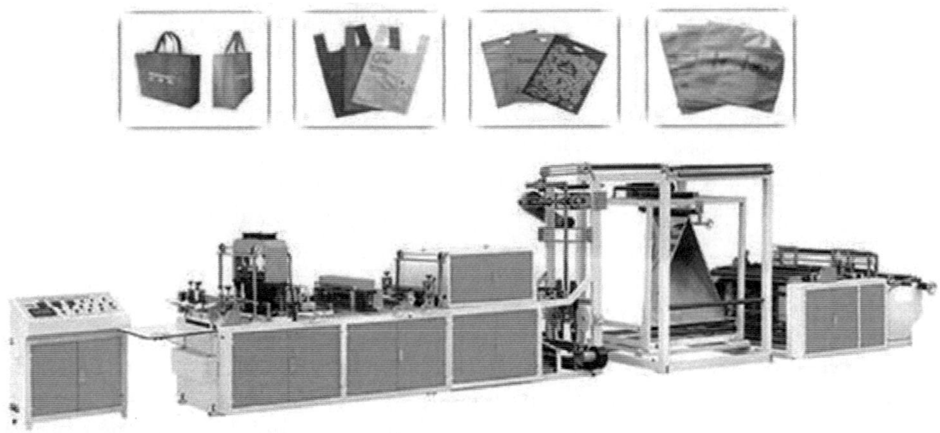

Figure 1. Multifunctional Non-Woven Bag Making Machine [1].

In past few decades, not much research has been noticed in similar application of bag making machine industry, but the most relevant related application used for pick-and-place machines are highlighted in this paper.

However, in every industry a pick and place applications are commonly used for different type of products. In recent years many researchers have studied the pick-and-place mechanisms for different applications such as RFID tags and smart card pick-and-place operations, in which the main task is to run the system at high speed while the pick-and-place die keeps on fabricating the RFID tags and smart card. In it the main factors controlled during the process of picking includes tape adhesion, wafer back side condition, machine hardware and programming [2].

Whereby another research on vacuum suction cup units for pick and place of aircraft flexible assembly design suggests a sufficient idea for vacuum suction pick-and-place module that can be achieved by introducing high flexibility to the end-effector (gripper), which can be obtained by three parallel used principles i.e. human and computer based analysis of the gripping object as well as mechanical adaptation of the gripper to the object with the help of servomotors and vacuum suction cups. It can be designed according to required specification, for example introducing three gripping vacuum suctions at the end-effector of the gripper [3]

The latest technology used to pick-up small items is used in Surface Mount Technology (SMT), it is a technology commonly used in electronic assembly. It is a method where components are mounted directly onto the surface of printed circuit board [4]. SMT uses visual location technology in its pick-and-place process, which offers a high speed and precision of component placement. Vision-Pro based system have been developed for SMT machine techniques; one of it is to solve the accurate image location by pattern-based location algorithms and the other is the calibration of camera, which is achieved by image wrapping technology through the checkerboard plate method [5], [6].

However, food is high-growth industry all over the globe. International food trading is demanding more raw materials every day. The current market demands more shelf life, improved quality, environmentally friendly materials and innovative packaging technology. The world's packaging machinery demands is increasing 5% per year. Many companies are adapting agile manufacturing practices. Agile manufacturing refers to the capability to quickly go from a set of novel customer requirements to a quality, finished product therefore to meet the market demand of packaging, companies are trying to work fast to increase their efficiency with their consumers to deliver the packaging material with in the time limit [7]. The main aims of this project are to develop an economical, fast, accurate and automated pick-and-place system, and to reduce the raw material wastage induced by manual workers during bag making machine process.

The main objectives achieved in this project are to, (i) differentiate between two colours of bags using pick-and-place system, (ii) distinguish the correct dimension of the bags, and (iii) pick-and-place the bag at defined positions.

2. Highlights on the technology for the development of pick-and-place equipment

The most difficult part in pick-and-place packaging of bags is the overlapping of the bags to other bags while pick-and-place process. The pick-and-place of plastic packaging bag system can be designed in a way that each plastic bag is picked and placed at one time only; meaning one bag per lift. The mechanism of such machine may possess a sensing technology that is able to recognize the presence of the plastic bag underneath the pickup lifter and sends the signals to the control system which is programmed to send signals further to various components of the packaging system. This system may also consists of number of servomotors, which are used for transmit signals at precise movement at a specific time to control the variables such as package placement and displacement points [8].

In order to detect the image underneath the pickup lifter image processing is a good option to be used at this stage to differentiate the colour difference between two packages and the correct size detection. A CCD camera with 32-bit microcontroller operated by computer software can be installed at a point of Image processing unit, which is low cost and suitable for small research [9]. An image processing method "Hue Histogram Algorithm" can be used in this invention to detect the plastic bag placed under pickup lifter which will determine the number of objects in a column by identifying the location of object by the difference in its length and width i.e. the plastic bag placed under it is on the right path to be picked up or not and the main aim of this technique is to calculate the time which is suitable for the pickup lifter to determine the distance between the object (plastic bag) and the pickup lifter [10]. But the restrictions in this method is that the target light interference with object colour i.e. the position of each object in the column cannot be identified correctly [9].

The limitations of detecting an image with a camera are always influenced by environmental illumination changes. It can be modified by using HSV (Hue, Saturation and Value), which is somehow similar to "Hue Histogram Algorithm" method. However due to illuminations effects on camera, the image processing unit cannot adapt sudden changes frequently. The illumination changes can be minimized by RGB colour space method which is very weak to adapt illumination and brightness changes of an image. The advantage of using HSV is that the illumination changes the value of the detected image only, whereas this value change does not affect the HS-Histogram colour of the image [11]. A MATLAB M-file script can be written to process the image automatically to be used as an image processing computer interference software to detect the size of the image by ruler and colour techniques [12].

Pneumatic system of actuator can be used at lift arm to create a push force at the vacuum suction end-effector. A novel pneumatic control system can be implemented as multiplex pneumatic control system which consists of an oscillator, a resonant valve and air tube; where the oscillator causes the air vibrations in air supply tube which is connected with the resonant valves. Therefore, the resonant valves are selectively controlled by air vibrations from the oscillator [13]. Limitations in pneumatic system are in compressor and many electrical wires forming the control valves, would complicates the entire pneumatic control system and the leaking of the valves.

A high-speed pick and place of the plastic film can be achieved by providing high adhesion strength and extremely low peel strength at the end-effector of the pick-and-place arm. For this reason, the plastic film roughness verses smoothness conditions can be measured to understand the critical properties required by the pick-and-place process [2]. The pick-and-place mechanism can be controlled by the number of servomotors which requires a perfect intelligent servomechanism of precise point-to-point positioning system. The idea suggested by [14] is that it can be used in pick-and-place control process by Flip-Chip technology and Iterative learning control (ILC) system, which can be used to enhance the point-to-point positioning accuracy of the control circuit.

Surface Mount Technology (SMT) pick & place machine designed by Autoronik (a German company) is a kind of machine that performs the pick-and-placing of electronic ICs on the electronic chip board. Mechanically machine is designed in such a way to provide high speed, repeatability and stability of the pick and place process which is achieved by using preloaded ball screws with AC-Servo motors. Such kind of mechanical system can be used in the development of any pick-an-place invention. The Autotronik pick-and-place machine uses Cognex Alignment Vision-Pro system to run its image processing capabilities which are synchronized with its control mechanism of pick-and-place. The machine also uses a bottom vision system to assure high-precision placement of pickup components. The machine also possesses flip-chip technology for point-to-point placement of the components over the electronic chip [15].

Another SMT pick-and-place machine by Manncorp, an America company, uses preloaded ball screws system to move the pick-up lifter at an accurate point with the use of closed-loop servo control system to achieve high performance of the machine but this machine has a dual gantry four heads which performs a pick-and-place task in two directions such as if one head is placing a component another will be picking up a component from other side, and the cycle continues hence increasing the reliability and speed of the machine. Perhaps all kind of SMT machines nowadays are using Cognex Alignment system "Vision on Fly" to achieve the image processing process with a bottom reference camera or vision system to achieve much easier and reliable fast performance tasks by the machine [16].

3. Concept Development and Evaluation

3.1. Concept of Process & System

The development of the mechanism for this invention is inspired by the SMTmax QM-2100 PCB pick & place machine which is operated by Cognex Visual-Pro Image processing software [17]. This invention system must identify the difference of two coloured bags and their dimensions during pick-and-place process while placing the correct bag into correct area.

The context diagram of pick-and-place processing system is given in Figure 2, in which the proposed setup for system idea is illustrated. In the process flow, the plastic bag dimensions, and colour is detected by the system meanwhile sending signals to the control system to adjust the lifter arm position before its pickup. The control system differentiates between two colours and the selected size of plastic bag by sensors, it further sends the signals to the actuator of the vacuum sucker at pickup arm to perform pickup function of the machine. After the bag is picked-up, the motion-control system is initiated, the decision processing system differentiates between the differences in colours and dimensions of the lifted bag and decides the placements position.

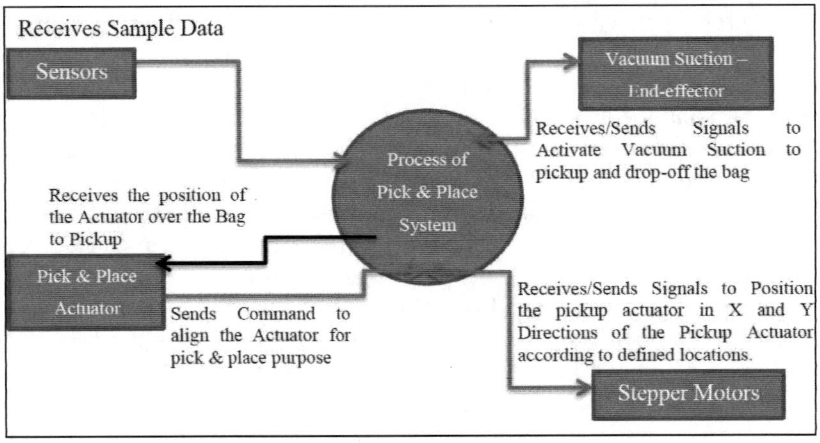

Figure 2. Process of Pick & Place System.

3.2. Design and System Development

The working area of Automatic Pick & Place Packaging System with Vacuum Lifter machine is set to be of approximately 45 cm x 45 cm, it is selected due to the small size of plastic bags which would be of about maximum 10 cm x 10 cm (length x width). Due to the testing of machine at its initial stage, the machine speed for pick-and-place is predicted to be set at five bags per minute.

Two belt driven stepper motor linear guides at perpendicular to each other are attached in order to achieve the two-way linear motion at X-direction and at Y-direction. And the Z-direction pickup task is controlled by a small DC motor. All the motion together would provide degree of freedoms to the vacuum actuator end-effector. An IR sensor mounted with a load cell to an external-baseplate is be used for Fiducial-mark recognition on the test sample of plastic film and its dimensions, respectively. The load cell is used to inspect the size of the bag during its pickup period. The placement flow chart with IR sensor and load cell for the concept of location training process can be given as in Figure 3.

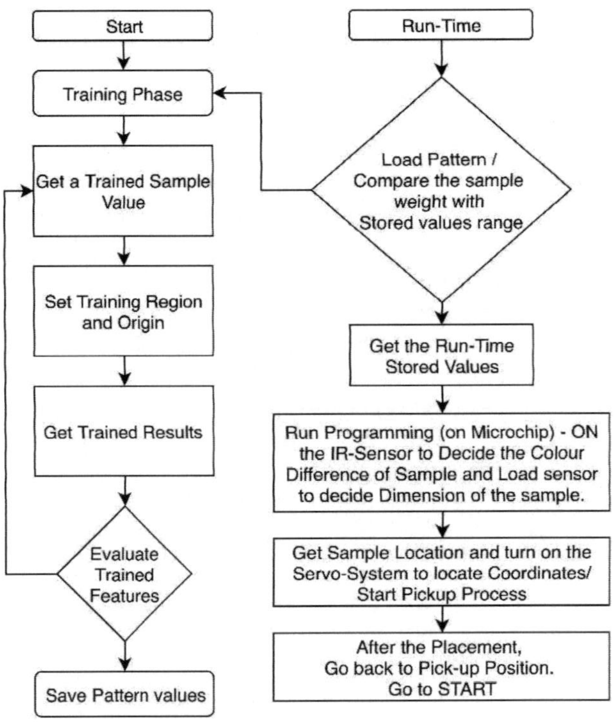

Figure 3. The Flow Chart of the Machine System with Location.

4. Methodology

4.1. Design and System Development

In the X and Y axis motion, the stepper-motor control system is selected due to its accuracy and precision of perfection in motion and they are inexpensive as compared to servomotors. The Microchip PIC16F877A is used to program the selected objectives of the project due to its best feature in the industrial research and development and inspection in the field of Automation.

To get a successful pickup by vacuum suction more studies and research are required to develop the model of pick-and-place end-effector to have an exact pressure of vacuum and the pressing power of pickup-actuator when the plastic bag is lifted. If the pressing power of vacuum lifter is exceeded, it might cause the lifter to brake or damage the material of the bag.

Fiducial Marks are coloured blocks presented at the edge of printed plastic film and by the use of optic, ultrasonic, colour and piezoelectric sensors the bag making machine detects for cutting edge or for punching. These coloured marks on each bag are sampled to help the film in the printed frame to be aligned and to detect the size or position of each bag on a printed film during running machine process. As far as there are no cameras attached into the bag making machine to understand the complete image whiles there are several sensors used in the machine during the bag making process to keep the running film aligned.

For example, an IR sensor attached to the machine at a set value to turn ON the next motor when plastic bag bypass across its wave-light, the changes in its voltage generates a signal to turn-on the next motor. But this phenomenon of mechanism is not suitable for this invention because a colour sensor itself cost about US$300 or above. And fast speed camera technology is even much more expensive then these sensors.

Therefore, after careful analysis, a simple IR-proximity sensor (of range 2cm) is used in the development process of this invention to detect the difference between the two colours of plastic film bags.

To achieve the dimension analysis of the plastic bag film, a sample is weighted by the use of load cell to achieve the different voltage output ratings of each sample weight and the changes in the range of micro voltages is programmed into the memory of microprocessor, that is, each sample voltage reading represents the specific Length and Width of the sample, for example a 6 cm x 6 cm sample gives a voltage difference of 0.03mV (as shown in Figure 4). This value range is fed into the microprocessor control algorithm to understand that each time this film is placed over the load cell.

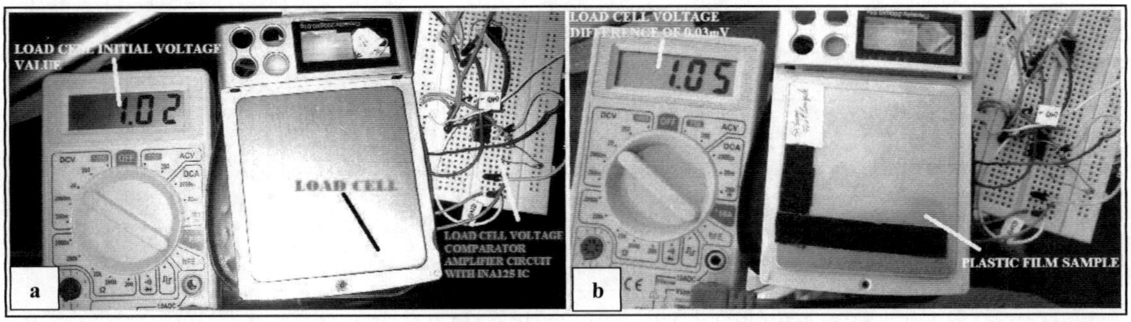

Figure 4. a) Load Cell Voltage output without Sample, b) Load Cell Voltage output with Sample.

4.2. Working Process

To meet the objectives of this project test samples used are of two colours. To specify the difference in two types of sample bags, a transparent plastic film sample is used. Each sample is introduced with a Fiducial Mark on its corner (see Figure 5). Therefore, the system will accept the bags of correct dimension and a bag with a Fiducial-mark only to consider it as an Accepted Bag. In case if the fiducial mark is missing and the sample bag is transparent, the system understands it as rejected bag operation.

There are three samples used in our test. The dimensions used are of (length x width) cm = (6 x 6) cm, (8 x 8) cm & (10 x 10) cm. These three different dimensions are made as 'Mode Selection' input parameter of the embedded system design.

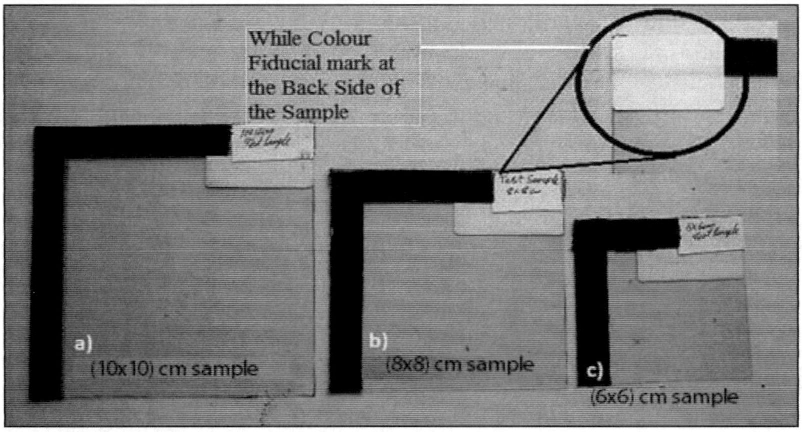

Figure 5. Test Samples of Transparent Plastic Films of various dimensions with Fiducial-Mark.

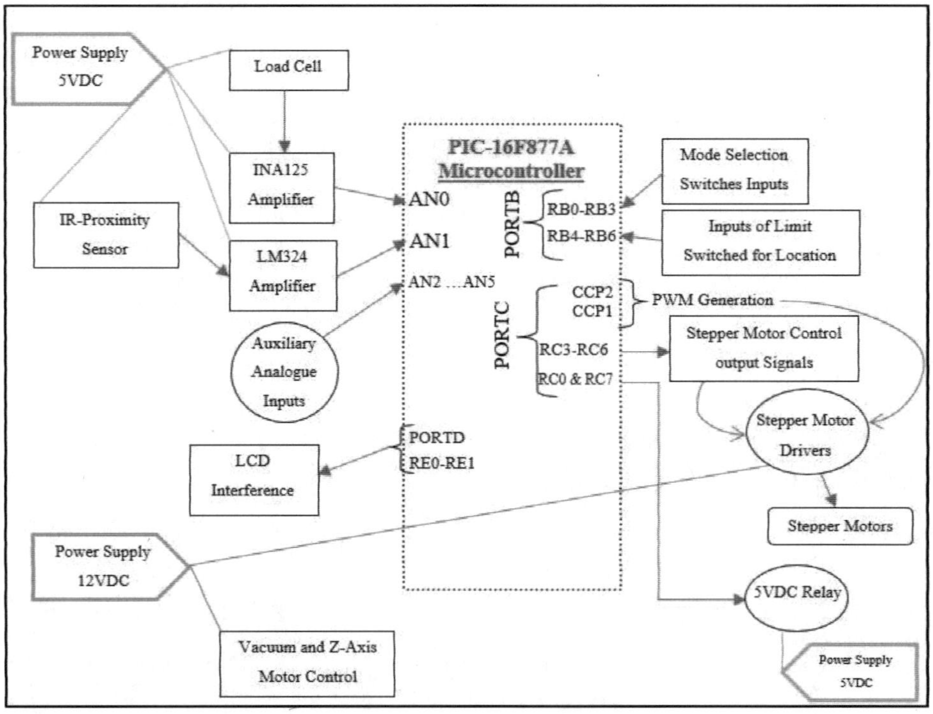

Figure 6. Block Diagram of Drive Unit.

4.3. System Block Diagram

The system block diagram is illustrated in Figure 6, showing the hardware architecture. On the block diagram, the inputs to the microcontroller are named by the hardware function of the input. It clearly describes that how each type of measurement is taken and what kind of the microcontroller is required. It is noticed from the drive unit block diagram that there are total 26 input/output ports (I/O) connections to the microcontroller, and due to many input/output connections PIC16F877A is selected to be used in this project [18]. The others most obvious points are the A/D converters, two PWMs, LCD Interference, enough discrete I/O for switches and other digital signal requirements, the embedded system references for hardware development can be found in [19], [20], [10]. And The system flow chart is given in the Figure 7.

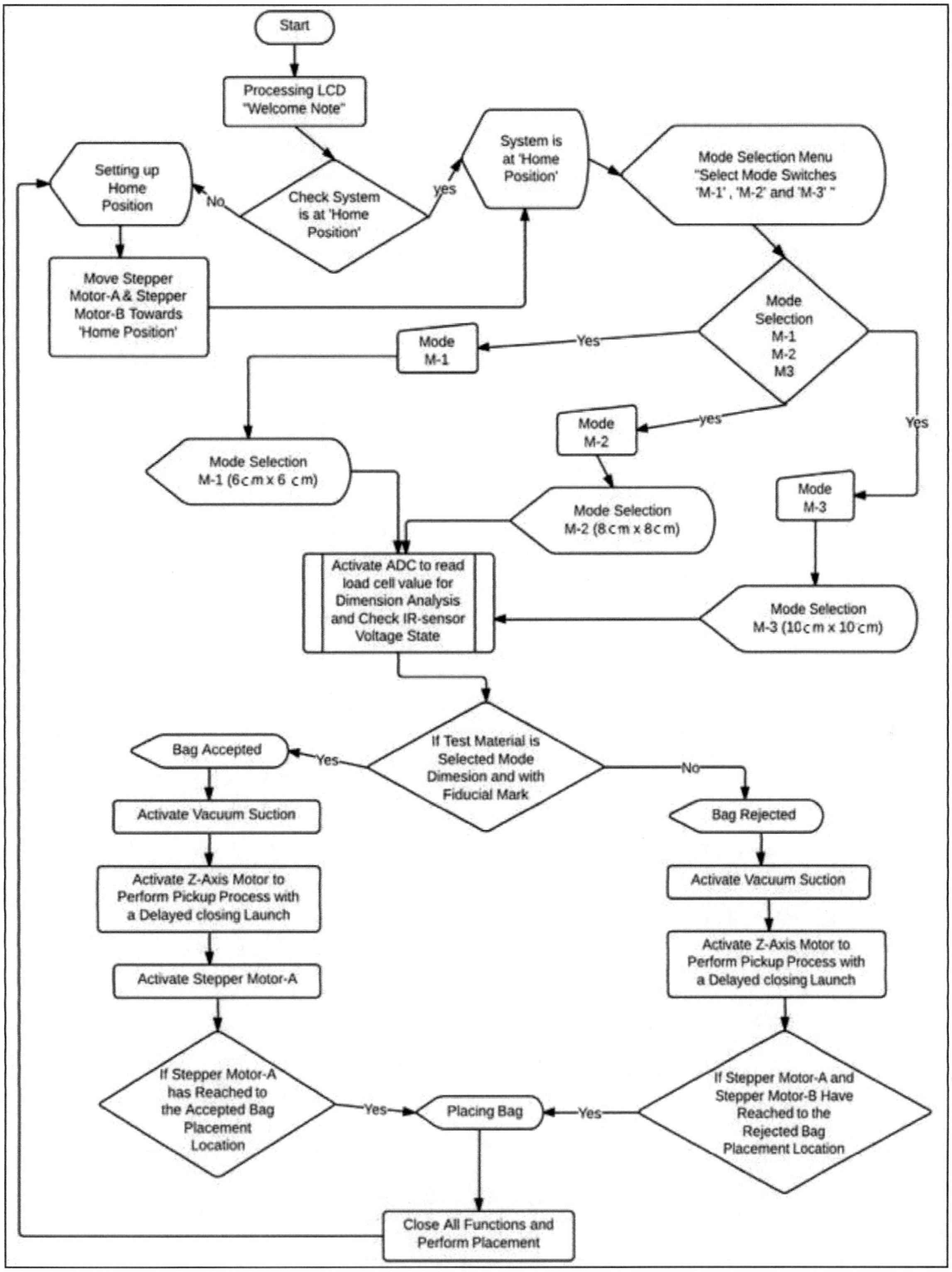

Figure 7. System Flow Chart

5. Simulation and Data Analysis

5.1. Experimental Setup
In this section the experimental setup is shown in both firmware and hardware forms of prototype to demonstrate the working of the project.

5.1.1. Dimension Analysis Monitoring System with the use of a Load Cell. The load cell used in this project is taken out from a precision weighing scale of resolution 0.01grams and can measures up to maximum of 200 grams. INA125 IC [21] is used as operational amplifier for loadcell system, as the output voltage change of a precision (milligram) load cell is in micro (μ) volts therefore a voltage amplification bridge connection is used to increase the gain of the change in voltage from μ volts to millivolts. Load cell used in this project have a cable with four wires; having +/- excitation and +/- signal lines. The basic wire construction of a load cell is given in Figure 8.

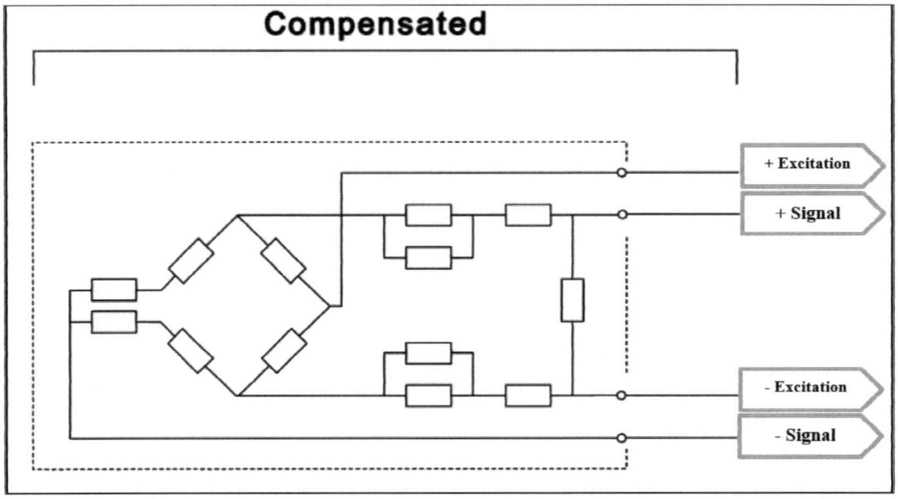

Figure 8. Four Wire Load Cell Compensated Diagram [22], [23]

The hardware design of load cell to measuring output voltage range, block diagram is shown in Figure 9, where it states that on what stage the A/D conversion is required while programing the complete setup process.

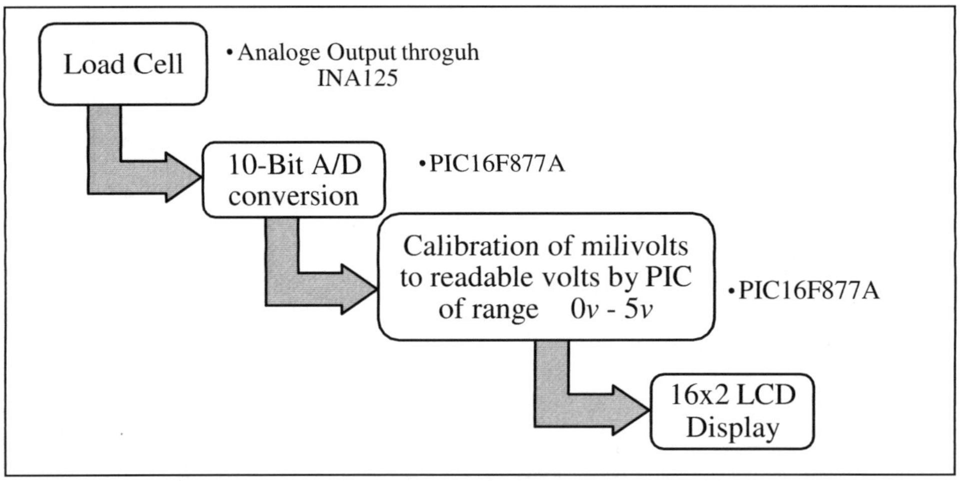

Figure 9. Block diagram of PIC16F877A-based weight monitoring system

5.1.2. Colour Difference Monitoring System with the use of IR-Proximity Sensor. The task of proximity sensor in this project is to differentiate between an opaque material and a transparent material (i.e. plastic bag film). The output quantity of this sensor is analog voltage, and the output voltage range in its setup is measured as voltage change which is set in range from 0v to 5v for microchip input pin. In general, the direct output voltage of this sensor is 3v. To trigger digital input signal of microcontroller LM324 operational amplifier is used. The amplification is set to achieve maximum output of 4 volts from 3 volts output of the proximity sensor. However, the gain Av, of 1.33 is set by using the non-inverting amplifier configuration.

5.2. System Hardware Circuit

The complete system circuit schematic of drive unit & display units is given in Figure 10, and the hardware setup in shown in Figure 11.

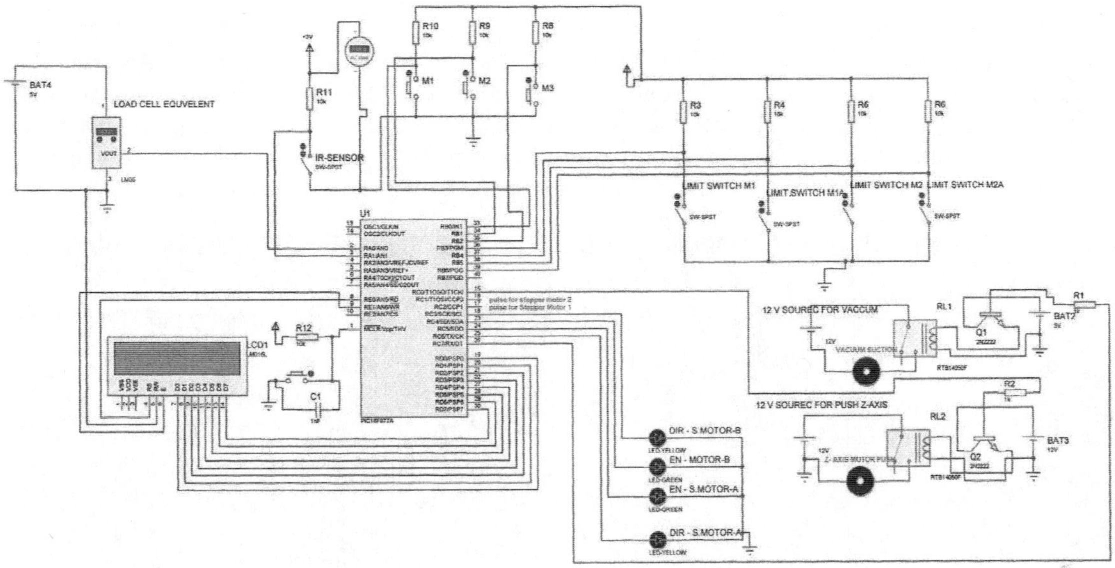

Figure 10. Complete System Circuit Schematic.

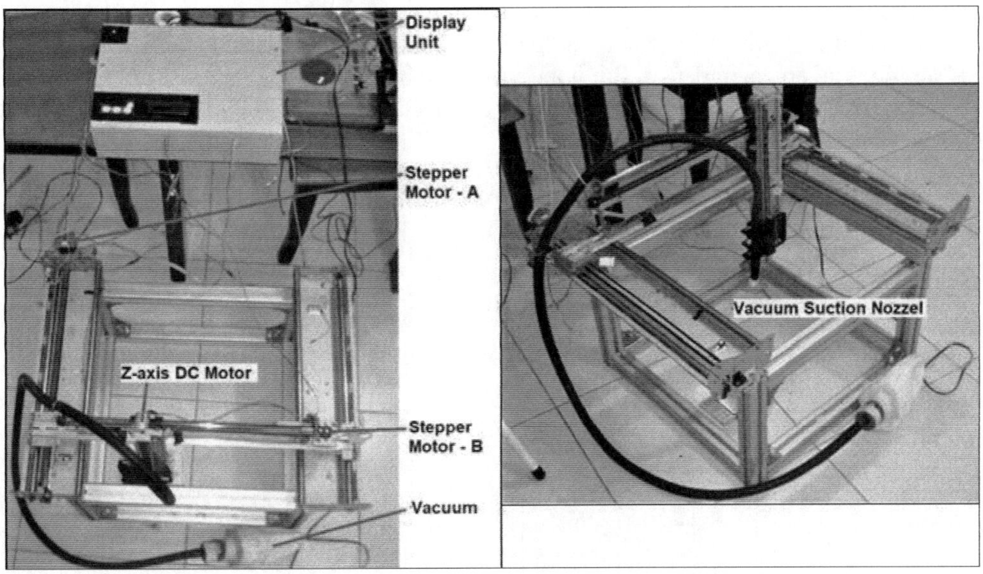

Figure 11. Complete System Mechanical Design Setup.

5.3. Data Collection

The main code of implementation of this project involves the analogue-to-digital conversion of load cell voltage to a readable voltage in microcontroller by ADC to set the values of desired selected dimension or size of the plastic film samples. Table 1 shows the dimensions of selected test samples and their weights in grams and change in voltage by load cell, tested by a voltmeter.

Table 1. Test Samples Details.

No. Samples	Sample Dimension [*w* x *l*] (*cm*)	Weight in grams (*g*)	Voltage Output Volts (V)
0	-	0.00	1.56
1	6 x 6	1.65	1.61
2	8 x 8	3.31	1.66
3	10 x 10	4.97	1.71

While testing the sample, are placed on the load cell test pan, before the mode selection button. The pickup arm is at home position that is above the loading pan. Once the mode selection button is pressed, for example, mode M-1 is selected to test sample of dimension (6 x 6) cm, and the bag is supposed to be accepted so that the pickup arm drops the accepted bag at location of accepted material or else if the bag is rejected it drops it at a location of rejected material, as shown in Figure 12.

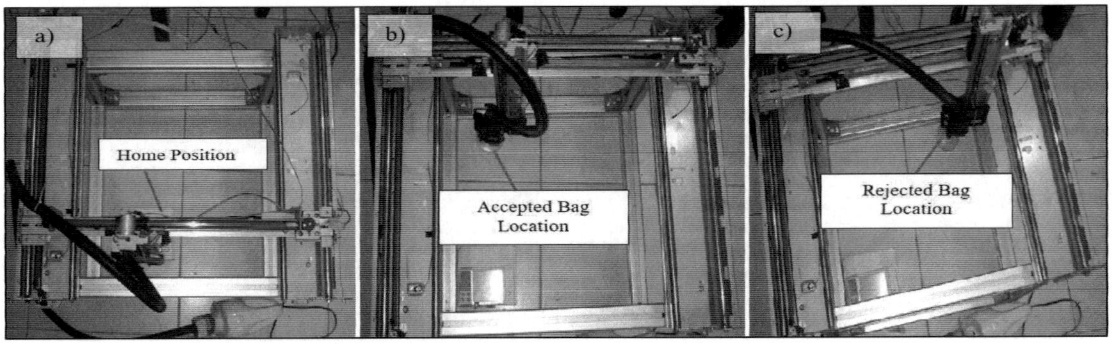

Figure 12. Test Samples Experimental Placement Position of Locations.

5.4. Data Analysis

Referring to the chart in Figure 13, output voltage versus distance of proximity sensor graph is achieved by keeping a refractive object across it. But if a transparent material is kept at 0 - 1 cm across the sensor, its output voltage does not change. And if an opaque object is kept at 0 - 1 cm, the voltage drops from 1.45v to 0.11v - 0.13v at 0cm to 5cm.

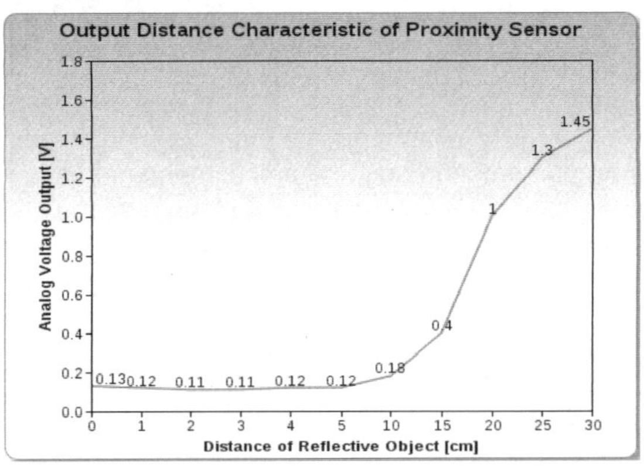

Figure 13. Proximity Sensor Voltage vs. Distance Characteristic.

5.5. Test Results

The main facts of the experiments are reported in Table 2 & 3. A total of three samples were tested for dimension analysis and colour difference to inspect all the modes of the system for pick-and-place process. The tick marks show the successful results achieved in testing. Nearly 87.5% correct samples were analyzed by the system. The average cycle time of the machine was much closer to the expected (i.e. the speed of machine during pick-and-place cycle).

Table 2. Summary of Test Results of Accepted Bags.

Test by Modes of Samples	Test of pick-and-place of Accepted Bag (T_a)				Time Taken for Complete Cycle (t/s)				Average Time ($T_{a,avg}/s$)
$(l \times w)$ cm	T_{a1}	T_{a2}	T_{a3}	T_{a4}	t_1	t_2	t_3	t_4	$T_{a,avg}$
6 x6	✓	✓	✓	✓	17.0	16.4	15.0	16.1	16.1
8 x 8	✓	✓	✓	✓	16.1	16.1	16.3	16.2	16.2
10 x 10	✓	✓	✓	✓	16.3	16.8	16.1	16.4	16.4
Total Average time of Machine of Machine to Pick-and-Place									16.2
Test by Modes of Samples	Machine Returning to it Home Position after Placing Accepted Bags ($T_{a.h}$)				Time Taken for Complete Cycle (t/s)				Average Time ($T_{a.h,avg}/s$)
$(l \times w)$ cm	$T_{a.h1}$	$T_{a.h2}$	$T_{a.h3}$	$T_{a.h4}$	t_1	t_2	t_3	t_4	$T_{a.h,avg}$
6 x6	✓	✓	✓	✓	11.8	11.4	11.9	11.9	11.8
8 x 8	✓	✓	✓	✓	11.0	12.3	12.1	12.4	12.0
10 x 10	✓	✓	✓	✓	12.1	12.4	12.0	12.2	12.2
Total Average time of Machine to Return Back to Home Position									12.0

Table 3. Summary of Test Results of Rejected Bags.

Test by Modes of Samples	Test of pick-and-place of Rejected Bag (T_r)				Time Taken for Complete Cycle (t/s)				Average Time ($T_{r,avg}/s$)
$(l \times w)$ cm	T_{a1}	T_{a2}	T_{a3}	T_{a4}	t_1	t_2	t_3	t_4	$T_{r,avg}$
6 x6	✓	✓	✓	✓	16.6	16.4	16.0	16.1	16.3
8 x 8	✓	✓	✓	✓	16.2	16.1	16.3	16.2	16.2
10 x 10	✓	✓	✓	✓	16.4	16.6	16.2	16.4	16.4
Total Average time of Machine of Machine to Pick-and-Place									16.3
Test by Modes of Samples	Machine Returning to it Home Position after Placing Rejected Bags ($T_{r.h}$)				Time Taken for Complete Cycle (t/s)				Average Time ($T_{r.h,avg}/s$)
$(l \times w)$ cm	$T_{r.h1}$	$T_{r.h2}$	$T_{r.h3}$	$T_{r.h4}$	t_1	t_2	t_3	t_4	$T_{r.h,avg}$
6 x6	✓	✓	✓	✓	12.1	12.4	11.9	12.4	12.2
8 x 8	✓	✓	✓	✓	12.1	12.3	12.1	12.4	12.2
10 x 10	✓	✓	✓	✓	12.0	12.0	12.0	12.1	12.0
Total Average time of Machine to Return Back to Home Position									12.1

From Table. 2 and Table. 3:

The average time taken by machine to perform a random pick-and-place, T_{avg} is:

$$T_{avg} = \frac{T_{a,avg} + T_{r,avg}}{2} \quad (1)$$

$$T_{avg} = 16.25s \approx 16.3s$$

The average time taken by machine to return to its Home Position after successful pick-and-place process, $T_{Home,avg}$ is:

$$T_{Home,avg} = \frac{T_{a.h,avg} + T_{r.h,avg}}{2} \quad (2)$$

$$T_{Home,avg} = 12.05s \approx 12.1s$$

There were three jams occurred during the test running of the machine. The achieved 87.5% of successful results can be calculated as:

Total number of tests conducted on samples from Table 2 & 3 are;

$$T_{test} = 24$$

Total Number of Jams Occurred during Test Running Cycle, $J_{test} = 3$

Percentage of unsuccessful results, α

$$\alpha = \frac{J_{test}}{T_{test}} \times 100 \quad (3)$$

$$\alpha = 12.5\%$$

Percentage of successful results, R_s;

$$R_s = 100 - \alpha \quad (4)$$

$$R_s = 87.5\%$$

5.6. Machine Test

Random test running of machine process was conducted to investigate the distribution of recovery time (i.e. idle time to fix the jam and restart the system completely). The first 3 jams occurred while random testing of machine is recorded; the time of recovery was noted as shown in Table 4 and the graphical distribution of recovery time is given in Figure 14.

Table 4. The counted Recovery Time of the Machine when it was Jammed.

No. Samples Running During Jam (*l x w) cm*	Total Number of Jams occurred during Random Test Phase	Time (in seconds) of Machine Recover after Reset Button is Pressed, T(s)			
		T_1	T_2	T_3	$T_{t,\ average}$
6 x 6	3	12.4	17.6	19.5	16.5
8 x 8	3	10.0	15.4	20.4	22.9
10 x 10	3	13.4	15.9	16.8	15.4
Maximum Average Time of Recovery for Machine					18.3

Based on the results achieved, the machine time of operation for each pick-and-place task is tremendously accurate as expected. The average time taken by the machine to perform one pick-and-place operation is about 16.3 seconds and if added the time of machine to return to home position, which is about 12.1 second, a total of about 28.4 seconds are achieved for one complete pick-and-place process.

Therefore, as tested that one complete pick-and-place cycle with machine returning to its home position would take about 28.4 second, the pick-and-place (of bag) speed of machine can be calculated as:

One complete cycle of pick-and-place, X = 16.3 sec

Time for machine to return to its home position, Y = 12.1 sec

For each bag pick-and-place operation: Total performance time, T_p;

$$T_p = X + Y \quad (5)$$

$$T_p = 28.4s$$

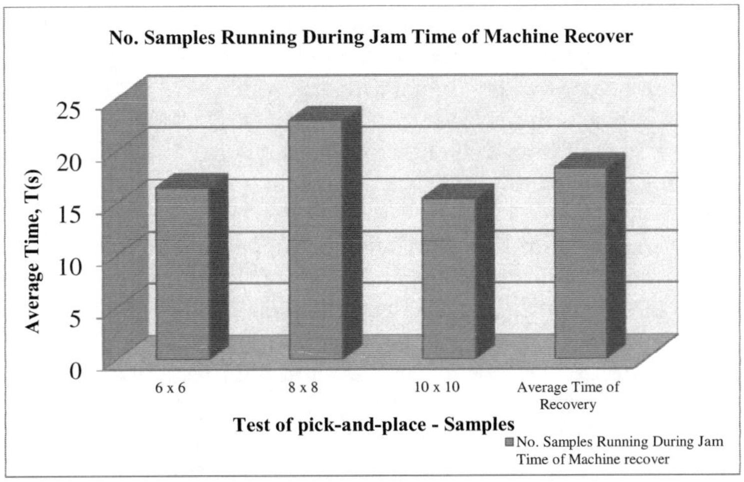

Figure 14 Measured Recovery Histogram.

Speed of machine to pick-and-place numbers of Bags per minutes (60 Seconds), S_{Bags};

$$S_{Bags} = \frac{60s}{T_p} \quad (6)$$

$$S_{Bags} = 2.1 \approx 2\ bags/min$$

Therefore, practically the machine can perform about 2 bags pick-and-place operations per minute, however the expected speed of machine was about 5 bags/minutes. The achieved speed of machine is vastly lesser then the actual bag making machine in the factory, which makes about 70 bags/minutes. The current invention speed is about 97% lesser then the actual bag making machine, but it is 60% accurate to the expected speed of the machine, given as:

Actual speed of Bag making Machine, S_{Actual} = 70 bags/min Expected

Speed of the current system to pick-an-place, $S_{Expected}$ = 5 bags/ min

Achieved speed of pick-and-place, S_{Bags} = 2 bags/min

Percentage of Limitation in the current speed of system from reality, S_{Lim} from reality:

$$S_{Lim}\ from\ reality = 100 - \left[\left(\frac{S_{Bags}}{S_{Actual}}\right) \times 100\right] \quad (7)$$

$$S_{Lim}\ from\ reality = 97.1\%$$

Percentage of Limitation in the current speed of system from expected speed, $S_{Expected}$:

$$S_{Lim}\ from\ expected = 100 - \left[\left(\frac{S_{Bags}}{S_{Expected}}\right) \times 100\right] \quad (8)$$

$$S_{Lim}\ from\ expected = 60\%$$

6. Conclusion and Recommendation

This project presents a pick-and-place process of plastic film bags after their bag making process to be picked up and placed at a desired location by the method of vacuum suction and the dimension analysis by a newly developed concept of load cell to monitor the dimensions of the film.

According to the experiments carried out on the prototype of the presented project, the approach allows the pick-and-place of plastic bags by a method of using vacuum suction with lower cycle of time as compared to the manual process, and required the presence of the only one human operator to only look after the each pick-and-place process performed by this machine and troubleshoot the machine jamming, if occurred during machine running cycles. Jams in pick-up end-effector do not affect the machine running cycle time significantly, as system can be restored shortly.

The only important factor that can be considered as the basis of this project is the costing related to the manufacturing and operation of this system. And such automation system can be predicted to be used widely by industries and consumers are willing to compromise the quality of the product in lower ratios and big profits. Therefore, in order to increase the accuracy of this project a better vacuum suction with the pressure sensor can be used with a digital pressure controller by the interference of microchip (of 32-bit or higher). To improve the accurate position sensing of the system a visual aid can be used, but all of these sensors are of high cost and need extra income to build them but once the system is built using perfect sensors to achieve more accurate results more the demand of the system will be increased in the market.

By using programmable logic controllers (PLCs) to the proposed system it can easily reduce the use of many sensors, especially used in the positioning of the system. Also, a PLC based system reduces the use of many other electronic components and reduces the size of the circuitry and makes the system reliable and more cost effective. PLC's are easy to be programed then a microchip and is user friendly plug-and-play device to control the automation of a system with lesser errors and more accurate and precise in its time delays for its functions and intrusions [24].

The project Automatic Pick & Place Packaging System with Vacuum Lifter can also be made to able to differentiate between not only two colours but various coloured bags of various dimensions to locate them accordingly to their desired location of more than two points of location as described in this project, which will benefit the industry to minimize the work load. The different sizes and colours of bags can be stored in the visual memory of system to demonstrate different bags accordingly and keep them in their desire set locations. Whereas, this picks-and-place project machines can be built in the bigger scale to support the pick-and-place purpose not only in food packaging as well as in heavy industries like, automobile parts, shipping industries and aircraft manufacturing plants.

7. References

[1] "Flexible Packaging Converting Machine," 4th April 2019. [Online]. Available: http://witymachinery.com/3-7-bag-making.html.

[2] R. Koepp, T. Allen and J. Fassett, "Achieving high speed RFID die pick and place operation," 33rd IEEE/CPMT International Electronics Manufacturing Technology Conference (IEMT), pp. 1-8, 2008.

[3] Q. J. Yi, Z. Guolei, Z. Hongyong and L. Fang, "Intelligent Design Approach for Vacuum Suction Cup Units in Aircraft Flexible Assembly," Interneational Forum on Computer Scence - Technology and Application (IEEE), pp. 146-149, 2009.

[4] J. Jirsa and K. Dusek, "Studies of surface mount technology and its risk analysis," 34th International Spring Seminar on Electronics Technology, pp. 216-219, 2011.

[5] L. Wei and Z. Jiao, "Visual Location System for Placement Machine based on Machine Vision," 5th IEEE International Symposium on Embadded Computing, p. 2008, 141-146.

[6] DREXEL University, "PRISM," 8 April 2018. [Online]. Available: http://www.pages.drexel.edu/~pyc23/IR_proximity_sensor.jpg. [Accessed 14 January 2013].

[7] Food & Beverage News, "Excellent scope for packaging machinery mftrs," http://www.fnbnews.com, Mumbai, 2007.

[8] J. A. Goodman, "Universal Packaging System with Vacuum Lifter," United States Patent, Vols. Patent no. 09/139,955, 1999.

[9] P. Palungsuntikul and W. Premchaiswadi, "Object Detection and Keep on a Mobile Robot by using a Low Cost Embaded Color Vision System," 8th International Conference on ICT and Knowledge Engineering, pp. 70-76, 2011.

[10] T. Braunl, Embedded Robotics - Mobile Robot Design and Application with Embedded Systems, New York: Springer-Verlag, 2008.

[11] Y. Kimura, H. Takemura and H. Mizoguchi, "Quantative Evaluation Methodes of Robust Detecting and Trackng with Color Cemera under Illumination Changes," IEEE International Conference on Systems, Man and Cybernetics, pp. 2263-2266, 2010.

IOP Conf. Series: Materials Science and Engineering **697** (2019) 012008 doi:10.1088/1757-899X/697/1/012008

[12] R. Ekawita, E. Widiatmoko, Suprijadi, H. Nawir, E. Susila and Khairurrijal, "Color image Processing for Measuring Lenght Deformation in Compresson Test," 2nd International Conference on Instrumentation, Communications, Information Technology, and Biomedical Engineering, 2011.

[13] Y. Nishioka and K. Suzumori, " A New Control Method Utilizing Multiplex Air Vibration for Multi-DOF Pneumatic Mechatronics Systems," IEEE/RSJ International conference on Robots and Systems, no. 978-1-4244-6679-4, pp. 3037-3041, 2010.

[14] B. Peng, J. Quan, Z. Yin and Y. Xiong, "The Analysis and Control of Pick-and-Place Process in Flip-Chip," 2nd Interneational Conference on Mechanical and Electronics Engineering (ICMEE), p. 2012, 404-408.

[15] Autotronik, "Model:BS385V2 SMT Pick & Place Machine," www.autotronik-smt.de, Amberg, 2012.

[16] Manncorp, "Flexible & Precise Automatic SMT Pick & Place Machine," https://www.manncorp.com/, California, 2012.

[17] SMTmax, "SMTmax," 2019. [Online]. Available: http://www.smtmax.com/detail.php?id=67. [Accessed 4 April 2019].

[18] Microchip Technology Inc., "PIC16F877A Datasheet," Microchip Technology Incorporated, USA, 2003.

[19] T. Wilmshurst, Designing Embedded Systems with PIC Microcontrollers: Principles and Applications, 2nd, Ed., UK: Newnes, 2009.

[20] J. Iovine, PIC Robotics - A Beginner's Guide to Robotics Projects Using the PICmicro, New York: McGraw-Hill Education TAB, 2004.

[21] Texas Instruments Incorporated, Instrumentation Apmlifier With Precision Voltage Reference, Dallas: www.ti.com, 2009.

[22] Interface Advance Force Measurment, Load Cells - A Primer on the Design and Use of Strain GageForce Sensors, User Guide ed., Scottsdale: http://www.interfaceforce.com, 2009.

[23] Vishay Revere Transducers B.V., "Load Cell Cabeling," www.vishaymg.com, The Netherlands, 2006.

[24] W. Bolton, Programmable Logic Controllers, UK: Newnes, 2009.

Nexus between Green Intellectual Capital and Organizational Learning Capability. Evidence from Malaysian Manufacturing Sector

Yusmazida Mohd Yusoff[1*], Muhamad Khalil Omar[1], Maliza Delima Kamarul Zaman[1]
[1]Faculty of Business and Management, Universiti Teknologi MARA, 42300 Bandar Puncak Alam, Selangor, Malaysia
*E-mail: yusmazida@gmail.com

Abstract. The environmental degradation have been concerned by many previous scholars claimed that the high demand for sustainable practices particularly in manufacturing sector due to the large amount and wide distribution of energy consumption. It is an urgent task for organizations to implement the innovation to cope with this issue. Apart from that, most scholars have suggested various possible solutions. Regrettably, very few studies have examined the importance of intellectual capital related to green protection known as green intellectual capital to achieve better performance. Furthermore, as the manufacturing is the main sector that focus on the transformation of low-carbon economy required organizations to keep themselves learned and updated about their business challenges. In spite of that, the role of management that facilitate learning to each member which under the concept of organizational learning capability is very important to obtain superior performance. Data were gathered from 168 manufacturing SMEs in Malaysia, and analysed using Partial Least Square approach. Results show that green relational capital has positive relationship with organizational learning capability while green human capital and green structural did not. Finally, the implications, limitations and future studies are discussed.

1. Introduction

Manufacturing industry is one of the biggest contributors to the Malaysia economy; however, it is also the highest contributor to the environmental degradations [1]. If nothing is done to solve this, environmental crises will continue to increase and it will worsen. In response to these concerns, many "green" strategies has been conceived due to greater environmental awareness [2]. Apart from that, Chen [3] suggested that investment in intellectual capital that is related to environmental protection known as green intellectual capital (GIC) not only meet environmental management, but also competitive advantage.

GIC is defined as all tangible assets or knowledge related to the green innovation or protection [3-5]. Three main components of GIC namely green human capital (GHC), green structural capital (GSC) and green relational capital (GRC). Additionally, the implement of innovation to fulfil the environmental pressure particularly in manufacturing sector towards green transformation is an urgent task due to the large amount and wide distribution of energy consumption [6].

However, the efforts and investment of the management towards GIC would be useless in the organizations if learning is not take place. Furthermore, the manufacturing is the main sector that focus on the transformation of low-carbon economy required organizations to keep themselves learned and updated about their business challenges [7]. Thus, it requires organizations to continuously changing, learn how to sustain and grow [8]. One the effective mechanism to achieve this goal is through learning.

In addition, previous researchers noted that it is hard for organizations to use knowledge exploitation to produce innovation without a substantial capability [9] which under the concept of organizational learning capability (OLC).

OLC can be describes as the set of management practices that help in facilitating the process of learning within organisations in aiming to increase performance of the organization [10, 11]. Hence, this study aims to examine the relationship between green intellectual capital dimensions and organizational learning capability.

2. Hypotheses Development

2.1 Green human capital and organizational learning capability

Baker [12] noted that high skilled employees are utmost important to the organisation. Human capital that comprised of all knowledge, experience and skills is invisible form and depends on organizational learning capability (OLC) to transform it into visible form. Human capital of the organization and absorptive capacity of individuals can be improved by investment in development and training [13]. The knowledge acquired can be embedded into OLC. Rotherberg [14] claimed that the more knowledge acquired by the employees, the more the better organizational learning capability will be generated. Similarly, previous scholars suggested that human capital is highly connected to OLC [15]. Based on these discussions, there hardly seems in the literature on the relationship between green human capital (GHC) and OLC. Therefore, this study offers the following hypothesis.

H1: There is a positive relationship between green human capital and organizational learning capability.

2.2 Green structural capital and organizational learning capability

Structural capital (SC) will help the development idea across the organisation and participation of the employees in decision-making process [14]. The implementation of good information technology system will help organization to acquire organizational learning capability [16]. Furthermore, organizations tend to absorb external knowledge related the negative consequences of their business operations on the environment such as pollution and waste and applied the new knowledge to facilitates the green innovation practices into their business activities [17]. Based on these discussions, there hardly seems in the literature on the relationship between green structural capital (GSC) and OLC. Therefore, this study offers the following hypothesis.

H2: There is a positive relationship between green structural capital and organizational learning capability

2.3 Green relational capital and organizational learning capability

Employees that have strong relationship and networks tends to have better information and resource allocation compared who are not [18]. A good relationship provides greater possibilities in knowledge sharing from their partners. The collaboration with other organizations can enhance organizational learning capability [19]. Thus, the organisations gain more knowledge that is new and used it to improve their OLC. Based on the above fact, there hardly seems in the literature on the relationship between green relational capital (GRC) and OLC. Therefore, this study offers the following hypothesis.

H3: There is a positive relationship between green relational capital and organizational learning capability.

3. Methodology

This study adopts quantitative research method. A seven-point Likert scale was used for all items in GIC constructs ranging from '1' for 'strongly disagree' to '7' for 'strongly agree'. The questionnaire was adopted from Huang and Kung [5] with 18 items. Meanwhile, the questionnaire for OLC was adopted from Gomez et al. [13] with 17 items in the measure and was measured using a 5 point Likert scale measurement. The summary of the key constructs is given in Table 1. The targeted respondents

IOP Conf. Series: Materials Science and Engineering **697** (2019) 012009 doi:10.1088/1757-899X/697/1/012009

consisting of the most knowledgeable individuals, including directors, human resource managers, production managers, research and development (R&D) managers, as well as assistant managers from SMEs manufacturing organisations in Malaysia. A total of 168 usable questionnaires were returned. All usable data were subsequently coded and analysed using Smart PLS 3.0 Structural Equation Modelling statistically techniques.

Table 1. Summary of Key Constructs, Sources of Questions and the Number of Items

Variable	Dimension	No of Items	Source
Green Intellectual Capital (GIC)	Green Human Capital (GHC)	5	Huang and Kung (2011)
	Green Structural Capital (GSC)	8	
	Green Relational Capital (GRC)	5	
Organisational Learning Capability (OLC)	Managerial Commitment	5	Gomez et al. (2005)
	System Perspective	3	
	Openness and Experimentation	5	
	Knowledge Transfer and Integration	4	

4. Results

4.1 Demographic profile of companies
Majority of the manufacturing SMEs (51.8%) have more than 150 employees, 51 organizations (30.4%) have employees from 51 to 150, followed by 23 organizations (13.57%) have employees from 5-50, as well as only 7 organizations (4.2%) having 5 employees. In term of years operating, 52 organizations (31.0%) have started a business in more than 20 years, 45 organizations (26.8%) started their operation during last 16 to 20 years followed by 37 organizations (22.0%) were operated in last 11 to 15 years, 17 organizations (10.1%) were operated in less 5 years to 10 years each.

4.2 Measurement Model
The model developed included a second order factor for organizational learning capability (OLC). Thus, the validity and reliability of first order, followed by second order were tested. Factor loadings, average variance extracted (AVE) and composite reliability (CR) were used to examine the convergent validity followed by discriminant validity. Table 1 showed all factors loading for first and second order. The AVE values of all constructs exceeded 0.5 and the CR values of all constructs exceeded 0.7, indicates that the measures were valid and reliable. The three criteria's also passed for second order factor; hence, convergent validity has been achieved. for both first and second order factors. One item was deleted from organizational learning capability construct (MC2) to ensure the AVE value above 0.5.

Table 2. Convergent validity

First Order	Second Order	Item	Loadings	CR	AVE
Green Human Capital		GH1	0.858	0.943	0.768
		GH2	0.912		
		GH3	0.931		
		GH4	0.891		
		GH5	0.780		
Green Structural Capital		GSC1	0.844	0.963	0.766
		GSC2	0.795		
		GSC3	0.897		
		GSC4	0.891		
		GSC5	0.918		
		GSC6	0.918		

		GSC7	0.936		
		GSC8	0.790		
Green Relational Capital		GRC1	0.879	0.964	0.844
		GRC2	0.894		
		GRC3	0.928		
		GRC4	0.947		
		GRC5	0.944		
Managerial Commitment (MC)		MC1	0.745	0.787	0.582
		MC3	0.743		
		MC4	0.849		
		MC5	0.705		
System Perspective (SP)		SP1	0.926	0.933	0.822
		SP2	0.940		
		SP3	0.851		
Openness and Experimentation (OE)		OE1	0.800	0.912	0.675
		OE2	0.825		
		OE3	0.833		
		OE4	0.767		
		OE5	0.877		
Knowledge Transfer and Integration (KT)		KT1	0.877	0.868	0.624
		KT2	0.758		
		KT3	0.802		
		KT4	0.713		
	Organizational Learning Capability	MC	0.661	0.930	0.769
		SP	0.891		
		OE	0.879		
		KT	0.797		

Note: CR = Composite Reliability; AVE = Average Variance Extracted

Next, using Fornel and Larker [19], the discriminant validity was verified in comparing the correlations between constructs and the average variance extracted from the construct. Table 4 below shows that all values of the square root of average is greater indicating that the measures were discriminant.

Table 3. Discriminant Validity.

	GHC	GRC	GSC	OLC
GHC	**0.879**			
GRC	0.748	**0.914**		
GSC	0.821	0.808	**0.872**	
OLC	0.550	0.625	0.511	**0.713**

Note: The diagonals represent the square root of the AVE, while the off diagonals represent the correlations. GHC = Green Human Capital; GRC = Green Relational Capital; GSC = Green Structural Capital; OLC = Organizational Learning Capability

4.3. Structural Model

Structural model represents the hypothesized relationship among the constructs. The standardized value of structural model relationship (path coefficients) is between -1 and +1. The closer the value to 1, means the strong positive relationship. Results shows that GRC ($\beta = 0.659$, $p < 0.05$) was positively related to organizational learning capability while GHC ($\beta = 0.175$ $p > 0.05$) and GSC ($\beta = -0.246$, $p < 0.05$) were positively associated to organizational learning capability. Thus, H3 was supported while H1 and H2 were not supported.

Table 4. Structural Model

Hypothesis		Path Coefficient	std Beta	Std error	t-value	p-value	Decision
H1	GHC → OLC	0.175	0.168	0.109	0.804	0.108	Not Supported
H2	GSC → OLC	-0.246	-0.239	0.118	1.041	0.037	Not Supported
H3	GRC → OLC	0.658	0.659	0.093	3.523	0.00	Supported

5. Discussion

The main objective of this study was to gain insight into the role played by GIC dimensions towards OLC. Three conclusions have been highlighted. Firstly, the empirical findings did not found an evidence to support the relationship between green human capital (GHC) and organizational learning capability (OLC). Thus, H1 was not supported. The results is similar with the study done by Omar, Mohd Yusoff and Kamarul Zaman [20] revealed that GHC-OLC link is not significant. However, the results is contradicted with other previous studies that acknowledge human capital as an important feature in intellectual capital. These can be seen from the study by [22,22] highlighted that human capital is the most critical factor of organisation's success. Most prior researches noted that SME organizations lack resources and low interest in environmental management [23]. Moreover, the new concept of GHC that is usually ignored by most organizations, especially among SMEs. In this case, manufacturing SMEs have less competent with environmental knowledge and experience due to lack of adequate training for upgrading skills and competencies. These training are necessary to empower their human capital to be embedded with environmental knowledge. In fact, human capital is highly connected to OLC [15] and the more knowledge acquired by the employees, the more the better organizational learning capability will be generated [14].

Secondly, green structural capital (GSC) was found to have no relationship on organizational learning capability (OLC). Thus, H2 was not supported. This finding is in contradicted with a previous study [16]. One main problem faced by organisations to invest in structure and system is due to higher cost [24]. It was believed that manufacturing SMEs have poor information technology system and knowledge repositories related to green protection. In fact, in the high-technology era, an update technology and system is required to ease the business process. The way knowledge created and transferred through all members in the organization will improve organizational learning in the organization. This limitation prevents the employees in the organisation to share their useful green knowledge to be embedded into OLC.

Thirdly, the finding of the study show that GRC has positive influence on OLC. Thus, H3 was supported. The result is similar with the study done by Omar et al. [20] revealed that the GRC has positive relationship with OLC. In addition, the result is consistent with Hsu and Fang [20] found that relational capital is the greatest factor among three types of intellectual capital contribute to better OLC. In this context, realizing that manufacturing SMEs have little resources, knowledge and expertise, they seek knowledge from their green relationships with their stakeholder. It is believed that the rich green network connections enabled the knowledge sharing on environmentalism and enhance the knowledge development in the organization. In addition, business models nowadays are transforming in which product and services is designed to be environmental-friendly. By having environmental knowledge, through a stable green relationship, it will increase their OLC.

6. Conclusion

The topic of green intellectual capital (GIC) become plausible solution to solve the environmental issues nowadays. However, GIC alone is not enough without intervention of organizational learning capability (OLC). However, the connection of two links are very limited. Hence, this study intends to better understanding the relationship between GIC and OLC in the context of manufacturing SMEs in Malaysia.

References

[1] Yong, J.Y., Yusliza, M-Y., Ramayah, T. & Fawehinmi, O. (2019). Nexus between green intellectual capital and green human resource management. *Journal of Cleaner Production*, ***215***, 364–374.

[2] Cheng, C., Yang, C., & Sheu, C. (2014). The link between eco-innovation, business performance: a Taiwanese industry context. *J. Clean. Prod.*, ***64***, 81–90.

[3] Chen, Y.-S. (2008b). The positive effect of green intellectual capital on competitive advantages of firms. *Journal of Business Ethics*, ***77*(3),** 271–286.

[4] Chang, C. H., & Chen, Y. S. (2012). The determinants of green intellectual capital. *Management Decision*, ***50*(1),** 74–94. https://doi.org/10.1108/00251741211194886

[5] Huang, C.-L. L., & Kung, F.-H. H. (2011). Environmental consciousness and intellectual capital management: Evidence from Taiwan's manufacturing industry. *Management Decision*, ***49*(9),** 1405–1425. https://doi.org/10.1108/00251741111173916

[6] Cai, W., Lai, K., Liu, C., Wei, F., Ma, M., Jia, S., … Lv, L. (2019). Promoting sustainability of manufacturing industry through the lean energy-saving and emission-reduction strategy. *Science of the Total Environment*, ***665***, 23–32. https://doi.org/10.1016/j.scitotenv.2019.02.069

[7] Camison, C., & Villar-Lopez, A. (2014). Organizational innovation as an enabler of technological innovation capabilities and firm performance. *Journal of Business Research*, ***67*(1),** 2891–2902.

[8] Bhaskar, A. U., & Mishra, B. (2017). Exploring relationship between learning organizations dimensions and organizational performance. *International Journal of Emerging Markets*, ***12*(3),** 593–609. https://doi.org/10.1108/IJoEM-01-2016-0026

[9] Leal-Rodríguez, A., Ariza-Montes, J., Rold an, J., & Leal-Mill an, A. (2014). Absorptive capacity, innovation, cultural barriers: a conditional mediation model. *J. Bus. Res.*, ***67*(5),** 763 768.

[10] Alegre, J., & Chiva, R. (2008). Assessing the impact of organizational learning capability on product innovation performance: An empirical test. *Technovation*, ***28*(6),** 315–326.

[11] Mbengue, A., & Sané, S. (2013). Capacité d'apprentissage organisationnel: analyse théorique et étude empirique dans le contexte des équipes de projets d'aide publique au développement. *Canadian Journal of Administrative Sciences/Revue Canadienne Des Sciences de l'Administration*, *30*(1).
[12] Baker, H. E. (1992). Employee socialization strategies and the presence of union representation. *Labor Studies Journal*, ***17***, 5–17.

[13] Gomez, P.J., Lorente J.C., & Cabrera, R. V. (2005). Organizational learning capability: A proposal of measurement. *Journal of Business Research*, ***58***, 715–725.

[14] Rothberg, G. S. (2009). Intellectual capital in business-to-business markets. *Industrial Marketing Management*, **2(1),** 19–34.

[15] Sachitra, V., & Chong, S. (2018). Resources , capabilities and competitive advantage of minor export crops farms in Sri Lanka An empirical investigation. https://doi.org/10.1108/CR 01-2017-0004

[16] Curado, C., Muñoz Pascual, L., & Galende, J. (2018). Antecedents to innovation performance in

SMEs: A mixed methods approach. *Journal of Business Research*, ***89***(January), 206–215. https://doi.org/10.1016/j.jbusres.2017.12.056

[17] Albort-Morant, G., Leal-Rodríguez, A., & De Marchi, V. (2018). Absorptive capacity, relationship learning mechanisms as complementary drivers of green innovation performance. *J. Knowl. Manag.*, ***22*(2),** 432–452.

[18] Hill, C.W.L., & Jones, G. R. (2001). *Strategic Management: An Integrated Approach*. Houghton Mifflin, Boston.

[19] Fornell, C., & Larcker, D. F. (1981). Evaluating structural equation models with unobservable variables and measurement error. *Journal of Marketing Research*, ***18*(1),** 39–50.

[20] Omar, M. K., Mohd Yusoff, Y., & Kamarul Zaman, M. D. (2019). The Effect of Organizational Learning Capability as a Mediating Variable in the Relationship between Green Intellectual Capital and Business Sustainability: Evidence from the Manufacturing Sector. *International Journal of Academic Research in Business and Social Sciences*, **9(6),** 337–352. https://doi.org/10.6007/IJARBSS/v9-i6/5974

[21] Hsu, YY. & Fang, W. (2009). Intellectual capital and new product development performance: The mediating role of organizational learning capability. *Technological Forecasting and Social Change*, ***76*(5)**, 664–677. https://doi.org/10.1016/j.techfore.2008.03.012

[22] Hsu, L.-C., & Wang, C.-H. (2010). Clarifying the Effect of Intellectual Capital on Performance: The Mediating Role of Dynamic Capability. *British Journal of Management*.

[23] Yacob, P. & Moorthy, M. (2012). Green Practices: Perception of Malaysian SME Owners/Managers. *International Journal of Academic Research in Economics and Management Sciences*, ***1*(3)**, 103–111.

[24] Vaghefi, N., Siwar, C., Aziz, S., & Ghani, A. (2015). Green Economy : Issues , Approach and Challenges in Muslim Countries. *Theoretical Economics Letters*, *5*(February), 28–35.

IOP Conf. Series: Materials Science and Engineering **697** (2019) 012010 doi:10.1088/1757-899X/697/1/012010

Supply Chain Performance Measurement Using Supply Chain Operation Reference (SCOR) in Sugar Company in Indonesia

E Kusrini[1], V N Helia[1*], M P Maharani[1]

[1]Industrial Engineering Department, Faculty of Industrial Technology, Universitas Islam Indonesia, Yogyakarta, Indonesia
*Email : vembri@uii.ac.id

Abstract. This study aims to measure supply chain performance by using SCOR (Supply Chain Operation Reference), a process-based model. It focuses on five business processes namely Plan, Source, Make, Deliver, and Return. There are 45 KPIs grouped into 5 main KPI attributes: Reliability, Responsiveness, Agility, Cost, and Asset Management. In order to get a more precise performance value, then weighted by using Analytical Hierarchy Process (AHP) method. The final value is obtained from the multiplication of the value of performance by the weight of importance. This study was conducted by taking a case study at one of the sugar company in Indonesia. The results of this study indicate that the supply chain performance of the sugar company is in a good category with the final grade of 70.94 from the 0-100 scale. Value of each process consecutively as follows: Plan 18.47, Source 16.12, Make 10.73, Deliver 10.79, and Return 7.1. Some recommendations are given to the company for evaluation and refinement focused on low business process values.

1. Introduction

XYZ Ltd is one sugar-cane based agroindustry company. The company has a sugar factory and a spiritus factory. Industry in the field of sugar production is one promising industry because sugar is one of the staple food needs used by humans every day. At present, the sugar-producing industry is increasing in number. In addition, demand for products, especially sugar, continues to increase. This is evidenced by the comparison between total production and total consumption. According to the Executive Director of the Indonesian Sugar Association (AGI), the need for sugar for both household and industrial will continue to increase in line with the increasing population. This phenomenon requires companies to rearrange their business strategies and tactics. Therefore, companies need to think about how to implement processes in producing products that are better, faster, and cheaper than other competitors.

Based on observations there are problems in the supply chain XYZ Ltd. These problems come from suppliers and consumers. Judging from the aspect of suppliers, the quality of sugar cane supplied is quite a lot which has not been in accordance with the standards for the production process. Usually, the sugar cane is damaged like wilting, too young, and the condition is dirty, causing low water content and sugar cane is not good. In addition, the weather factor is also a determinant of the quality of sugarcane, when it is in the rainy season it will usually result in increased sugarcane growth rates and affect the quality

IOP Conf. Series: Materials Science and Engineering **697** (2019) 012010 doi:10.1088/1757-899X/697/1/012010

of sugarcane due to decreased yield levels. Furthermore, from the aspect of the consumer where there are still products returned by consumers due to damage or return products.

From the problems found in the aspects of supply chain performance, it can be concluded that supply chain performance is less efficient. It is necessary to evaluate the supply chain system at XYZ Ltd as a whole so that the company know the supply chain conditions and improve the supply chain process going forward. Some companies carry out supply chain analysis in improving their performance to gain competitive advantages [1]. Performance measurement is also carried out for many studies [2]. The Hellenic Sugar Industry (HSI) conducts a combination of supply chain analysis with analysis of customer demand, logistics costs and the development of suitable transportation models [3]. Many methods of supply chain performance measurement system (SCPMS). One method of measuring supply chain performance is using processes, where the activities of business processes are used to identify performance measures and metrics [4]. SCOR (Supply Chain Operation Reference) provides standard process definitions, including terminology and metrics [5]. The Supply Chain Council develops SCOR that aims to evaluate supply chains. SCOR also provides usual performance framework, standard terminology, and optimum practices [6]. SCOR also has advantages in terms of process details and can be used for benchmarks. Some SCOR applications are widely used in many industries, one of them in Taiwan [7]. The SCOR model is a useful tool and is widely followed by the manufacturing industry [8], but it is not easy to apply it. For example, the application of SCOR in the construction industry [9].

One process and other processes in the supply chain has a different critical level [10]. To be able to distinguish priorities between processes, one weighting method is used, namely the Analytical Hierarchy Process [11]. Thus, the supply chain can also be combined with the Analytical Hierarchy Process (AHP) to overcome various problems, such as in the poultry supply chain [12]. SCOR with AHP (Analytical Hierarchy Process) and fuzzy AHP to create a valuation model in risk uncertainty in the supply chain of body manufacturing bus [13]. SCOR is also quite extensive in its application if it is associated with environmental issues, namely green SCOR [14]. Therefore, in this study using SCOR and AHP in measuring performance in sugar companies and providing recommendations on improvements in order to improve the efficiency of supply chain performance at XYZ Ltd.

2. Methodology

The focus of the study is the measurement of supply chain performance using SCOR 11.0 [15] and AHP methods [11]. The K-Chart research explains in detail the research focus and state of the art from this research. The measurement of supply chain performance using the SCOR and AHP method. K-Chart describes that the type of manufacturing or the object of research is the food industry with products, namely sugar. In the next sequence is the type of production where the company uses the type of production of MTS (Make To Stock). Then, the aspects to be analyzed are economic aspects with the method chosen is SCOR 11.0.

The research stage is as follow in figure 1. Firstly, mapping the business process with SCOR approach. The SCOR approach has a plan, source, make, deliver, return, and enable activities. Then, the next stage is designing performance metrics. In evaluating using SCOR there are several parameters including reliability, responsiveness, agility, cost and asset management. These parameters needed many data. The reliability data related to fulfilling the perfect order needed, such as demand data, forecasting data, on-time delivery, timelines of order receipt, verification of the number and quality of products from suppliers, verify the number and quality of return products from the customer. The responsiveness data in this paper means cycle time in completing orders from consumers. It is obtained directly from production, marketing, purchasing, and installation. There are procurement cycle time, production cycle time, delivery cycle time, and return cycle time. The agility data is data added time that needed by the company in completing orders from consumers. The cost data is data on total expenditure cost incurred every period. Assets management data has to do with cash to cash cycle time, which is the time required in its business cycle. Supporting data are pending sales day, inventory supply data, etc.

The AHP method in this study is used for weighting wherein the results of SCOR and AHP will be integrated so that the results of the supply chain performance from the company are obtained. To be able to understand the meaning of the calculation results, an analysis is performed for each of the performance attributes.

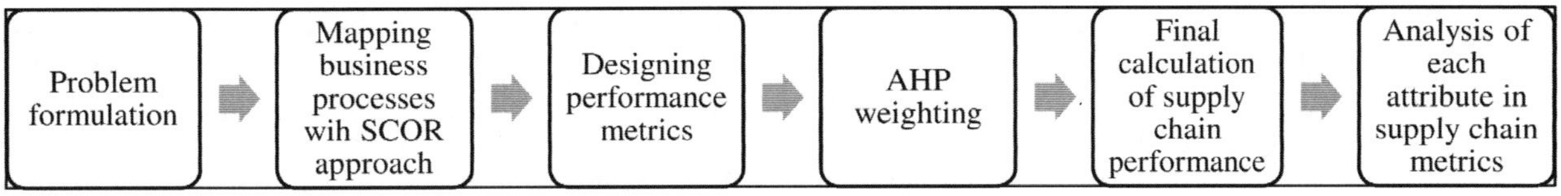

Figure 1. Research stage

3. Result and Discussion

3.1. Mapping Business Processes with the SCOR Approach

This study uses the SCOR approach to supply chain measurement. Therefore, it is necessary to map business processes first. Starting from level one and level two to the third level accompanied by business metrics. Table 1 shows the mapping of XYZ Ltd business processes at level 1 and level 2 using SCOR. Then, the whole process and sub-process are summarized into one in the supply chain flow pattern using the SCOR approach (figure 2).

Table 1. Mapping Business Processes XYZ Ltd

No.	Business Process	SCOR Level 1	SCOR Level 2	Implementing
1	Plan	Plan	Plan Make	Director, Department
2	Supply	Source	Source Stocked Product	Farming Facility & Planting Section, Warehouse Section, Accounting & Finance Section
3	Production	Make	Make to stock	Fabrication Section, Installation Section
4	Distribution	Deliver	Deliver Stocked Product	Marketing Section, Sugar Warehouse Section
5	Return	Return	Deliver Return Defective Product	Marketing Section, Sugar Warehouse Section
6	Management	Enable	Manage Performance	Director, Department

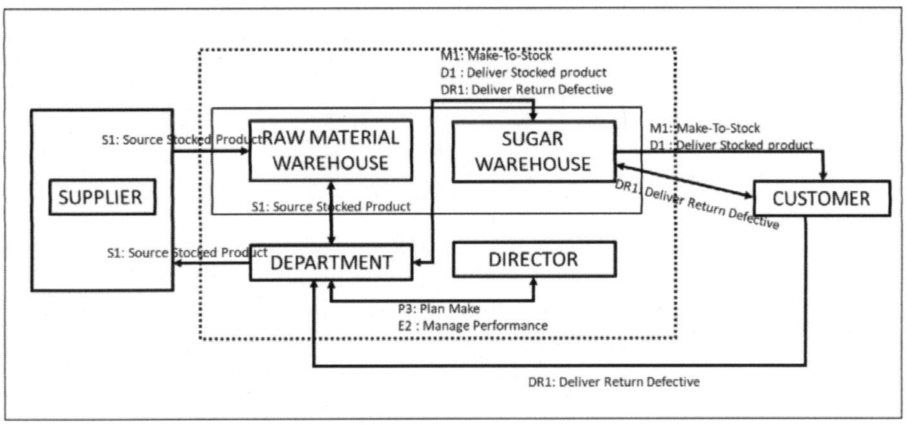

Figure 2. Supply Chain Flow Pattern With SCOR Approach

3.2. Designing Performance Metrics

Each process has a performance metric obtained from the SCOR approach. In addition, based on the mapping of previous business processes, it can be identified as the factors that influence the supply chain performance assessment of XYZ Ltd. This is defined as the performance attributes of each metric. The specific description of each process and the classification of metrics into performance attributes can be explained as follows (Table 2).

Table 2. Designing Performance Metrics

NO.	Level 1 SCOR	Level 2 SCOR	METRIC	PERFORMANCE ATTRIBUTE
1	PLAN	Plan Make	RL 3.37 Forecast Accuracy	RELIABILITY
2			RS 3.42 Identify, prioritize, and aggregate production requirement cycle time	RESPONSIVENESS
3			RS 3.36 Identify, assess, and aggregate production resources cycle time	
4			RS 3.13 Balance production resource with production requirement cycle time	
5			RS 3.28 Establish production plans cycle time	
6	SOURCE	Source Stocked Product	RL 3.18 %Order/lines processed complete	RELIABILITY
7			RL 3.20 %Order/lines received on-time to demand requirement	
8			RL 3.24 %Order/lines received damage free	
9			RL 3.21 %Order/lines received with correct content	
10			RL 3.25 %Product transferred on-time to demand requirement	
11			RS 3.14 Verify product cycle time	RESPONSIVENESS
12			RS 3.8 Authorize supplier payment cycle time	
13			CO 3.009 Purchased material cost	COST
14			AM 3.16 Inventory days of supply-raw material	ASSET MANAGEMENT
15	MAKE	Make To Stock	RL 3.49 Schedule achievement	RELIABILITY
16			RL 3.58 Yield	
17			RS 3.49 Issue material cycle time	RESPONSIVENESS
18			RS 3.101 Produce & test cycle time	
19			RS 3.142 Package cycle time	
20			RS 3.114 Release finished product to deliver cycle time	
21			CO 3.014 Production labor cost	COST
22			CO 3.016 Production property, plant, & equipment cost	
23			AM 3.9 Capacity utilization	ASSET MANAGEMENT
24	DELIVE R	Deliver Stocked Product	RL 3.33 Delivery item accuracy	RELIABILITY
25			RL 3.34 Delivery location accuracy	
26			RL 3.35 Delivery quantity accuracy	
27			RL 3.41 Orders delivered damage free conformance	
28			RL 3.50 Shipping documentation accuracy	
29			RS 3.100 Process inquiry & quote cycle time	RESPONSIVENESS
30			RS 3.112 Receive, enter, & validate order cycle time	
31			RS 3.116 Reserve resources & determine delivery date cycle time	
32			RS 3.18 Consolidate orders cycle time	
33			RS 3.16 Build loads cycle time	
34			RS 3.126 Ship product cycle time	
35			RS 3.102 Receive & verify product by customer cycle time	
36			AG 3.32 Current delivery volume	AGILITY
37			CO 3.022 Transportation cost	COST
38			AM 3.45 Inventory days of supply-finished goods	ASSET MANAGEMENT
39	RETURN	Deliver Return Defective Product	RS 3.5 Authorize defective product return cycle time	RESPONSIVENESS
40			RS 3.19 Current customer return order cycle time	
41			RS 3.104 Receive defective product cycle time	
42			RS 3.136 Transfer defective product cycle time	
43			AG 3.31 Current deliver return volume	AGILITY
44	ENABLE	Manage Performance	RS 3.78 Manage production performance cycle time	RESPONSIVENESS
45			CO 3.017 Production GRC, inventory & overhead cost	COST

The Plan process has a subprocess namely Plan Make. The metrics in this sub-process are : (1) Forecast accuracy with the attributes of Reliability; (2) Identify, prioritize, and aggregate cycle time production requirements; (3) Identify, assess and aggregate cycle time production resources; (4) Balance production resources with cycle time production requirements, and (5) Establish cycle time production plans with Responsiveness attributes. The Source process has subprocesses there are Source Stocked Products, the metrics in this subprocess are : (6) % Orders/lines processed complete; (7) % Orders/lines received on-time to demand requirements, (8) % Orders/lines received defect free; (9) % Orders/lines received with correct content; (10) % Product transferred on-time to demand requirements with

Reliability attributes; (11) Verify cycle time products; (12)Authorize cycle time payment suppliers with Responsiveness attributes; (13)Purchased material costs with Cost attributes; (14) Inventory days of supply-raw material with Asset Management attributes.

The Make process has subprocess namely Make-to-Stock. The metrics in this subprocess are: (15) Schedule Achievement; (16)Yield with Reliability attribute; (17) Issue material cycle time; (18) Produce and test cycle time; (19) Package cycle time; (20) Release finished product to deliver cycle time with Responsiveness attribute; (21) Production labor cost; (22) Production property, plant, and equipment cost with the attribute Cost; (23) Capacity utilization with the Asset Management attribute. Deliver process has sub-processes, namely Deliver Stocked Product. The metrics in this subprocess are: (24) Delivery item accuracy; (25) Delivery location accuracy; (26)Delivery quantity accuracy; (27) Orders delivered defect-free conformance; (28) Shipping documentation accuracy with Reliability; (29) Process inquiry and quote cycle time; (30) Receive, enter, and validate orders cycle time; (31) Reserve resources & determine delivery date cycle time; (32) Consolidate orders cycle time; (33) Build loads cycle time; (34) Ship products cycle time; (35)Receive & verify products by customer cycle time with Responsiveness attributes; (36) Current delivery volumes with Agility attributes; (37)Transportation costs with Cost; (38) Inventory days of supply-finished goods with attribute Asset Management.

Table 3. Results of XYZ Ltd Score Calculation

L1	Metric Name	Final Score	Weight	Score	Total
PLAN	*Reliability*	91,69	0,67	61,43	92,37
	Responsiveness	93,75	0,33	30,94	
SOURCE	*Reliability*	99,83	0,33	32,94	76,79
	Responsiveness	63,77	0,22	14,03	
	Cost	55,28	0,25	13,82	
	Asset Management	80	0,20	16	
MAKE	*Reliability*	53,24	0,36	19,17	63,14
	Responsiveness	70,1	0,22	15,42	
	Cost	38,89	0,22	8,55	
	Asset Management	100	0,20	20	
DELIVER	*Reliability*	99,98	0,30	30	71,94
	Responsiveness	69,86	0,29	20,26	
	Agility	66,67	0,12	8	
	Cost	55,39	0,20	11,08	
	Asset Management	28,95	0,09	2,6	
RETURN	*Responsiveness*	86,8	0,70	60,76	70,76
	Agility	33,33	0,30	10	
ENABLE	*Responsiveness*	50	0,47	23,5	42,94
	Cost	36,69	0,53	19,44	

IOP Conf. Series: Materials Science and Engineering **697** (2019) 012010 doi:10.1088/1757-899X/697/1/012010

The Return process has subprocesses namely Deliver Return Defective Product. The metrics in this subprocess are: (39) Authorize defective product return cycle time; (40) Current customer return order cycle time; (41) Receive defective product cycle time; (42) Transfer defective product cycle time with Responsiveness attribute; (43) Current deliver return volume with Agility attribute. The Enable process has subprocesses that are Manage Performance. The metrics in this subprocess are: (44) Manage production performance cycle time with Responsiveness; (45) Production GRC attributes, inventory, and overhead costs with Cost attributes. So, there are 45 metrics from 6 processes in the implementation of SCOR at XYZ Ltd.

3.3. AHP weighting

The score results for each attribute in table 3 obtained from the previous calculation. The weight used is the result of weighting calculations on level 2 using the AHP method. AHP calculations are performed on levels 1,2 and 3, where weighting at level 3 is given the assumption that each metric has the same important weight. The data used amounted to 4 people, therefore in the previous calculation also used Geometric Mean method to find the average value of the paired comparison. After that, a consistency test is also carried out where if the value of Consistency Ratio (CR) less than 0.1, then the data is declared true or consistent. Based on the results of this attribute score calculation, a score can be obtained for each process, where the score of this process will be recalculated to determine the final measurement of the supply chain performance.

3.4. Final Calculation of Supply Chain Performance

The final calculation of supply chain performance is obtained from the calculation of the results of SCOR calculations with AHP. Based on the calculation, it can be obtained the final score for the performance of the XYZ Ltd supply chain is 70.94 that can be seen in table 4. The data used is the total score for each process multiplied by the weight of each process. The weight is obtained from the weighting results at level 1 using the AHP method.

Table 4. Results of Final Score of Supply Chain Performance of XYZ Ltd

Process	Total Score	Weight	Final Score
Plan	92,37	0,20	18,47
Source	76,79	0,21	16,12
Make	63,14	0,17	10,73
Deliver	71,94	0,15	10,79
Return	70,76	0,10	7,1
Enable	42,94	0,18	7,73
Total			70,94

3.5. Analysis of Each Attribute in Supply Chain Metrics

Table 4 indicates the value of attribute scores in each process. The Plan process has two attributes. There is Reliability with a score of 91.69 and Responsiveness 93.75. This shows that for planning XYZ Ltd is good in terms of accuracy and time efficiency. However, there are still some attributes that still score below 70 including the Source process with the attributes of Responsiveness and Cost. Then, Make process with attributes Reliability and Cost, Deliver process with attributes of Responsiveness, Agility, Cost, and Asset Management, Return process with attribute Agility, and finally, Enable process with the attributes of Responsiveness and Cost.

Therefore, the company needs to evaluate and improve so that the company's supply chain performance can be better. Suggestions for improvements are improvements focused first on sugar cane farmers because to produce sugar the main raw material used is sugar cane. Thus, sugar cane is the most important thing in the continuity of production at XYZ Ltd. It is important to always maintain the quality and continuity of sugarcane from sugar cane farmers so that in the future XYZ Ltd can maintain the quality of sugar produced with the maximum production target according to achieve. It is also necessary to improve the quality of the Source process. Identification of key performance indicators and best practices that conducted an analysis of level 3 (metric) needs to be done. In modeling the Source process a Design Chain Operation Reference (DCOR) can be made before integrating it with the SCOR model [16].

Table 4 shows that the Return process has the lowest value. One way to manage the Return process is to use functional integration in the company [17]. The focus of that research is on the integration of marketing and logistics which can affect the effectiveness of the return management process. The results of the identification of external factors that can influence it, such as the customer market, competitive environment, and regulatory environment. Therefore, repairs also need to be done for the product delivery process, because every month there are still products returned by consumers. It happened because they are damaged during the shipping process. The company needs to add a protector to the product that is sent to minimize damage to the packaging. In terms of the speed of time in shipping also need to be considered to increase customer satisfaction to the company.

4. Conclusions and Recommendations

Supply chain measurements are carried out using SCOR 11.0. The conclusion of the research is as a priority weight determinant which includes 5 processes, namely Plan, Source, Make, Deliver, and Return. It is known that the level of supply chain performance XYZ Ltd is 70.94 on a scale of 0-100. This can be said that the level of performance of the XYZ Ltd supply chain is good. In addition, the weighting results using AHP according to 4 experts are for the highest weight process in the Source process with a weight of 0.21 and the lowest weight in the Return process with a weight of 0.10. Suggestions that can be given to companies are companies can hold training for farmers and company employees, tighten the process of quality control for raw materials and finished products, provide additional protection for products shipped, provide subsidies for superior seeds to farmers to improve product quality, holding forums with farmers to establish good relationships with companies, tighten selection supplier and prioritize suppliers loyal to maintain contract sustainability.

Companies need to establish methods used to assess supply chain performance so that companies can improve performance from all sides, such as product quality and company resources. So, it is expected that the company will get more profit and overall company performance will increase as well. For further research, it is expected to be more completed in determining the performance measurement using SCOR. For example, we can add skills to people in using SCOR and make sustainability to implementing SCOR in industry. It is hoped that this method of SCOR 11.0 can be developed to measure supply chain performance by integrating supply chain performance measurement with the environment, ICT, etc.

REFERENCES

[1] Delipinar GE, Kocaoglu B 2016 Using SCOR Model to Gain Competitive Advantage: A Literature Review *Procedia - Social and Behavioral Sciences* **229** pp 398-406.

[2] David R, Josep J 2014 Study on performance measurement systems-Measures and Metrics *Int. J of Scientific and Research Publications* Vol. **4**, Issue 9, 1.

[3] Ioannou G 2005 Streamlining the supply chain of the Hellenic sugar industry, J. of Food Eng. **70** (3) pp 323-332.

[4] Chan FTS, Qi HJ 2003 Feasibility of performance measurement system for supply chain: A process-based approach and measures *Integrated Manufacturing Systems* **14**(3) pp. 179–190.

[5] Stewart G 1997 'Supply-chain operations reference model (SCOR): The first cross-industry framework for integrated supply-chain management', *Logistics Information Management,* **10**(2) pp. 62–67.

[6] Huan SH, Sheoran SK, Wang G 2004 A review and analysis of Supply Chain Operations Reference (SCOR) model *Supply Chain Management: An International Journal* **9**(1) pp 23-29.

[7] Lee JS, Tzong-Ru, Shiu, Yi-Siang, Sivakumar P 2012 The applications of SCOR in manufacturing: two cases in Taiwan *Procedia Engineering* **38** pp 2548-2563.

[8] Prakash S, Sandeep Gunjan S, Rathore A, Supply Chain Operations Reference (SCOR) model: an overview and a structured literature review of its application, 2013.

[9] Persson F, Jonas B, Orjan G 2010 Construction logistics improvements using the SCOR model-Tornet case, Advances in Production Management Systems. New Challenges, New Approaches, Springer Berlin Heidelberg pp 211-218.

[10] Palma-Mendoza JA, Neailey K, Roy R 2014 Business process re-design methodology to support supply chain integration *Int. J. of Inf. Mgt* **34** (2) pp 167-176.

[11] Saaty TL, Vargas IG 2012 Models, methods, concepts, & applications of the analytic hierarchy process (2nd. Ed.), New York : Springer.

[12] Bukhori IB, Widodo KH, Ismoyowati D 2015 Evaluation of Poultry Supply Chain Performance in XYZ Slaughtering House Yogyakarta Using SCOR and AHP Method *Agriculture and Agricultural Science Procedia* **3** 221-225.

[13] Butdee S, Phuangsalee P 2019 Uncertain risk assessment modelling for bus body manufacturing supply chain using AHP and fuzzy AHP *Procedia Manufacturing* **30** pp 663-670.

[14] Ntabe EN, LeBel L, Munson AD, Santa-Eulalia LA 2015 A systematic literature review of the supply chain operations reference (SCOR) model application with special attention to environmental issues *Int. J. of Production Economics* **169** pp 310-332.

[15] SCC 2012 Supply Chain Operations Reference Model Revision 11.0. Pittsburg, PA :Supply Chain Council Inc.

[16] Zuniga R, Marcus S, Klaus-Dieter T 2013 Study on the Application of DCOR and SCOR models for the sourcing Process in the Mineral, Raw Material Industry Supply Chain. Dynamics in Logistics, Springer Berlin Heidelberg pp 211-220.

[17] Mollenkopf D, Russo I, Frankel R 2007 The returns management process in supply chain strategy *Int. J. of Physic. Dist. & Log. Mgt* **37** (7) pp 568-592.

Acknowledgments

This research was supported by Directorate of Research and Community Service or DPPM (Direktorat Penelitian dan Pengabdian Masyarakat) Universitas Islam Indonesia, Yogyakarta, Indonesia and Industrial Engineering Department, Faculty of Technology Industry, Universitas Islam Indonesia, Yogyakarta, Indonesia.

Integration of UI Greenmetric performance measurement on ISO 14001 implementation in higher education

R Nurcahyo[1*], F S Handika[1], D S Gabriel[1], and M Habiburrahman[1]

[1]Industrial Engineering Department, Faculty of Engineering, Universitas Indonesia, Depok, Indonesia
[*]E-mail: rahmat@eng.ui.ac.id

Abstract. Increasing sustainability concern in the world, encouraging the growth of higher education participation. Many higher education institutions adopt Environmental Management System (EMS) ISO 14001 and UI GreenMetrcis performance measurement separately to achieve sustainability. The two systems have similar goal but different approach related to sustainability. The objective of this research is to integrate UI Greenmetric performance measurement on ISO 14001 implementation in higher education. The similarity goals of them, making the analysis of the using of UI GreenMetric performance measurement in the implementation of ISO 14001 EMS in higher education needs to be done. Based on that analysis, it is known the agreement level of relationship between EMS ISO 14001 and UI GreenMetric by using frequency analysis, as well as the level of their implementations in higher education through mixed method. The result is UI GreenMetric performance measurement can be used to implement ISO 14001 EMS, so that it can support the achievement of environmental performance in higher education.

1. Introduction

Higher education activities and operations to fulfill its function has impact on the environment. Green campus is an effort by higher education institution to participate and promote in reduction of the negative impact on the environment [1]. Higher education have commitment for continuos improvement to protect environment [1, 2, 3, 4]. Environmental Management System (EMS) ISO 14001 have been implemented by many higher education institutions around the world. EMS ISO 14001 objectives are improvement of environmental performance and reduction of negative impact on the environment [5]. Moreover, higher education make efforts towards continuous improvement of environmental performance through EMS [1]. It indicates the importance of EMS implementation as an effort to achieve sustainability in higher education.

Previous studies proposed evaluation for environmental performance in higher education. The UI GreenMetric World University Ranking is an initiative of Universitas Indonesia which is being launched in 2010 for green campus assessment. There are 6 criteria with 39 indicators to measure the environmental performance in UI GreenMetric

Alshuwaikhat and Abubakar (2008) stated that ISO 14001 EMS is an effective tool for controlling processes in a system regarding to the environment, because it is considered to contribute in achieving the goals as green campus [1]. Thus, it can be said that the concept of a green campus and ISO 14001 EMS related each other.

2. Literature Review

2.1. Sustainability Concept

According to Jorge, Madueño, Cejas, and Peña (2014), the concept of sustainability is part of the merger between the economic, social, and environment dimension, published in the Brundtland Report by World Commission on Environmental and Development (WCED) in 1987 [6]. The concept of sustainability is often represented as a combination of the three dimensions of social, economic, and environment, in which sustainability is an area where an organization can operate and maintain the harmony of these values [7].

The economic dimension describes a system of production, distribution, and consumption of wealth or well-being, which can be generally defined as a sense of someone's needs satisfaction. Furthermore, the social dimension is described as a system of group life that considers the interests in maintaining and improving the standard of human life. Meanwhile, the environment dimension describes a system that provides integrity and protection of the ecosystems as well as pays attention on productivity and sustainable functioning of these ecosystems [7].

One of the effort to implement sustainability is the adoption of international standard such as ISO 14001 and ISO 9001. The adoption of these standards significantly improving several aspects, not only creates suatainable economic but also company competitiveness and operational performances such as production volume, production efficiency, and time efficiency [8].

2.2. Green Campus

Alsuwaikhat and Abubakar (2008) stated that the beginning of the sustainability concept in higher education is the existence of the Stockholm Declaration in 1972. The declaration described the relationship of interdependence between humans and the environment and proposed several ways to achieve sustainability. Then, the higher education continues to find ways to engage and deploy their resources in order to respond the challenges of a balance between economic, technological developments, and environmental protection. Thus, the higher education compelled to incorporate the concept of sustainability into the system being operated by them, through the implementation of green campus [1].

According to Alsuwaikhat and Abubakar (2008), green campus is a form of management for the higher education environment by considering the construction of environmentally buildings and transportation facilities. Environmentally building (green building) can help to reduce energy usage and improve comfort for the university. Then, the green on the transportation system (green transportation) in higher education can be achieved through the using of bicycles in the campus as well as the providing of public transportation for all staffs, students, and visitors. Thus, it can reduce emissions and congestion [1].

2.3. UI GreenMetric

According to Lauder, Sari, Suwartha, and Tjahjono (2015), UI GreenMetric is a tool developed by Universitas Indonesia to help the higher education measure sustainability efforts within the university environment [9]. UI GreenMetric's advantage is its rating system and it can be used as a framework and standards guidelines for realizing the green campus, so higher education can achieve better rankings in the next evaluation. UI GreenMetric can help higher education to better understand what are the things should be measured by them. There are six categories on UI GreenMetric, namely setting and infrastructure, energy and climate change, waste, water, transportation, and education.

2.4. ISO 14001 Environmental Management System

Environmental Management System (EMS) is part of the organization's management system that aims to control the environmental aspects associated with activities, products, and services produced by them. ISO (2004) defines EMS as a set of elements that are interrelated and a part of the organization's management system used to develop and implement environmental policies as well as to manage the environmental aspects [10, 11]. There is sustainability process in EMS, in which the sustainability concept is defined by the Environmental Protection Agencies (2007) as an ability to achieve economic prosperity continuously to protect nature and provide a high quality of life for the people [12, 13].

Sustainability processes in EMS referenced to ISO 14001. ISO 14001 is an international environmental standards related to EMS which allows an organization to plan policies and objectives based on information about environmental influences [14]. Meanwhile, ISO 14001 as defined as an international standard based on the concept that environmental performance will be achieved when the environmental aspects are identified, so it contributes in improving the continuous improvement. Then, they stated that ISO 14001 EMS is not only as a system of environmental performance, but also can be used to control the sustainable development and creating value in the process of quality improvement [15]. Furthermore, according to ISO (2004), ISO 14001 guidelines consist of four clauses, which in the forth clause consists of clause 4.2 environmental policy, 4.3 planning, 4.4 implementation and operation, 4.5 checking, and 4.6 management review [10].

2.5. Frequency Analysis

In a research, researcher needs to make the description of the data presentation in tables and graphs. In table, the data are arranged according to a rule or a specific category so that, an overview of the data can be read easily and systematically. Then in graph, the data are visualized in images. By using frequency tables, it can be seen a general description of the data. The table is made with data sets that have been grouped by category or specific rules. Frequency table can be presented in the some columns, for example column variable categories, frequency, cumulative frequency, and cumulative percent [16].

2.6. Mixed Method

Sugiyono (2013) stated that a comprehensive data are complete data which is a combination of quantitative and qualitative data. With the combination of these methods, the data would be more valid. Thus, the data accuracy that can not be validated with quantitative methods, can be validated using qualitative methods, or conversely. Moreover, with mixed method, the reliability of the data that can not be tested using quantitative data, can be tested using qualitative data, or conversely [17].

3. Methods

This study use quantitative and qualitative data. For collecting quantitative data, the study use a research instrument, questionnaire. There are two questionnaires that should be designed, the first questionnaire is aim to determine the level of the relationship agreement between the requirements of ISO 14001 EMS with UI GreenMetric categories, and the second questionnaire is aim to determine their implementation level in higher education. Every statement in the questionnaire regard to the requirements of ISO 14001 EMS, clauses 4.2 to clause 4.6, and also six categories of UI GreenMetric. Then, for collecting qualitative data, the study use secondary data related to the implementation of ISO 14001 EMS and UI GreenMetric in higher education.

In designing of the first questionnaire, researcher was brainstorming with experts of UI GreenMetric. Next, determining the target respondents, the respondents are experts on sustainability and participant of UI GreenMetric. After that, questionnaire can be given to the respondents. In the process of collecting data, the questionnaire also gets revised from expert on sustainability, so it needs to be improved in some parts of the questionnaire. And then, in designing of the second questionnaire, the researcher used indicators in the implementation of ISO 14001 EMS based on Environmental Management Guide for the University from Environmental Protection Agency. For the assessment associated with the UI GreenMetric categories, it uses indicators of UI GreenMetric as statements in the questionnaire. The target respondents in second questionnaire are experts on sustainability and higher education that have been certified ISO 14001. After that, the questionnaire can be given to the respondents.

From both of questionnaires, quantitative data were obtained, and then processed using frequency analysis. In the second questionnaire, for supporting the result of quantitative data, qualitative data will be used through the case studies. That data is about information related to the implementation of ISO 14001 EMS requirements and UI GreenMetric categories in respondents' higher education. Furthermore, quantitative and qualitative data will be conducted a comparative analysis using a mixed method to determine the compatibility between quantitative and qualitative data that were obtained.

4. Results and Discussion

4.1. The Agreement Level Identification of Relationship between ISO 14001 EMS and UI GreenMetric

Table 1. Survey result.

Clause \ Category	Setting and infrastructure	Energy and climate change	Waste	Water	Transportation	Education
4.2	73	81	82	79	81	73
4.3.1	71	84	84	82	80	68
4.3.2	66	70	73	66	64	60
4.3.3	70	75	78	78	74	74
4.4.1	68	70	70	70	71	71
4.4.2	64	69	68	62	70	76
4.4.3	60	72	71	69	70	72
4.4.4	64	72	68	61	61	62
4.4.5	59	64	66	59	61	61
4.4.6	65	67	71	71	69	64
4.4.7	56	66	65	58	85	51
4.5.1	78	78	77	77	67	64
4.5.2	64	72	70	68	63	55
4.5.3	57	65	63	64	62	57
4.5.4	65	64	66	61	64	64
4.5.5	66	66	64	64	69	60
4.6	68	66	65	66	81	64

Based on the first questionnaire, the agreement level of relationship between ISO 14001 EMS and UI GreenMetric is obtained by using frequency analysis. The result is presented in Table 1.

4.2. The Implementation Level Identification of ISO 14001 EMS and UI GreenMetric in Higher Education

In this stage, quantitative and qualitative data is used to identify the implementation level of ISO 14001 EMS and UI GreenMetric in higher education. Based on the second questionnaire, quantitative data is obtained by frequency analysis, as presented in Table 2. Furthermore, qualitative data is obtained by case studies in respondents' higher education.

Table 2. Calculation result of the implementation ratings of ISO 14001 ems and UI Greenmetric in higher education.

	Ratings Mean	
	University of Cadiz	Vytautas Magnus University
ISO 14001 EMS Requirements		
Environmental Policy	05,00	02,06
Planning	04,00	02,52
Implementation and Operation	04,53	04,37
Checking	04,23	02,83
Management Review	05,00	03,42
Total	22,76	15,21
UI GreenMetric Categories		
Setting and Infrastructure	04,00	03,31

Energy and Climate Change	04,76	03,88
Waste	04,82	03,05
Water	05,00	03,98
Transportation	04,00	02,83
Education	05,00	04,31
Total	27,58	21,36

4.3. Case Study in Higher Education: University of Cadiz

The explanation related to the implementation of ISO 14001 EMS and UI GreenMetric can be outlined as follows:

A. ISO 14001 EMS Implementation

- Environmental policy: The higher education ensures that its policy includes a commitment to comply with the rules related to environment.
- Planning: The higher education implements the law and the current environmental regulation as well as ensures its compliance with the rule in future to manage the environment.
- Implementation and operation: The higher education implements the updating principle which aims to ensure the formation of human resources continuously to address environmental challenges in higher education and also implements external communication with suppliers.
- Checking: The higher education ensures the compliance of legal rules and procedures that have been designed and implemented for environmental management.
- Management review: The higher education evaluates its commitment to ensure the compliance with environmental problems in the future.

B. UI GreenMetric Implementation

- Setting and infrastructure: The higher education has did infrastructure planning effort that pay attention between the design and the using of open spaces and buildings with environmental impact.
- Energy and climate change: The higher education tries to pay attention in resource management by the using of renewable energy to reduce environmental impact.
- Waste: The higher education tries to do residue management effort to optimize the management of residues or waste materials generated from various activities, as well as improves reuse and recycling activities.
- Water: The higher education makes efforts to pay attention in resource management through water usage.
- Transportation: The higher education have paid attention for the suitability of infrastructure planning between the designing and using of the road with environmental criteria in their area.
- Education: The higher education provides environmental education to promote environmental practices and raise awareness contributing to solve environmental problems. [17].

4.4. Case Study in Higher Education: Vytautas Magnus University

The explanation related to the implementation of ISO 14001 EMS and UI GreenMetric can be outlined as follows:

A. ISO 14001 EMS Implementation

- Environmental policy: The higher education needs to develop environmental policy consistently and implements them efficiently.
- Planning: The higher education has arranged their environmental goals. However, between short-term and long-term goals have not been indicated properly.
- Implementation and operation: The higher education cooperates with stakeholders from the outside. In other words, they have established an external communication well.
- Checking: The higher education has implemented the measurement and reporting properly.
- Management review: The higher education makes the changes to environmental policy, but it still needs attention to keep it running consistently

IOP Conf. Series: Materials Science and Engineering **697** (2019) 012011 doi:10.1088/1757-899X/697/1/012011

B. UI GreenMetric Implementation

- Setting and infrastructure: The higher education allocates financial resources for infrastructure development and renovation as the sustainability efforts.
- Energy and climate change: The higher education tries to do energy saving and reduction of greenhouse gas emissions.
- Waste: The higher education has Go Green program that invites to reduce the using of paper and promotes recycling. However, this program has not been implemented in all areas.
- Water: The higher education pays attention in water usage and saving water.
- Transportation: The higher education executes Go Green program promoting the using of public transportation. But, it has not been implemented in all areas.
- Education: The higher education has provided several courses for their students that related to environmental and sustainability topics.

5. Discussion

5.1. The Agreement Level Analysis of Relationship between ISO 14001 EMS and UI GreenMetric

Referring to Table 1, it can be seen that the respondents agreed if each ISO 14001 EMS requirements associated with each UI GreenMetric category. The explanation as follows:

1. Setting and Infrastructure Category

Based on Table 1, it can be seen that all requirements of ISO 14001 EMS are assessed agree related to setting and infrastructure category. It shows that the implementation of them need to be considered in this category. However, the implementation of monitoring and measurement requirements obtain the highest relationship rating with setting and infrastructure category. Higher education concentrate on infrastructure development activities that pay attention in the green area to protect the environment for achieving sustainability through this category. Thus, it is important for higher education to monitor and measure the infrastructure development activities in order to keep attention to environmental impact. Therefore, the implementation of these requirements should be considered to support the achievement of the performance in this category. Top of Form Bottom of Form

2. Energy and Climate Change Category

Based on Table 1, it can be seen that all requirements of ISO 14001 EMS are assessed agree related to the category of energy and climate change. This shows that the implementation of them is also worth noting in this category. But, the implementation of environmental aspects requirement obtain the highest relationship rating with energy and climate change category. Higher education concentrate to achieve sustainability in this category through the using of energy efficient and environmental friendly. Thus, it is important for higher education to know the negative impact of energy sources used. Therefore, the implementation of this requirement should be considered to support the achievement of the performance in energy and climate change category.

3. Waste Category

Based on Table 1, it can be seen that all requirements of ISO 14001 EMS get agree assessment related to the category of waste. It shows that the implementation of all requirements of ISO 14001 EMS need to be considered in this category. However, implementation of the environmental aspects requirement is given the highest relationship rating with waste category. To achieve sustainability in the category of waste, higher education concentrate on management and recycling of waste. Thus, it is important for higher education to know the negative impact caused by the waste management that is not good. So that, the implementation of the requirement should be considered to support the achievement of the performance in this category.

4. Water Category

Based on Table 1, it can be seen that all requirements of ISO 14001 EMS are assessed agree related to the category of water. This shows that the implementation of all requirements of ISO 14001 EMS need to be considered in water category. But, implementation of environmental aspects requirement get the highest relationship rating with water category. Higher education concentrate to achieve sustainability in this category through water management activities in order to use more efficient. Thus, it is important for higher education know the negative effects caused by the uncontrolled using of water. Therefore, the implementation of that requirement should be considered to support the achievement of the performance in water category.

5. Transportation Category

Based on Table 1, note that all requirements of ISO 14001 EMS are assessed related to the category of transportation. It shows that the implementation of all requirements of ISO 14001 EMS need to be considered in this category. However, the implementation of environmental requirements objective and management review obtain the highest relationship rating with transportation category. Higher education is focus to achieve sustainability in this category through the transportation system management. Thus, it is important for higher education to arrange the environmental policy and management review related to the activities of transportation in their areas. It aims to reduce the negative impact on the environment due to air pollution generated by vehicles in it. Thus, the implementation of the requirements should be considered to support the achievement of the performance in the category of transportation.

6. Education Category

Based on Table 1, it can be seen that all requirements of ISO 14001 EMS are assessed agree relate to the category of education. This shows that the implementation of all requirements need to be considered in this category. But, the implementation of goals, objectives, and program requirement is scored the most related to education category. Higher education is focus to achieve sustainability in this category by providing educational activities related to the environment. Thus, it is important for higher education to determine the goals and objectives of the activities. Through the appropriate program, the higher education can achieve the goals and objectives that have been determined previously. So that, the implementation of that requirement should be considered to support the achievement of the performance on education category.follows:

5.2. The Implementation Level Analysis of ISO 14001 EMS and UI GreenMetric in Higher Education

Analysis of the implementation level from ISO 14001 EMS and UI GreenMetric can be done by comparing the quantitative data, based on the assessment of respondents on the second questionnaire that is presented in Table 2, as well as qualitative data, based on information about the real implementation of ISO 14001 EMS and UI GreenMetric in higher education.

Based on Table 2, it can be seen that the average assessments of the implementation of ISO 14001 EMS requirements and UI GreenMetric categories at University of Cadiz are respectively 22,76 and 27,58. Meanwhile, Vytautas Magnus University get average assessments of implementation of ISO 14001 EMS requirements and UI GreenMetric categories respectively are 15,21 and 21,36. The ideal total scoring for the entire statements in ISO 14001 EMS if all respondents answered strongly agree is 25. Then, for the entire statement on the UI GreenMetric if all respondents is strongly agreed, it will obtain 30 an ideal total scoring of 30.

Based on the above explanation, University of Cadiz is considered to have implemented ISO 14001 EMS and UI GreenMetric better than Vytautas Magnus University. Meanwhile, from the case studies, overall University of Cadiz have implemented ISO 14001 EMS and UI GreenMetric well. Then, at Vytautas Magnus University, although the implementation of ISO 14001 EMS requirements and UI GreenMetric categories are good, but there are still lack that must be improved in order to increase their achievement. This indicates that there is no difference between quantitative and qualitative data generated from both higher education. Thus, qualitative data can support quantitative data.

Overall, it can be said that the implementation of ISO 14001 EMS and UI GreenMetric interrelated. If higher education implements the ISO 14001 EMS very well, then the result of environmental performance, which is measured by using of UI GreenMetric, is also very good. However, if the implementation of ISO 14001 EMS in higher education is not good, then the result of environmental performance is also not good. So that, environmental performance in higher education can be achieved as expected by implementing both concepts properly.

IOP Conf. Series: Materials Science and Engineering **697** (2019) 012011 doi:10.1088/1757-899X/697/1/012011

6. Conclusion

Both of UI GreenMetric and ISO 14001 EMS are aim to achieve sustainability regard to the environmental aspects of the organization, in this case the higher education. Thus, it is important to analyse the using of UI Green Metric performance measure in the implementation of ISO 14001 EMS in higher education. By doing this analysis, it will be knew the requirements of ISO 14001 EMS are considered related to the UI GreenMetric categories and the implementation of the ISO 14001 EMS requirements for environmental performance in higher education that measured by UI GreenMetric categories. Therefore, it is expected to improve environmental performance at the higher education.

Based on the analysis of the relationship agreement level between the requirements of EMS ISO 14001 and category UI GreenMetric, it obtains the conclusion that each ISO 14001 EMS requirement have relationship with one of the categories in UI GreenMetric. So that, the implementation of these requirements is considered supporting the achievement of environmental performance based on categories of UI GreenMetric. Furthermore, from the analysis of their implementation level, it can be concluded that the implementation of ISO 14001 EMS can support the achievement of environmental performance measured by UI GreenMetric. Through the implementation of ISO 14001 EMS, the environmental management process can be controlled, so that the environmental performance can be improved. Therefore, in the implementation of ISO 14001 EMS in higher education, UI GreenMetric performance measure can be used to improve the environmental performance achievement.

Time limitations of the study cause the data that are collected has not met expectations. It is because the lack of the number of respondents who give feedback on the study. Thus, this study still needs to be further investigated with more number of respondents in order to obtain a more valid research results. For further research, it can integrate ISO 14001 EMS and UI GreenMetric as an environmental performance measurement method in higher education that more comprehensive.

7. References

[1] Alshuwaikhat H M and Abubakar I 2008 An integrated approach to achieving campus sustainability: assessment of the current campus environmental management practices *Journal of Cleaner Production.* **16** 1777-1785

[2] Faghihi V, Hessami A R and Ford D N 2014 Sustainable Campus Improvement Program Design Using Energy Efficiency and Conservation *Journal of Cleaner Production.* **30** 1-10

[3] Tan H, Chen S, Shi Q and Wang L 2014 Development of green campus in China *Journal of Cleaner Production.* **64** 646-653

[4] Nejati, Mostafa, Nejati, Mehran 2013 Assessment of sustainable university factors from the perspective of university students *Journal of Cleaner Production.* **48** 101-107

[5] Disterheft et al 2012 Environmental management systems (EMS) implementation processes and practices in european higher education institutions – top-down versus participatory approaches *Journal of Cleaner Production.* **31** 80-90

[6] Jorge ML, Madueño JH, Cejaz MYC, Peña FJA An approach to te implementation of sustainability practices in Spanish higher education *Journal of Cleaner Production.* **106** 34-44

[7] Herremans IM, Reid RE Developing awareness of the sustainability concept *Journal of Environmental Education.* **34** 16-20

[8] Alfredo, Elfianus I, Nurcahyo Rahmat 2018 The Impact of ISO 9001, ISO 14001, and OHSAS 18001 Certification on manufacturing industry operational performance. *Proc. Int. Conf. on Industrial Engineering and Operations Management (Bandung)* pp 1862-1866

[9] Lauder A, Sari RF, Suwartha N, Tjahjono G. Critical review of a global campus sustainability ranking: GreenMetric *Journal of Cleaner Production.* **xxx** 1-12

[10] ISO 2004 *Sistem manajemen lingkungan – Persyaratan dan panduan penggunaan* vol 2 (Jakarta: Badan Standardisasi Nasional)

[11] Campos LMS, Heizen DADM, Verdinelli MA, Miguel PAC 2015 Environmental performance indicators: a study on ISO 14001 certified companies *Journal of Cleaner Production.* **99** 286-296

IOP Conf. Series: Materials Science and Engineering **697** (2019) 012011 doi:10.1088/1757-899X/697/1/012011

[12] Testa F, Rizzi F, Daddi T, Gusmerotti NM, Frey M, Iraldo F 2014 EMAS and ISO 14001: the differences in effectively improving environmental performance *Journal of Cleaner Production.* **68** 165-173
[13] Environmental Protection Agency 2007 *Environmental management guide for higher educations and higher education (*New England: Author)
[14] Comoglio C, Botta S 2012 The use of indicators and the role of environmental management systems for environmental performances improvement: A survey on ISO 14001 certified companies in the automotive sector *Journal of Cleaner Production.* **20** 92-102
[15] Da Fonseca LMCM 2015 ISO 14001:2015: An improved tool for sustainability *Journal of Industrial Engineering and Management.* **8** 37-50
[16] Wahana Komputer 2009 *Panduan praktis: SPSS 17 untuk pengolahan data statistik* (Yogyakarta: Andi Offset)
[17] Sugiyono 2013 *Metode penelitian kombinasi (Mixed methods)* (Bandung: Alfabeta)

Acknowledgments
This work was supported by the HIBAH PIT Q1Q2 under contract number NKB-0320/UN2.R3.1/HKP.05.00/2019.

IOP Conf. Series: Materials Science and Engineering **697** (2019) 012012 doi:10.1088/1757-899X/697/1/012012

Modeling and simulation of manufacture sector data in Malaysia with detection of outliers: An ARMA-GARCH approach

I M Md Ghani[1] and H A Rahim[1]

[1] Faculty of Ocean Engineering Technology and Informatics, Universiti Malaysia Terengganu, 21030 Kuala Nerus, Terengganu Darul Iman, Malaysia

E-mail: intanmartina23@gmail.com

Abstract. Manufacturing is one of the major sectors that contribute to the economy of Malaysia. The situation of manufacturing sales, especially rubber gloves received the attention of investors to forecast. However, the pattern of economics exposed to unexpected changes, which called outlier which occurred because of internal or external factors. Consequently, give shock to time series data. The main approach used is the hybrid of Autoregressive Moving Average (ARMA) model and Generalized Autoregressive Conditional Heteroscedasticity (GARCH) model. The effectiveness of the volatility model with the existence of outliers are measured by Monte Carlo method. Besides, to find the best ARMA-GARCH model there are four models produced from different specifications in ARMA(a,b) and GARCH(g,h) models. In this paper, we used 216 monthly price data of Standard Malaysian Rubber Grade 20 (SMR 20) in Malaysia. The validity comparison of diagnostic checking is measured on AIC, AICc, SIC and HQIC. While the forecasting performance evaluated using MSE, RMSE and MAPE. The results of the empirical analysis indicate that the ARMA(2,0)-GARCH(1,2) model is the appropriate model to forecast the price of SMR 20 that were used to the manufacturing of rubber gloves in Malaysia.

1. Introduction

According to [1], about 23% from Gross Domestic Product are contributed by manufacturing sectors. One of the industries that contributed to manufacturing sectors are the rubber glove industry. The domestic consumption of natural rubber used in 2015 is 73.5% or 349,031 tonnes. However, the domestic consumption of natural rubber shows fluctuations from the period from 2008 to 2010; see Figure 1. These indicates that most of the time series data are influenced by heavy-tailed distribution and volatility clustering. The existing of both characteristics can give negative affect on volatility modelling as well as forecasting. Based on the previous research, Generalized Autoregressive Conditional Heteroscedasticity (GARCH) model is used to accomplish the accuracy of modelling as well forecasting.

The GARCH model is highly recognized compared to Autoregressive Conditional Heteroscedasticity (ARCH) model. In recent years, there have been various amount of study on GARCH(1,1) model to analyze time series data such as [2, 3, 4, 5, 6, 7, 8, 9]. Some references supports that GARCH(1,1) model is popular among others specifications because it is fits many data series well [10] and sufficient to capture the volatility clustering in the data [11]. However, the phenomenon of outliers can give impacts to GARCH model.

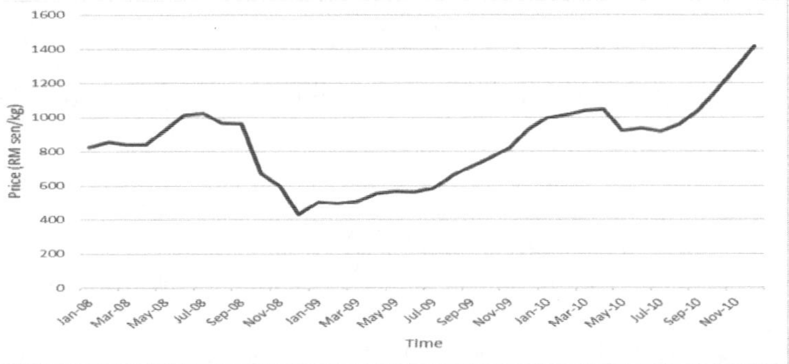

Figure 1. The trend of SMR 20 (January 2008 – December 2010).
Source: Malaysian Rubber Board.

The first serious discussion of outlier emerged during the 1970s by [12]. As highlighted by [13], the stock market volatility forecasts can be enhanced by revising the additive outlier. In 2005, [14] extended their investigation to innovative outlier. The GARCH model was preferred by both studies to forecast volatility and check out the effect of outlier. Several studies have revealed that the implications of outliers can tendency the parameters estimation of GARCH model [15, 16, 17], on estimation and identification of the GARCH-type models [18, 19], and also on forecasting [13, 17, 18]. Thus, previous research was tested the efficiency of the volatility model with hybrid ARMA(a,b)-GARCH(g,h) model.

Although the above investigations reported many interesting results, little work has examined on different specifications on ARMA(a,b)-GARCH(g,h) model. In particular, the effectiveness of volatility model in the presence of outliers via Monte Carlo method is less reported. The paper is planned as follows. The description of the ARMA(a,b) model, GARCH(g,h) model and additive outlier is explained in the next section. Then, the process of Monte Carlo simulation also presented. After that we applied to 216 monthly prices of SMR 20. Finally, the conclusion is summarized in the last section.

2. Materials and Methods

The time series models that will discussed in this section are Autoregressive Moving Average (ARMA) model and Generalized Autoregressive Conditional Heteroscedasticity (GARCH) model. The formation of ARMA(a,b) model is from the coupled of autoregressive (AR) model and moving-average (MA) model which can denoted in sigma notation as [20]

$$y_t = \mathrm{A} + \sum_{i=1}^{a} \varphi_i y_{t-i} + \sum_{j=1}^{b} \theta_j \varepsilon_{t-j} + \varepsilon_t \tag{1}$$

where A is a constant term, φ_i are the parameters of the AR term of order a , θ_j are the parameters of the MA term of order b , and ε_t is the error term at time t . The order a and b are positive integers.

In the recent decades, GARCH model which developed by [21] has been one of the major interesting in nonlinear time series model. This is due to the ARCH model utilize more parameters compared to GARCH model [22]. The conditional mean equation (r_t) and conditional variance equation (σ_t^2) in GARCH(g,h) model formulated as in Eq.(2) and Eq.(4), respectively.

$$r_t = \mathrm{A} + \varepsilon_t \tag{2}$$

where r_t is a return series $(\ln(R_t / R_{t-1}))$, A is a constant term and ε_t is residual term at time t.

$$\varepsilon_t = z_t \sigma_t \quad , z_t \sim N(0,1) \tag{3}$$

$$\sigma_t^2 = \eta + \sum_{i=1}^{g} \beta_i \sigma_{t-i}^2 + \sum_{j=1}^{h} \alpha_j \varepsilon_{t-j}^2 \tag{4}$$

where z_t is the standardized residual, σ_t^2 is the conditional variance at time t, η is the constant parameter, σ_t^2 is the forecast variance for the current period (GARCH term) and ε_t^2 is the recent information about volatility observed in the current period (ARCH term) with conditions $\eta > 0, \beta_i \geq 0; i = 1,\ldots, g$ and $\alpha_j \geq 0; j = 1,\ldots,h$.

In order to determine the persistence of GARCH(g,h) model, the non-negative constants can calculated as $\sum_{i=1}^{g} \beta_i + \sum_{j=1}^{h} \alpha_j$. When the $\sum_{i=1}^{g} \beta_i + \sum_{j=1}^{h} \alpha_j < 1$, hence GARCH(g,h)model is called covariance stationary. If the constant parameter, η combined with $\sum_{i=1}^{g} \beta_i + \sum_{j=1}^{h} \alpha_j$, the level of unconditional variance (known as long term volatility) can be determined by

$$\bar{\sigma}^2 = \mathrm{var}(\varepsilon_t) = \frac{\eta}{1 - \left(\sum_{i=1}^{g} \beta_i + \sum_{j=1}^{h} \alpha_j \right)} \tag{5}$$

Since the general mathematical form of GARCH(g,h) model in Eq.(4), the GARCH(1,2) model can be written as follows

$$\sigma_t^2 = \eta + \beta_1 \sigma_{t-1}^2 + \alpha_1 \varepsilon_{t-1}^2 + \alpha_2 \varepsilon_{t-2}^2 \tag{6}$$

where σ_t^2 is the conditional variance at time t, η is the constant parameter, σ_{t-1}^2 is the forecast variance in the preceding period (GARCH term), ε_{t-1}^2 is the information about volatility observed in the preceding period (ARCH term) and ε_{t-2}^2 is the information about volatility observed in the two preceding period (ARCH term) with condition $\eta > 0, \beta_1 \geq 0, \alpha_1 \geq 0$ and $\alpha_2 \geq 0$.

Although there are various types of outliers that effect to time series data yet this research only focused to additive outliers (AO). The past forty years, [12] described AO in AR model. AO is also known as an external or exogenous change [23]. The recording errors, measurement errors or external factors are among the aspects that caused an AO.

From Eq.(4), GARCH(g,h) model can be written as an ARMA(a,b) model for ε_t^2 [21] as

$$\varepsilon_t^2 = \eta + \sum_{i=1}^{a} (\alpha_j + \beta_i) \varepsilon_{t-i}^2 + \kappa_t - \sum_{i=1}^{b} \beta_i \kappa_{t-j} \tag{7}$$

with $a = \max\{g,h\}$ and $\kappa_t = \varepsilon_t^2 - \sigma_t^2; t = 1,2,\ldots,n$ where ε_t^2 known as outlier free time series while κ_t known as outlier-free residuals. Equation (7) can be written

$$\varepsilon_t^2 = \frac{\eta}{1-\alpha(G)-\beta(G)} + \frac{1-\beta(G)}{1-\alpha(G)-\beta(G)}\kappa_t = \frac{\eta}{1-\alpha(G)-\beta(G)} + \pi^{-1}(G)\kappa_t \tag{8}$$

with $\alpha(G) = \sum_{j=1}^{h}\alpha_j G^j, \beta(G) = \sum_{i=1}^{g}\beta_i G^i$ and $\pi(G) = \frac{1-\alpha_j(G)-\beta_i(G)}{1-\beta_i(G)}$.

According to [24], the presence of AO in GARCH model turn into

$$e_t^2 = \omega_{\mathrm{AO}}\xi_{\mathrm{AO}}(G)I_t(P) + \varepsilon_t^2 \tag{9}$$

From Eq.(9) can be interpreted as a regression model for ε_t^2 and rewrite as

$$e_t^2 = \omega_{\mathrm{AO}}x_t + \varepsilon_t^2 \tag{10}$$

where

e_t^2 is an observed series ε_t^2,

ω_{AO} is the magnitude effect of AO, which is $\omega_{\mathrm{AO}}(P) = \frac{\sum_{t=T}^{n} e_t^2 x_t}{\sum_{t=T}^{n}(x_t)^2}$,

x_t represent as $x_t = \begin{cases} 0 & ,t < P \\ 1 & ,t = P \\ -\pi_k & ,t = P+k, k \geq 1 \end{cases}$,

$\xi_{\mathrm{AO}}(G)$ is the dynamic pattern of AO effect, where $\xi_{\mathrm{AO}}(G) = 1$,

$I_t(P)$ is an indicator function that explained the effect of outliers as

$I_t(P) = \begin{cases} 1 & ,t = P \\ 0 & ,\text{otherwise} \end{cases}$, where P is the position of AO occurred.

2.1 Model Selection Criteria

The performance between different types of ARMA(a,b)-GARCH(g,h) models specification can be compared based on four model selection criteria: Akaike Information Criteria (AIC) [25], the corrected Akaike Information Criteria (AICc), Schwarz's Bayesian Information Criterion (SIC) [26] and the Hannan-Quinn Information Criterion (HQIC) [27]. The AIC, AICc, SIC and HQIC can be computed as

$$\mathrm{AIC} = -2\ln(L) + 2k \qquad \mathrm{AICc} = \mathrm{AIC} + 2\frac{k(k+1)}{N-k-1}$$

$$\mathrm{SIC} = -2\ln(L) + \ln(N)k \qquad \mathrm{HQIC} = -2\ln(L) + 2\ln(\ln(N))k$$

where L is the value of the likelihood function that evaluated at the parameter estimates, N is the number of observations, and k is the number of estimated parameters. When comparing among

ARMA(a,b)-GARCH(g,h) models, the smallest value of AIC, AICc, SIC and HQIC are chosen as the best model.

2.2 Model Goodness of Fit
In order to determine the model's goodness of fit, the performance of forecasting model is evaluated using three measures: Mean Square Error (MSE), Root Mean Square Error (RMSE) and Mean Absolute Percentage Error (MAPE).

$$\text{MSE} = \frac{1}{T}\sum_{t=T_1}^{T}(\sigma_t^2 - \hat{\sigma}_t^2)^2 \qquad \text{RMSE} = \sqrt{\frac{1}{T}\sum_{t=T_1}^{T}(\sigma_t^2 - \hat{\sigma}_t^2)^2}$$

$$\text{MAPE} = \frac{1}{T}\sum_{t=T_1}^{T}\left|\left(\frac{\sigma_t^2 - \hat{\sigma}_t^2}{\sigma_t^2}\right)\times 100\right|$$

where T is the number of total observations and T_1 is the first observation in out-of-sample part. The σ_t^2 and $\hat{\sigma}_t^2$ is the actual and predicted conditional variance at time t, respectively. The ARMA(a,b)-GARCH(g,h) model that have minimum value of MSE, RMSE and MAPE was selected as the best forecast model.

2.3 Simulation Study
The GARCH model was generated by using *tseries* package [29] and *fGarch* package [30] in the statistical software environment R [28]. The effectiveness of the ARMA(a,b)-GARCH(g,h) models were compared in two scenarios: contaminated with 0% AO and contaminated with 10% AO. For each scenario, there are three types of time series length (T) which are 120, 360 and 480. The simulation process carried out in this study is summarized as the following algorithm:

1. The true value of coefficients were fixed as $A = 1.0265, \varphi_1 = -0.3581, \varphi_2 = 0.0198$, $\eta = 31.0877, \alpha_1 = 0.0967, \alpha_2 = 0.3322, \beta_1 = 0.001821$ during specified ARMA(2,0)-GARCH(1,2) model in *garchSpec* function.
2. The GARCH process were simulated 120 series length with normal innovations.
3. The coefficients of the ARMA(2,0)-GARCH(1,2) model were estimated with normal innovations in *garchFit*function.
4. In the scenario of 0% AO, the efficiency of ARMA(2,0)-GARCH(1,2) model were calculated.
5. About 10% from time series length contaminated as AO. The positions and magnitudes of AO are recognized.
6. The coefficients of the ARMA(2,0)-GARCH(1,2) model were estimated with normal innovations during contamination 10% AO.
7. In the scenario of 10% AO, the efficiency of ARMA(2,0)-GARCH(1,2) model were calculated.
8. Step (1) to (7) repeated by setting time series length to 360 and 480.

3. Results and Discussion

3.1 Simulation Results
Table 1 presents the summary of the descriptive statistics during 0% AO and 10% AO. The kurtosis values in the scenario of 10% AO are 8.3639, 17.5987 and 17.0654 for time series length 120, 360 and 480, respectively. From this data we can revealed that there is leptokurtic distribution during contaminated.

IOP Conf. Series: Materials Science and Engineering **697** (2019) 012012 doi:10.1088/1757-899X/697/1/012012

Table 1. Summary of descriptive statistics for 0% and 10% additive outlier.

Criteria	0% AO			10% AO		
	T=120	T=360	T=480	T=120	T=360	T=480
Mean	-0.0044	0.0140	0.0789	0.3328	-0.1252	0.0018
Variance	1.0909	0.9922	1.0003	6.3115	6.9388	10.3543
Standard deviation	1.0445	0.9961	1.0001	2.5123	2.6342	3.2178
Kurtosis	-0.3047	-0.2124	-0.1377	8.3639	17.5987	17.0654
Skewness	-0.0062	0.0202	-0.0310	1.9662	-0.8984	-1.1096

As shown in Table 2, the two scenarios were compared based on selection criteria. When simulated 120 observations with contaminated 10% AO, there were an increased of 47.12%, 47.21% and 46.09%, respectively in AIC, SIC and HQIC criteria. In addition, the AIC, SIC and HQIC criteria showed increased when the time series length increased to 360 and 480. These finding highlights that the value of three criteria's during contamination AO is larger than without contamination AO.

Table 2. Selection criteria for 0% and 10% additive outlier.

Percentage contamination	Criteria	T=120	T=360	T=480
	AIC	2.9782	2.8587	2.8393
0%	SIC	2.9719	2.8580	2.8389
	HQIC	3.0442	2.8888	2.8633
	AIC	4.3814	4.8003	5.1723
10%	SIC	4.3751	4.7996	5.1719
	HQIC	4.4474	4.8304	5.1962

Table 3 provides the comparison goodness of fit among contaminated 0% and 10% AO. The MSE value in time series length 120 during situation 0% AO and 10% AO displayed 1.0240 and 6.5103, respectively. This showed the percentage increment of 535.77%. While there was an increased of 152.15% in RMSE criteria. When the percentage of contamination increased, both criteria's shows larger value. This situation is consistent when the time series length increased.

Table 3. Goodness of fit for 0% and 10% additive outlier.

Percentage contamination	Criteria	T=120	T=360	T=480
0%	MSE	1.0240	0.9836	0.9731
	RMSE	1.0119	0.9918	0.9865
10%	MSE	6.5103	6.8454	10.2186
	RMSE	2.5515	2.6164	3.1967

3.2 Empirical Results

In this section, the monthly price data of Standard Malaysian Rubber Grade 20 (SMR 20) are investigated. These data consist of period from January 2000 to December 2017. All observations were taken from official portal of Malaysia Rubber Board. From 216 observations, about 151 observations were selected into in-sample part as model estimation while 65 observations were selected into out-of-sample part as model evaluation in forecasting.

There is a clear trend of 151 monthly observations prices of the SMR 20 in Malaysia from January 2000 to July 2012. When monthly prices converted to log returns, the plot in Figure 2 (right) illustrates there are large negative values especially on October 2008 and December 2008. Both figures can explained there exist volatility clustering in monthly returns on SMR 20.

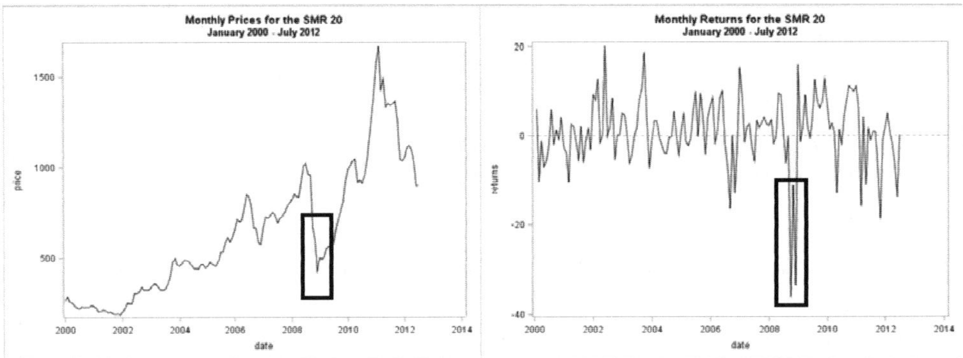

Figure 2. In-sample results of monthly prices (left) and returns (right) of SMR 20.

The descriptive statistics of the monthly returns for SMR 20 are shown in Table 4. The kurtosis value display 5.1872, which is greater value than normal distribution. From this value, we can state that there is leptokurtic distribution during model estimation part.

Table 4. Summary of descriptive statistics of monthly returns of SMR 20.

Mean	Variance	Standard deviation	Kurtosis	Skewness
0.8070	60.6026	7.7848	5.1872	-1.2846

The first step in time series data is to test the stationarity. There are three tests that considered in this paper: Phillips-Perron (PP) test [31], Augmented Dickey-Fuller (ADF) test [32] and Kwiatkowski-Phillips-Schmidt-Shin (KPSS) test [33]. The finding of ADF test is consistent with the PP test, which we reject the null hypothesis of a unit root at the 5 percent level of significance. These indicates that the series is stationary where there is no unit root. Meanwhile we do not reject the null hypothesis of a unit root in KPSS test. These indicates that the series is stationary and there is no unit root. All three tests proved that the series is stationary and has no unit root, therefore we proceed the next step.

After confirmed heteroscedastic exist in residuals, then we proceed to specify the model. We generates four models from different conditional mean and conditional variance specification in ARMA(a,b)-GARCH(g,h) models where a and b were 2 and 0, respectively. While g and h were either 1 or 2. The four different specification models was compared based on criteria such as AIC, AICc, SIC and HQIC to select the best model. The comparison among four models are highlighted in Table 5. The ARMA(2,0)-GARCH(1,2) model presents the minimum value for all selection criteria (AIC, AICc, SIC and HQIC). Therefore, we decided ARMA(2,0)-GARCH(1,2) model as the best model in-sample part.

Table 5. Comparison of selection criteria for ARMA(a,b)-GARCH(g,h) models.

Models	AIC	AICc	SIC	HQIC
ARMA(2,0)-GARCH(1,1)	1030.8990	1031.4864	1048.9628	1038.2377
ARMA(2,0)-GARCH(1,2)	1022.2898	1023.0786	1043.3643	1030.8517
ARMA(2,0)-GARCH(2,1)	1034.8354	1035.1112	1046.8779	1039.7279
ARMA(2,0)-GARCH(2,2)	1034.8291	1035.1050	1046.8717	1039.7216

In Figure 3 (left) there is a clear trend of 65 monthly observations prices of the SMR 20 from August 2012 to December 2017. The plot in Figure 3 (right) illustrates there are large negative values especially on April 2017. Both figures can be explained that there exists volatility clustering in monthly prices and returns on SMR 20.

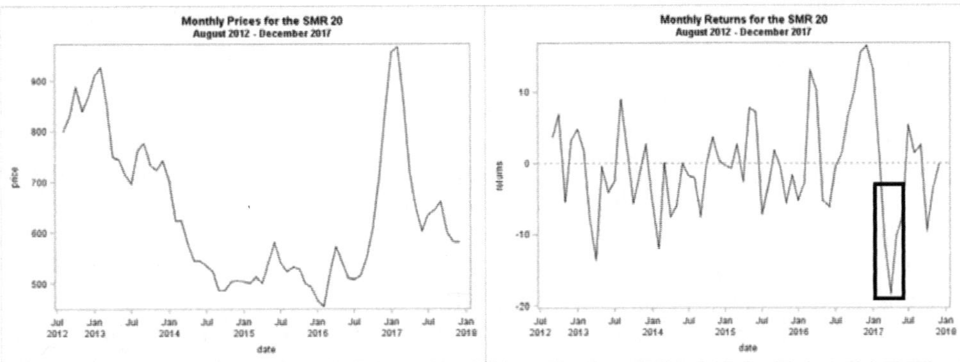

Figure 3. Out-of-sample results of monthly prices (left) and returns (right) of SMR 20.

In out-of-sample part, the performance of forecasting was evaluated. The evaluation of ARMA(a,b)-GARCH(g,h) models in out-of-sample are summarized in Table 6. As illustrated in Table 6, the accurate forecast for MSE and RMSE statistic is ARMA(2,0)-GARCH(1,1) model. While ARMA(2,0)-GARCH(1,2) model present more accurate in MAPE statistic. Therefore among four models, we decided ARMA(2,0)-GARCH(1,2) model as the best accurate model in out-of-sample part. This is supported by [34] which suggested that the selection of MAPE statistic is the best if there are different results between three statistics.

Table 6. Comparison of goodness of fit for ARMA(a,b)-GARCH(g,h) models.

Models	MSE	RMSE	MAPE
ARMA(2,0)-GARCH(1,1)	34.3860	5.8640	812.5882
ARMA(2,0)-GARCH(1,2)	34.3939	5.8646	800.5116
ARMA(2,0)-GARCH(2,1)	34.3862	5.8640	813.7431
ARMA(2,0)-GARCH(2,2)	34.3862	5.8640	813.7431

For overall part, there is a clear trend of 216 monthly observations prices of the rubber SMR 20 in Malaysia from January 2000 to December 2017 depicted in Figure 4. When monthly prices converted to log returns, plotted of monthly returns of rubber SMR 20 clearly show exhibit volatility clustering.

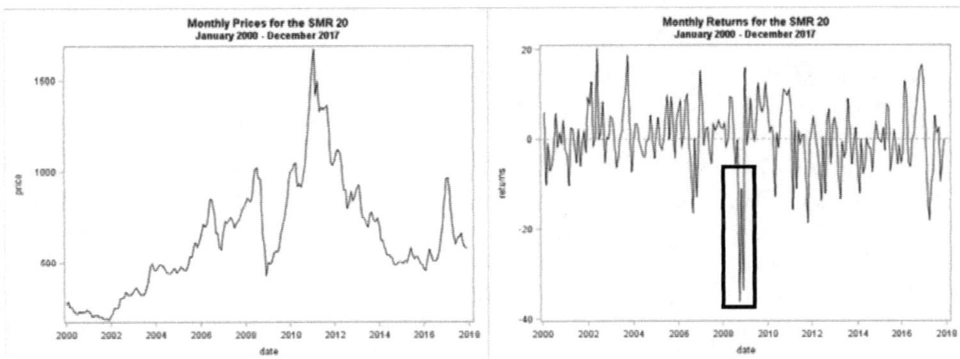

Figure 4. Overall results of monthly prices (left) and returns (right) of SMR 20.

Since the best model in estimation part and forecasting part are consistent, then the estimated conditional mean and conditional variance of ARMA(2,0)-GARCH(1,2) model can expressed as

$$\hat{y}_t = 1.0265 - 0.3581 y_{t-1} + 0.0198 y_{t-2}$$

$$\hat{\sigma}_t^2 = 31.0877 + 0.001821\sigma_{t-1}^2 + 0.0967\varepsilon_{t-1}^2 + 0.3322\varepsilon_{t-2}^2$$

The outliers in monthly price of SMR 20 were detected automatically by using X12 procedure. During this process, the best ARIMA model selection which is $(1,1,0)\times(0,0,0)_{12}$ was automatic identify based on [35]. The default initial critical values used for additive outlier (AO) and level change outlier (LC) detection is 3.971414. It appears that, the LC on June 2002 was identified as potential outliers. While there are two types of outliers that included in the regression model. There are LC and AO which located on October 2008 and December 2008, respectively. This is due to the fact that the global economic recession happened in the second half of the year. In addition, when the critical value decreases to 3.3, there are two additional outliers detected. These includes LC on June 2002 and AO on March 2011. As it happens on 11 March 2011, the Earthquake and Tsunami hit Japan which triggered panic selling.

4. Conclusion

This paper presents the modeling and numerical simulation of SMR 20 return price that used in manufacturing rubber gloves. The following conclusions were obtained. 1) Based on the modelling result shows that ARMA(2,0)-GARCH(1,2) model is the best model for the period of in-sample part and out-of-sample part compared to three different specifications of ARMA(a,b)-GARCH(g,h) model. These findings of this paper suggest that GARCH(1,2) model is suitable used to model the current SMR 20 rubber price compared to GARCH(1,1) model. 2) One of the more significant findings to emerge from this paper is that the efficiency of ARMA(2,0)-GARCH(1,2) model were decreased when outliers were present. The outliers that we have identified therefore assists in our understanding of the influence of outlier in time series data. more research is required on the modelling of rubber gloves by using latex price. Considerably more work will need to be done to determine the outlier types of outliers that effects on the behavioural time series data such as level change outliers and temporary change outliers based on different specification of ARMA(a,b)-GARCH(g,h) model.

References

[1] Department of Statistics Malaysia Retrieved from http://www.dosm.gov.my/v1/ 28 February 2019
[2] Pham H T and Yang B S 2010 Estimation and forecasting of machine health condition using ARMA/GARCH model*Mechanical Systems and Signal Processing* **24** 546–558
[3] Wang Y and Wu C 2012 Forecasting energy market volatility using GARCH models: Can multivariate models beatunivariate model *Energy Economics* **34** 2167–2181
[4] Hickey E, Loomis D G and Mohammadi H 2012 Forecasting hourly electricity prices using ARMAX-GARCH models: An application to MISO hubs *Energy Economics* **34** 307–315
[5] Salisu A A and Fasanya I O 2013 Modelling oil price volatility with structural breaks. *Energy Policy* **52** 554–562
[6] Huq M M, Rahman M M, Rahman M S, Shahin A and Ali A 2013 Analysis of Volatility and Forecasting General Index of Dhaka Stock Exchange *American Journal of Economics* **3(5)** 229–242
[7] Joukar A and Nahmens I 2015 Volatility Forecast of Construction Cost Index Using General Autoregressive Conditional Heteroskedastic Method *Journal of Construction Engineering and Management* **142(1)** 04015051
[8] Goh H H, Tan K L, Khor C Y and Ng S L 2016 Volatility and Market Risk of Rubber Price in Malaysia: Pre- and Post-Global Financial Crisis *Journal of Quantitative Economics* **14(2)** 323–344

[9] Epaphra M 2017 Modeling Exchange Rate Volatility: Application of the GARCH and EGARCH Models *Journal of Mathematical Finance* **7** 121–143
[10] Hill R C, Griffiths W E and Lim G C 2011 *Principles of Econometrics* ed Fourth (United States of America: John Wiley & Sons, Inc) chapter 14 pp 526
[11] Brooks C 2014 *Introductory Econometrics for Finance* ed Third (New York: Cambridge University Press) chapter 9 pp 430
[12] Fox A J 1972 Outlier in time series *Journal of the Royal Statistical Society. Series B (Methodological)* **34(3)** 350–363
[13] Franses P H and Ghijsels H 1999 Additive outliers, GARCH and forecasting volatility *International Journal of Forecasting* **15(1)** 1–9
[14] Charles A and Darné O 2005 Outliers and GARCH models in financial data *Economics Letters***86(3)** 347–352
[15] Sakata S and White H1998High breakdown point conditional dispersion estimation with application to S&P 500 daily returns volatility*Econometrica* **66(3)**529–567
[16] Melo Mendes BVD 2000 Assessing the bias of maximum likelihood estimates of contaminated GARCH models *Journal of Statistical Computation and Simulation* **67(4)** 359–376
[17] Charles A 2008 Forecasting volatility with outliers in GARCH models *Journal of Forecasting* **27(7)** 551–565
[18] Carnero M A, Peña D and Ruiz E 2007 Effects of outliers on the identification and estimation of GARCH models *Journal of Time Series Analysis* **28(4)** 471–497
[19] Carnero M A, Peña D and Ruiz E 2012 Estimating GARCH volatility in the presence of outliers *Economics Letters* **114(1)** 86–90
[20] Box G E P, Jenkins G M, Reinsel G C and Ljung G M 2015*Time series analysis: Forecasting and control* ed Fifth (Hoboken, New Jersey: John Wiley & Sons, Inc)
[21] Bollerslev T 1986 Generalized autoregressive conditional heteroskedasticity *Journal of Econometrics* **31(3)** 307–327
[22] Poon S H and Granger C W 2003 Forecasting volatility in financial markets: A review *Journal of Economic Literature* **41(2)** 478–539
[23] Urooj A and Asghar Z 2017 Analysis of the performance of test statistics for detection of outliers (additive, innovative, transient and level shift) in AR (1) processes *Communications in Statistics Simulation and Computation* **46(2)** 948–979
[24] Chen C and Liu L-M 1993 Joint estimation of model parameters and outlier effects in time series *Journal of the American Statistical Association* **88(421)** 284–297
[25] Akaike H 1974 A new look at the statistical model identification *IEEE Transactions on Automatic Control*AU-19 716–722
[26] Schwarz G 1978 Estimating the dimension of a model *The Annals of Statistics* **6(2)** 461–464
[27] Hannan E J and Quinn B G 1979 The determination of the order of an autoregression *Journal of the Royal Statistical Society. Series B (Methodological)* **41(2)** 190–195
[28] The R project for statistical computing Retrieved from http://www.r-project.org/ 11 June 2019
[29] tseries: Time series analysis and computational finance Retrieved from http://cran.r-project.org /web/packages/tseries 11 June 2019
[30] fGarch: Rmetrics – Autoregressive conditional heteroskedastic modelling Retrieved from http://cran.r-project.org/web/packages/fGarch 11 June 2019
[31] Phillips P C B and Perron P 1988 Testing for a unit root in time series regression *Biometrika* **75(2)** 335–346
[32] Dickey D A and Fuller W A 1979 Distribution of the estimators for autoregressive time series with a unit root *Journal of the American Statistical Association* **74(366a)** 427–431
[33] Kwiatkowski D, Phillips P C B, Schmidt P and Shin Y 1992 Testing the null hypothesis of stationarity against the alternative of a unit root *Journal of Econometrics* **54(1–3)** 159–178
[34] Makridakis S 1993 Accuracy measures: Theoretical and practical concerns *International Journal of Forecasting* **9(4)** 527–529

[35] Gómez V and Maravall A 2001 *Automatic modeling methods for univariate series* In: Peña D, Tiao GC and Tsay RS *A course in time series* (New York: Wiley) p 171-201

Modelling and simulation of three-roll bending process of bimetal Circular Billet

A S Alaboodi[1], N B Yusoff[2] and A A Alnasser[2]
[1] Mechanical Engineering Department, Qassim University, 51452 Buraydah, Saudi Arabia
[2] Department of Mechanical Engineering, University of Malaya, 50603 Kuala Lumpur, W. Persekutuan, Malaysia
*E-mail: alaboodi@qu.edu.sa

Abstract. Bending process is an important metal forming technique which is used on industrial scale. An asymmetric three-roll bending is sub-type bending process and has a simple configuration. The present study has focused on the investigation of the bonded bimetal billet. The two material consist of copper alloy rod inside an aluminium alloy (AA 6061-T6) tube with one common dimension between the rod and the tube which is the radius of rod and the inner radius of tube. The effect of varying the radius has considered on this study. The results shown that the increasing of copper alloy radius increased the reaction load and the mass of the bimetal billet with exponential increment.

1. Introduction

Alloys consist of two or more different parent materials. Because of the characteristic advantage in term of thermal, mechanical, physical, corrosion resistant and other properties, the composite material has been used in different industrial applications [1–4]. Fabrication of alloys can be attained through several existing processing methods including casting [5,6], diffusion bonding [7–9], roll bonding [10–15] and explosive welding [16–18]. The selection of fabrication technology and materials composition have influenced the mechanical properties of manufactured alloys [19].

Complex metal products for manufacturing, like large plastic forming, bending and hemming, the formability and flow behaviour of alloys are not the same with special characteristics relative to homogeneous material [20]. Verguts and Sowerby [21] have investigated the bending of bimetal laminated sheets. Majlessi et al. [22] have studied the bending behaviour of a linear stress-strain model for the bi- and tri-metal plates. Yoshida et al. [23] found that the bendability and residual stress of bimetal plates have influenced the relative position of the weak and the strong steel layers. Ryutaro et al [24] have testified that the thickness ratio and relative position of component layers were influenced by the bimetal spring-back in draw-bending operation. Xiao et al. [25] and Yilamu et al. [26] have applied the finite element (FE) modelling to study the behaviour and response of stainless steel cladded with aluminium bimetal sheets on bending and flexural. Advanced bending theory and Ludwik hardening law have utilized to investigate the bending of bimetal plates as explained elsewhere [20, 27]. The authors have compared their results of the bending characteristics with FE model under different material conditions. Other researchers were use a hybrid method of deformation and investigate it either experimentally or through finite element methods [28].

Most of researchers have investigated the total performance of bimetal laminate sheets / plates / rods under roller bending, such as bending angle, bending moment, thinning, and bimetal billet fabrication processes. The main objective of this paper is to study the effect of variables of combination of two different metals called bimetal, including the ratios of component, relative positions and bending characteristics of bimetallic materials. Numerical simulation using finite element modelling were also used to investigate the bending deformation process.

2. Methods and Materials

2.1. Geometric Modeling

A schematic of the asymmetric roll bending process is shown in Fig.1 where a billet is being processed by three identical rolls in the three-roll bending machine. The process would start when bottom roll goes up in the vertical direction without rotation. In the same time, the plate to be bent is passed between the rollers. In this manner, the plate is pressed against the top rolls. This is followed by the rotation of top rolls in the clockwise sense. Further, the center of the bottom roll describes the diameter of the bending.

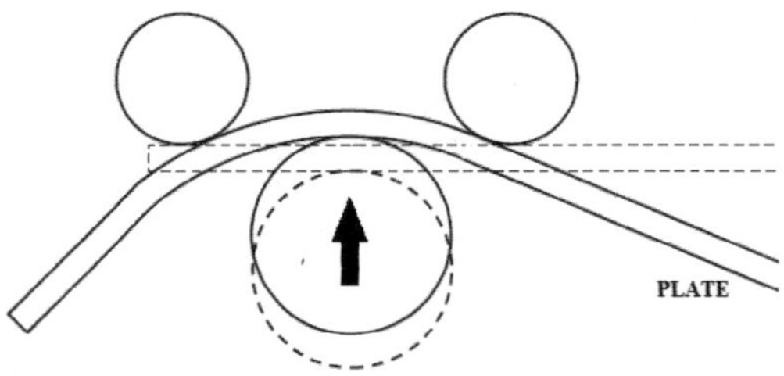

Figure 1. Schematic of asymmetrical three-roll bending process.

2.2. Material Modeling

In the present model, bimetal contains two type of material in which all of them are elasto-plastic material. Here, aluminium alloy (AA 6061-T6) tube and copper solid rod were used in the material model. The mechanical propertied fed in to the FE model is illustrated in Table 1. The Von-Mises yield criterion with isotropic hardening was used for matrix yield surface. The simulation was carried out at room temperature, hence all properties were provided at room temperature only.

Table 1. Copper and Aluminium Alloy (AA 6061-T6) mechanical properties.

Property	Copper	Aluminium (AA 6061-T6)
Density kg/m^3	8300	2770
Young's Modulus GPa	110	71
Poisson's Ratio	0.34	0.33
Yield Strength MPa	280	280
Tensile Ultimate Strength MPa	430	310

The circular billet was made from two materials as shown in Fig 2. The rod of cupper material with radius was varied from 25 to 45 mm. In addition, the second material is a tube of AA 6061-T6 aluminium alloy in which the inner radius was same as the copper rod radius and the outer radius was kept constant value of 50 mm.

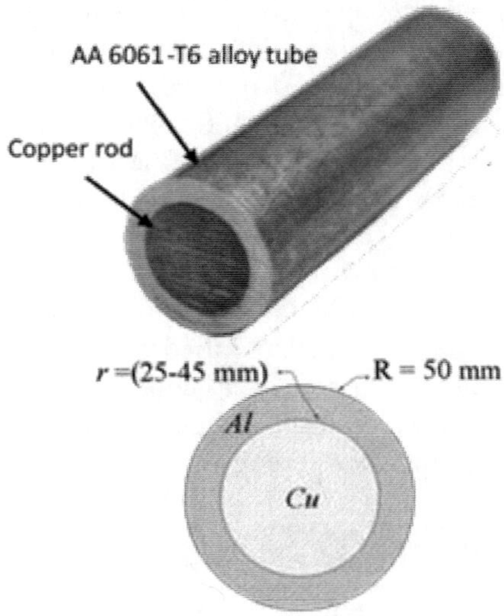

Figure 2. Billet with two materials (Cooper alloy and Aluminium Alloy).

3. Finite Element Modelling

The model mesh of bimetal billet consists of 60415 nodes and 48408 elements of quad/tringle combination for AA 6061T6 alloy and copper rod respectively which is shown in Fig. 3. The adaptive mesh was used to overcome an elemental distortion during the FE running. The three rollers were treated as a rigid body.

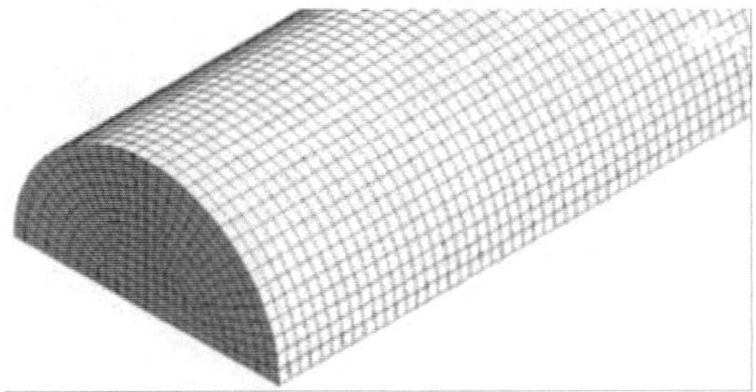

Figure 3. 3D Mesh of the biomaterial circular billet

Figure 4. Bending processes showing the circular bimetal billet and the three roller with arrow indicate the type and direction of its motion.

Figure 4 shows the three roller bending operation with the feeding of bimetal circular billet. The circular billet was model as a halve circle taking the consideration of the symmetry and to reduce the tone of running. The analysis step was done by two steps. The two rollers on the right side was kept idle on the first step then rotates clockwise with 180o. However, the roller on the right side was translated on the negative side of x-axis with 20 mm on the first step and 40 mm at the second step. Moreover, the same roller was kept idle of rotation on the first step and rotate with 180o at the second step.

4. Results and Discussion

The following definition and assumptions were applied in order to get acceptable results and to minimize the computational cost. Billet material was assumed isotropic. The elastic-plastic behavior of the material was defined by bilinear elastic-plastic which is having constant value of elastic modulus, yield strength and tangent modulus. The values are illustrated in Table I

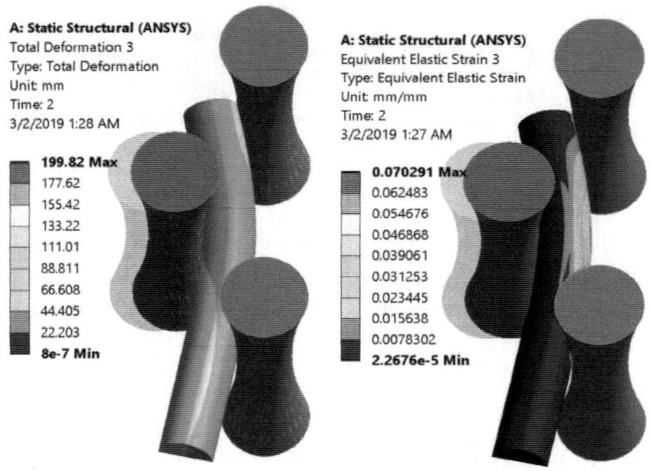

Figure 5. Total deformation of the billet at the end of lading (a) and equivalent elastic strain of billet (b).

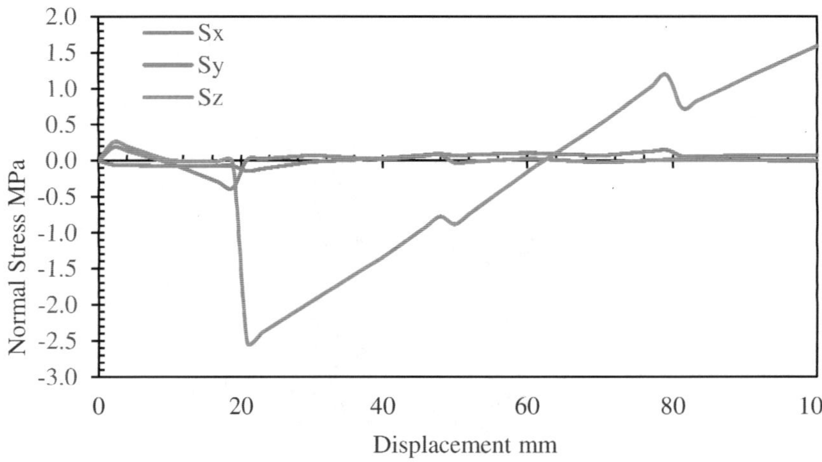

Figure 6. Normal stress in directions x, y and z through the centre of circular billet from (A to B) with r = 30 mm.

Table 2. Structural Steel NL and Aluminium Alloy NL mechanical proper.

#	r mm	Total mass kg	Reaction Force kN
1	25	11.41	926.33
2	30	13.09	1232.98
3	35	15.06	1644.34
4	40	17.34	2186.91
5	45	19.93	2903.35

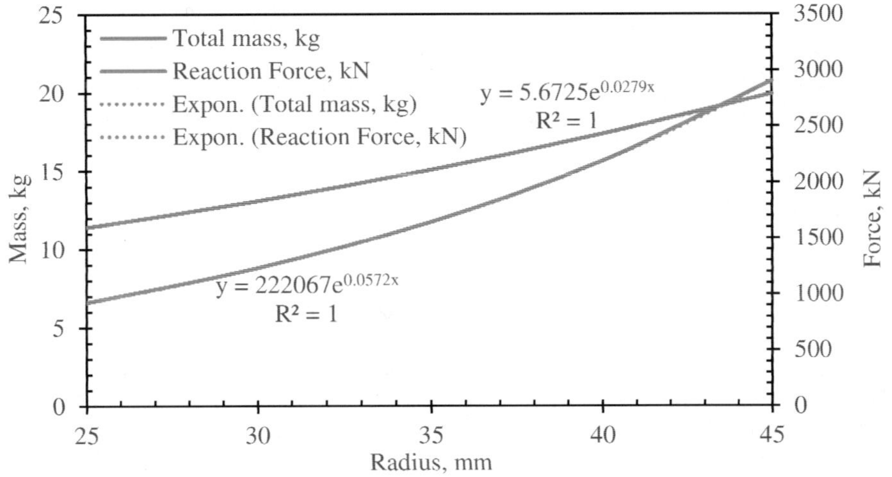

Figure 7. Displacement in y direction vs. reaction force. (Elastic and Plastic Portions).

5. Conclusion

The finite element analyses were carried out to study bimetal billet bended process of an asymmetrical three-roll with different material of circular cross section geometry successfully. The bilinear elastic/plastic material model was selected for the billet corresponding to force and tensile tests. The inner material is copper with higher density and strength while the aluminium is in the outer side of the billet. Results of the simulation has shown that the reaction force and the mass increased exponentially with the increase of inner radius. The exponential power constant is higher for the reaction force as about twice of the mass.

References

[1] Dhib Z, Guermazi N, Ktari A, Gasperini M, Haddar N. Mechanical bonding properties and interfacial morphologies of B austenitic stainless steel clad plates. Mater Sci Eng 2017;**696**:374–86.

[2] Khodadad Motarjemi A, Koc'ak M, Ventzke V. Mechanical and fracture characterization of a bi-material steel plate. Intern J Press Vessels Piping 2002;**79**:181–91.

[3] Inoue J, Nambu S, Ishimoto Y, Koseki T. Fracture elongation of brittle/ductile multi- layered steel composites with a strong interface. Scripta Materialia 2008;**59**:1055–8.

[4] Yoshida F, Urabe M, Hino R, Toropov VV. Inverse approach to identification of material parameters of cyclic elasto-plasticity for component layers of a bimetallic sheet. Intern J Plastic 2003;**19**:2149–70.

[5] Liu N, Jie J, Lu Y, Wu L, Fu Y, Li T. Characteristics of clad aluminum hollow billet prepared by horizontal continuous casting. J Mater Process Technol 2014;**214**:60–6.
[6] Xu J, Gao X, Jiang Z, Wei D, Jiao S. Microstructure and hot deformation behaviour of high-carbon steel/low-carbon steel bimetal prepared by cen- trifugal composite casting. Intern J Adv Manufact Technol 2016;**86**:817–27.
[7] Noh S, Kasada R, Kimura A. Solid-state diffusion bonding of high-Cr ODS ferritic steel. Acta Materialia 2011;**59**:3196–204.
[8] Deqing W, Ziyuan S, Ruobin Q. Cladding of stainless steel on aluminum and carbon steel by interlayer diffusion bonding. Scripta Materialia 2007;**56**:369–72.
[9] Gao X, Jiang Z, Wei D, Jiao S, Chen D, Xu J, Zhang X, Gong D. Effects of temperature and strain rate on microstructure and mechanical properties of high chromium cast iron/low carbon steel bimetal prepared by hot diffusion-compression bonding. Mater Design 2014;**63**:650–7.
[10] Jing YA, Qin Y, Zang X, Y. The bonding properties and interfacial morphologies of clad plate prepared by multiple passes hot rolling in a protective atmosphere. J Mater Process Technol 2014;**214**:1686–95.
[11] Knezevic M, Nizolek T, Ardeljan M, Beyerlein IJ, Mara NA, Pollock TM. Texture evolution in two-phase Zr/Nb lamellar composites during accumulative roll bonding. Intern J Plastic 2014;**57**:16–28.
[12] Kocich R, Macháčková A, Fojtík F. Comparison of strain and stress conditions in conventional and ARB rolling processes. Intern J Mechan Sci 2012;**64**:54–61.
[13] Wang HY, Li X, Sun J, Wang ZH, Zhao DW, Zhang DH. Analysis of sandwich rolling with two different thicknesses outer layers based on slab method. Intern J Mechan Sci 2016;**106**:194–208.
[14] Xie HB, Jiang ZY, Liu XH, Wang GD, Tieu AK, Yang M, Manabe K. Application of fuzzy control of laminar cooling for hot rolled strip. J Mater Process Technol 2007;**187-188**:715–19.
[15] Xie HB, Jiang ZY, Liu XH, Wang GD, Tieu AK. Prediction of coiling tempera- ture on run-out table of hot strip mill using data mining. J Mater Process Technol 2006;**177**:121–5.
[16] Akbari-Mousavi SaA, Barrett LM, Al-Hassani STS. Explosive weld- ing of metal plates. J Mater Process Technol 2008;**202**:224–39.
[17] Bataev IA, Lazurenko DV, Tanaka S, Hokamoto K, Bataev AA, Guo Y, Jorge AM. High cooling rates and metastable phases at the inter- faces of explosively welded materials. Acta Materialia 2017;**135**:277–89.
[18] Reid SR. A discussion of the mechanism of interface wave gener- ation in explosive welding. Intern J Mechan Sci 1974;**16**:399–413.
[19] Ha JS, Hong SI. Deformation and fracture of Ti/439 stainless steel clad composite at intermediate temperatures. Mater Sci Eng 2016;**651**:805–9.
[20] Govindasamy G, Jain MK. Modeling and experimental assessment of bending characteristics of laminated bilayer sheet materials. J Eng Mater Technol 2016;**138** 031014-031014-18.
[21] Verguts H, Sowerby R. The pure plastic bending of laminated sheet metals. Intern J Mechan Sci 1975;**17**:31–51.
[22] Majlessi SA, Dadras P. Pure plastic bending of sheet laminates under plane strain condition. Intern J Mechan Sci 1983;**25**:1–14.
[23] Yoshida F, Okada T, Itoh M, Harada Y, Ohmori M. Bendability of aluminium and steel-clad chromium plates. Metals Mater 1998;**4**:426–31.
[24] Hino R, Goto Y, Yoshida F. Springback of sheet metal lami- nates in draw-bending. J Mater Process Technol 2003;**139**:341–7.
[25] Xiao X, Hsiung C-K, Zhao Z. Analysis and modeling of flexural de- formation of laminated steel. Intern J Mechan Sci 2008;**50**:69–82.
[26] Yilamu K, Hino R, Hamasaki H, Yoshida F. Air bending and springback of stainless steel clad aluminum sheet. J Mater Process Technol 2010;**210**:272–8.
[27] Govindasamy G, Jain MK. Assessment of analytical models of pure bending of sheet materials using the digital image correlation method. Strain 2016;**52**:191–204.
[28] Alaboodi AS. Finite element study of the hybrid expansion of tube-to-tubesheet joints. Thin-Walled Structures. 2019 Apr 1;**137**:347-52.

IOP Conf. Series: Materials Science and Engineering **697** (2019) 012014 doi:10.1088/1757-899X/697/1/012014

Performance measurement using supply chain operation reference (SCOR) model: a case study in a small-medium enterprise (SME) in Indonesia

E Kusrini[1], M A B Rifai[1], S Miranda[1]

[1]Industrial Engineering Department, Faculty of Industrial Technology, Universitas Islam Indonesia, Yogyakarta, Indonesia

Abstract. Nowadays, measuring supply chain performance is a topic that attracts many researchers. Performance measurement can be used as a reference to improve performance in order to compete in the market. This study aims to measure the supply chain performance in a small and medium enterprise (SME) producing sports clothes in Yogyakarta. This research utilizes the performance attributes from the Supply Chain Operation Reference Model (SCOR). The business process is identified as the baseline to determine performance metrics on each process (plan, source, make, deliver, return and enable) and performance attributes, i.e. reliability, responsiveness, agility, cost, and asset management efficiency. According to the experts within the company, there are only 27 of 40 performance metrics obtained valid. The overall performance score is at a good level with a value of 77.89. Among the metrics, it is found 9 metrics in marginal and average level while the remaining metrics gained the value of more than 70. This supply chain performance analysis can support the company decision making in order to improve its performance at an excellent level.
Keywords: Performance metric, Supply chain performance measurement, SCOR, SME

1. Introduction

Supply chain management system is a set of approaches utilized to efficiently integrate suppliers, manufacturing, warehouse, and stores, so that the product is produced and distributed at the right quantities, to the right location, and the right time in order to minimize system-wide cost while satisfying service level requirements. [1]. Supply chain performance measurement is a management action that needs to be done the company can continue to grow and excel in competition. Performance measurement is systematic learning to find out the company's potential capabilities in creating management, products, or services. Performance measurement provides results on whether companies achieve their goals and mission [2]. Regarding the goals and mission of a company, revenue and consumer have a correlation. The company will gain high revenue if the company has many customers, this becomes challenging among company to acquire many customers. According to [3] the company has to give full attention to this concept because it is very important in order to meet customer satisfaction.

Customer satisfaction is a concept applied by the company to gain many customers. A common customer can change to become a loyal customer. A loyal customer is important for the success and continuity of the company [4]. The analysis performance and measurement is an assessment of activities, then the result used to know the performance whether still effective and efficient to fulfill or increase customer satisfaction [5]. Analysis performance of supply chain management is an important aspect for the successful [6]. Thus, it is required to measure the supply chain performance related to determining a business strategy to achieve the company's goals.

IOP Conf. Series: Materials Science and Engineering **697** (2019) 012014 doi:10.1088/1757-899X/697/1/012014

There are various models in measuring supply chain performance, but there is no agreement on which model is most appropriate to be used. Some of the most widely used models include performance-based measurement, Balanced Scorecard (BSC), Total Cost of Ownership and Life Cycle Assessments (TCO/LCA), Maturity Assessments, SCOR model [7,8,9,10,11]. In this research, the SCOR model is employed to measure the performance of the supply chain. SCOR is a model developed by the supply chain council [12]. SCOR has the characteristics to manage business activities to meet customer demand. SCOR is updated regularly to adapt changes to supply chain business processes. This makes SCOR more suitable for other products than other methods for measuring supply chain performance. According to [13] the SCOR model is mostly used in manufacturing companies and also [14] which can be used by all companies including service companies.

Some research in supply chain performance measurement using the SCOR model has been conducted in different industries such as manufacturing, convection, construction [11,15,16,17] and provide many benefits for industrial development. All of the performance measurement was conducted to maintain competitiveness in the market and take the advantages [18]. Supply Chain Operation References (SCOR) is a powerful tool for evaluating, communicate supply chain management when taking a decision within the company, supplier, and customers, then SCOR link it all the business process, performance metrics, practices, and people skills into the unified structure. SCOR is the most promising model for supply chain management decision making [10,19]. SCOR is a process reference model that the purpose of the reference model is to define process architecture in a way that aligns with key business functions and goals. The scope of SCOR describes the business activities can be linked to each other with all phases of satisfying customer demand. The model support by six primary management processes that is Plan, Source, Make, Deliver, Return, Enable. By describing these processes, the model can be used to describe the level of complexity of the supply chain. Each industry can be known how the depth and breadth they are with all of the supply chain members. SCOR describe the process, not the function, the model focused on the activity involved, not the person or organization element that performs the activity.

The company where this study conducted produces sportswear such as Hoodie, T-shirt, and Jersey as their main product. As an SME enterprise established in 2014, it is necessary to perform well regarding customer demand fulfillment as well as business competition. Supply chain performance deals with expanded supply chain activities in meeting end-customer requirements, including product availability, on-time delivery, and all necessary inventory and capacity in the supply chain to deliver such performance responsively [20]. Improvement can not be applied without a measurement. The measurement is important to see how well the supply chain is doing started from the suppliers' suppliers until the product comes to the end-user. However, this company does not understand the term of supply chain nor the performance measurement. Therefore, for supporting this small enterprise, the objective of this research is to identify key performance indicators and measure them by utilizing the SCOR model. The result of the measurement is urgently needed as guidance to initiate an action of improvement.

2. Research methodology

This research consist of the following steps:

1. Identifying the business process. In order to measure performance, the business process needs to be identified. Each process will be measured according to the 5 activities, plan, source, make, deliver, and return.
2. Identifying and validating performance metrics. Identify performance metrics that are suitable for SME and validating performance metric aims to select the metric that is valid for SME based on expert judgment. The questionnaires are distributed to experts and ask them to rate 1 until5to show their level of agreement (from strongly disagree to strongly agree) with the given statement (items) on a metric scale [21]. Based on the score result, the performance metric that possesses a score of less than 4 will be rejected, and do not apply for the performance measurement process [22].

3. Measuring performance and propose improvement. Measuring the performance is conducted through observation and interview. The result will be normalized using SNORM de Boer. Calculation of normalization value that aims to equalize the unit of value (parameter) of each measurement matrix. Indicators normalized by SNORM, which valued from a range of 0 – 100 [23,24], with the interpretation of the result as presented in Table 1.

$$\text{Larger is better: SNORM} = \frac{(Si - Smin)}{(Smax - Smin)} x\ 100 \quad (1)$$

$$\text{Lower is better: SNORM} = \frac{(Smax - Si)}{(Smax - Smin)} x\ 100 \quad (2)$$

Si = Average value
Smax = The value of achieving the best performance indicators
Smin = The value of achieving the worst performance indicator

Table 1. Performance indicator

System Monitoring	Performance Indicator
<40	Poor
40 – 50	Marginal
50 – 70	Average
70 – 90	Good
>90	Excellent

Source:[25]

3. Results and discussion

The business process is shown in Figure 1 below. When orders come from customers both online and offline, the administration will check the fabric inventory in the warehouse. If the inventory is sufficient, the order will be continued to the design. Yet, if it is less than the safety stock (SS), the administration will order a certain amount to the fabric supplier. Then the fabric supplier will send fabric orders to the warehouse. The company design a logo for the clothes ordered in the design section to be sent to the third party called the printing section together with the fabric (material). The fabric that has been printed will be delivered to the confection for sewing. After completion, all order is delivered to the company's warehouse for the pressing process of name and number requested on Jersey, then the orders are packed. Finally, they are ready for shipping to the customers. Fabrics are sent from suppliers to administration, continue sending them to the printing process.

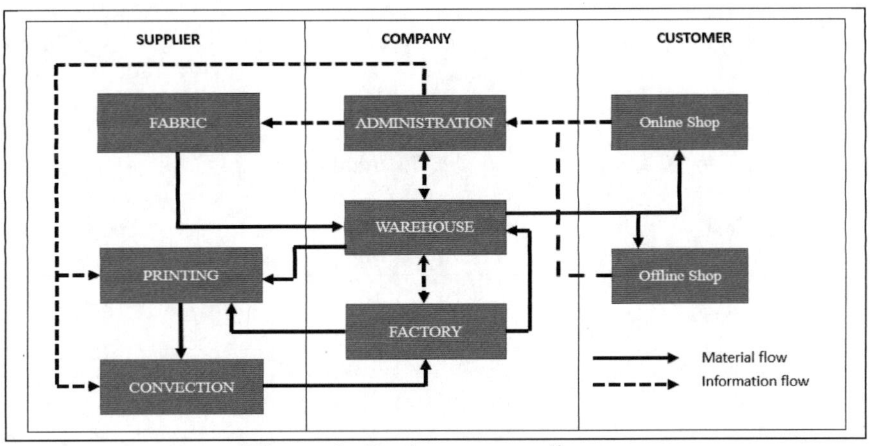

Figure 1. Business scope diagram

3.1. Identifying and validating performance metrics

The metrics were obtained by the sub-processes identified which they were measured by particular performance metrics stated on SCOR. For example, of 5 sub-processes within plan process (sP), it was found only plan source (sP2) due to the actual plan process conducted in the company. Performance metrics obtained, in Table 2, were linked to the sub-processes so identifying the sub-process involved becomes the most important part of quantifying the supply chain performance. Since this study focuses on make-to-order (MTO) sportswear products, all sub-process involved was related to the company's business process. The sub-processes are plan source (sP2), source make-to-order products (sS2), make-to-order (sM2), deliver make-to-order products (sD2), source return defective product (sSR1), deliver return defective product (sDR1), manage supply chain human resource (sE4), and manage supply chain assets (sE5).

There are 40 performance metrics identified through literature reviews. These metrics were validated by distributing a questionnaire to 3 experts who have been working for more than 3 years in the company. The validation example of the reliability performance attribute is presented in the following Table 1 applying Likert Scale. Performance metric scores more than 4 are necessary and will be quantified. There are only 27 performance metrics selected by the experts while the rest eliminated. The definition and formula of 27 performance metrics are taken from SCOR. According to the perceived scores provided in Table 1, inventory accuracy, the defect rate of raw material, meeting with a customer, and packaging error were not quantified due to the score of less than 4. Before scoring each metric, the experts have been explained the definition of the metrics in order to ensure the validity of those scores given. The definitions were taken from SCOR including the formula as shown in Table 2 below. Reliability becomes the most metrics gained in each sub-process following by responsiveness, cost, and agility. Yet, asset management was not measured since it did not meet the business process within the company.

Table 2. Sub-process, performance attributes, and performance metrics

Supply Chain Operation References Version 12.0					
Level 1	**Level 2**	**Performance Attributes**	**Performance Metrics**	**Definition**	**Formula**
(sP) Plan	(sP2) Plan Source	Responsiveness	Time Identify new product specification	The number of times to finish new product identification	Time used started from order/ customer's request received by the company until product design ready for production
		Responsiveness	Lead Time Product	The average time associated with the agreed time of product sent/ ready to pick	How long the product ready to pick by customer
		Cost	Cost for plan	Cost used to make a plan	The total cost to make a plan

Table 3. The example of reliability perceived importance scores

No	RELIABILITY PERFORMANCE METRIC	RESULT			
		Expert 1	Expert 2	Expert 3	Average
1.	Inventory Accuracy	3	3	5	3.67
2.	Source Fill rate	5	4	5	4.67
3.	Defect rate of raw material	4	2	4	3.33
4.	Delivery item accuracy	5	3	5	4.33
5.	Deviation of arrival schedule	4	5	5	4.67
6.	Yield/ product defect	4	4	5	4.33
7.	Schedule achievement of production	5	5	5	5.00
8.	Fill rate/ fulfillment customer order	5	5	5	5.00
9.	Perfect condition	5	5	5	5.00
10.	Delivery quantity accuracy	4	5	5	4.67
11.	Achievement of schedule delivery	3	4	5	4.00
12.	Customer complaint	4	4	4	4.00
13.	Warranty and return	4	3	5	4.00
14.	Meeting with customer	2	2	3	2.33
15.	Packaging error	3	2	2	2.33

Table 4. Normalization of Plan Processes (sP)

No	Performance metrics	Actual	Smin	Smax	Snorm
	Plan (sP)				
	Responsiveness				
1.	Time Identify new product specification	2 Days	3 days	1 ½ days	66
2.	Lead Time Product	14 Days	30 days	10 days	80

Based on the interview's result, it took 2 days for identifying the new product process. Thus, the actual score of this metric is 2. The owner of the company said that he gives a half days and the worst time is 3 days fo time to finish identifying. Both of them are affected by the condition of the job at that time. Thus the normalization calculation of time identify new product specification is:
Time Identify new product specification: (2-3)/(1.5-3) x 100 = 66.

3.2. Discussion
It was identified 27 selected metrics through the interview with production and supply chain manager. The final score was obtained by normalizing the actual score using SNORM shown in Table 4. Each process comprises different performance metrics related to the business process. The performance score of supply chain management reached 77.89. The value shows that it is at a good level of performance. According to the performance metric indicator in Table 1, there are 9 performance metrics reached marginal and poor levels while the remaining metrics are at a good level with the minimum value of 70 presented in Table 5.

Delivery item accuracy, for instance, means that supplier always delivers with the right quantity and never run out of stock. It is also supported by good communication. Other good level metrics are Additional Order to Supplier, Fill Rate/ Fulfillment customer order, and Manage Capital Asset Cycle Time. On the other hand, some lower-level performances require further investigation and improvement. Metric time identify new product specification got a low score because the design operator has more than one job simultaneously, thus it needs to be improved by proposing better scheduling. The cost for the plan metric is the lowest score because there is no cost control for overhead cost when the process plan takes place. Source flexibility can be improved by doing research in sourcing to find more vendors. Current manufacturing cycle time took a long route for distribution to the company's warehouse which can be optimized. Since customers tend to choose the cheaper price, it has affected the calculation of cost for finish product metrics. The company must persuade the customer when meeting them for the first time to choose the best offering.

Table 5. Supply chain performance

No	Performance metrics	Performance Score	No	Performance metrics	Performance Score
1.	Time Identify new product specification	66	15.	Cost for Finish Product	40
2.	Lead Time Product	80	16.	Fill Rate/ Fulfilment customer order	100
3.	Cost for Plan	40	17.	Perfect Condition	99
4.	Source Fill Rate	92	18.	Delivery Quantity Accuracy	86
5.	Delivery Item Accuracy	100	19.	Achievement of Delivery Schedule	93
6.	Deviation of Arrival Schedule	85	20.	Shipment Product Cycle Time	66
7.	Source Cycle Time	85	21.	Customer Complaint	93
8.	Additional Order to Supplier	100	22.	Warranty and Return	99
9.	Source Flexibility/ alternative supplier	50	23.	Warranty Cycle Time	66
10.	Cost for Order to Supplier	71	24.	Days Payable/ Customer Payment	55
11.	Yield/ Product Defect	99	25.	Warranty Cost	70
12.	Achievement of Production Schedule	81	26.	Manage Capital Asset Cycle Time	100
13.	Current manufacturing cycle time	50	27.	Man power capability	42
14.	Manufacturing Product Flexibility	75			
				Final Score	**77.89**

Long shipment product cycle time is caused by consolidating all orders before the delivery process. To improve the warranty cycle time, the company should increase communication with the customer regarding the warranty of the product. It takes on average of 14 days to do payment for Days Payable/ Customer payment because the customers are required to pay the Down Payment when ordering a product then they are allowed to finish it later before delivering the orders. Manpower capability happened due to the lack of focus by the operators that they must complete more than one job in time. These metrics require a depth investigation in order to identify the root cause which can lead an improvement strategy. This analysis can assist the company to determine proper action for each process. The source process has 7 metrics followed by delivery, make, return, delivery, plan, and enable. Of 7 processes, only 1 process obtained the marginal level while the others were good. This process performance is at a good level with an average value of 83.25. Even though delivery (sD) was measured by 5 metrics, it achieved a higher average value than the source process (sS). Return (sR) performance was better than the making process (sM) because of the performance was almost excellent. Overall, the process performance of the supply chain was good according to the performance indicator in Table 1 above since only 30% of metrics were less than 70. However, these performances were not compared to other SMEs for benchmarking. It can not yet be concluded whether the scores were the best in class. Moreover, it was found metrics at marginal, average, and poor levels for each process particularly performance attribute of cost in plan (sP) and make (sM) processes. The smaller the cost the better performance while the actual condition was slightly reaching the maximum limit (Smax).

Responsiveness, reliability, and agility refer to customer-focused. Responsiveness comprises 7 metrics to be measured and it was at a good level. Responsiveness is the speed at which tasks are performed regarding providing products to the customer by the supply chain. The speed is the cycle time. The smaller the cycle time the better performance achieved. The company's responsiveness performance have been conducted properly that it must be maintained for company sustainability. Reliability is the ability to perform tasks as expected focusing on the predictability of the outcome of a process. It relates 3 main points considered in reliability: to be on-time, the right quantity, the right quality. It was at an excellent performance that aligned with the customer-focused target. Meanwhile, there were only 2 metrics identified in agility attributes and they are also at a good level. It can be concluded that these performances were able to follow the roles of keeping customer satisfaction. The company must continue its performance in order to win the business competition.

Besides customer-focused, SCOR explains that cost and asset management are part of internal-focused. In order to conduct a well supply chain, it must be supported by a healthy financial represented by cost and asset management. In this study, cost consisted of only 2 metrics measured while asset management was quantified by a metric in enable process (sE). These 3 metrics were at marginal and poor levels. This condition was aligned with the metrics identified for both cost and asset management. It was difficult to determine the metrics involved since the company has not put its attention on controlling cost and asset management. It was concluded that they did not yet have knowledge about it. To increase the performance scores particularly for these attributes, knowledge transformation is highly required related to control and manage cost and asset management. Thus, more metrics can be added in the future. This situation will gain a better level if it can be compared to other companies running a similar business, sportswear confection. So, this company can learn, copy, and apply the best-in-class process from others.

4. Conclusion

Supply Chain Operation Reference Model (SCOR) is a comprehensive process-based model to measure the performance process of the supply chain. The measurement can inform the company position among other companies running a similar business as well as a benchmark to the best-in-class company. Utilizing the SCOR model, it is identified a list of 27 metrics spread into six performance attributes, i.e. reliability, responsiveness, agility, cost, and asset management efficiency to measure performance. According to the result, the supply chain performance is at a good level of 77.89. Further research can measure the supply chain performance considering the risk management and personnel capabilities stated in the latest SCOR 12. It can be combined with a depth interview to investigate the root cause of each metric calculated on the average, marginal, or poor level. Particularly internal-focused performance attributes, cost, and asset management, they can be explored to gain more information on how SMEs control and manage these attributes. By conducting benchmarking research of the supply chain performance score, the company can set a better strategy in order to achieve its business goals.

References

IOP Conf. Series: Materials Science and Engineering **697** (2019) 012014 doi:10.1088/1757-899X/697/1/012014

[1] Levi S D, Kaminsky P, and Levi E S 2008 *Designing and Managing the Supply Chain: Concepts Strategies and Case Studies* (New York: Mc Graw-Hill Education)
[2] Kulkarni S and Sharma A 2004 *Supply Chain Management* (New Delhi: McGraw-Hill Education Private Limited)
[3] Angelova B and Zekiri J 2011 *J. of Academic Research in Business and Social Sciences* 2222-6990
[4] Durukan T and Bozaci I 2011 *J. of Business and Social Science* **23**
[5] Liputra D T, Santoso, and Susanto N A 2018 Jurnal Rekayasa Sistem Industri 119-125
[6] Agami N, Saleh M, and Rasmy M 2012 *J. of Organizational Man. Studies* **20**
[7] Gunasekaran A, Patel A, and McGaughey R E 2004 *J. production economic* 333-347
[8] Liebetruth T 2017 *Procedia Engineering* 539-544
[9] Golparvar M and Seifbarghy M 2009 *J. of Industrial Engineering* 59-69
[10] Huan S M, Sheoran S K, and Wang G 2004 *Supply Chain Management: An International Journal* 23-29
[11] Hasibuan A, Arfah M, Parinduri L, Hernawati T, Suliawati, Harahap B, and Purwadi A 2018 *J. of Physics: Conference Series, IOP Publishing Ltd*
[12] Information on www.supplychain.org
[13] Delipinar G E and Kocaoglu B 2016 *Procedia - Social and Behavioral Sciences* 398 – 406
[14] Kazemkhanlou H and Ahadi H R 2014 *Proc. Int. Conference on Industrial Engineering and Operations Management (Bali)* p 274
[15] Wibowo M and Sholeha 2015 *Procedia Engineering* 25-31
[16] Wigati D T, Khoirano A D, Alsana S, and Utama D R 2017 *J. Industrial Services* 46-52
[17] Meliantika, Tanjung W N, and Nurhasanah N 2014 *Seminar National on Innovation and Application of Technology in Industry* p 2085
[18] Pourjavad E and Shahin A 2018 *Intern. J. of Fuzzy Systems* 901–912
[19] APICS 2017 *Supply Chain Operation Reference Model SCOR, Version 12.0* (Chicago: APICS)
[20] Hausman W H *The Practice of Supply Chain Management: Where Theory and Application Converge* (ISOR) pp 61-67
[21] Joshi A, Kale S, Chandel S, and Pal D 2015 *British Journal of Applied Science & Technology* 396-403
[22] Aramyan L H, Lansink A G, Van Der Vorst J, and Kooten O F 2007 *Supply Chain Management: An International Journal* 304–315
[23] Waaly A N, Ridwan A Y, and Akbar M D 2018 *MATEC Web of Conferences* **204** 1-15
[24] Kusrini E, Subagyo, Masruroh N A 2015 *Management Science and Applications* 551-562
[25] J.H. Trinekens and H.H. Hvolby 2000 *Proc. of the 3rd CINET Conference (Aalborg)* pp 399-409

IOP Conf. Series: Materials Science and Engineering **697** (2019) 012015 doi:10.1088/1757-899X/697/1/012015

Identify Key Success Factors Using Interpretive Structural Modeling (ISM) : A Case Study in Small and Medium Enterprise in Indonesia

E Kusrini[1*], W Safitri[1], V N Helia[1]

[1]Industrial Engineering Department, Faculty of Industrial Technology, Universitas Islam Indonesia, Yogyakarta, Indonesia
*Email : elisakusrini@uii.ac.id

Abstract. Small Medium Enterprises (SMEs) have a very significant role in the economy of the community. The large number of SMEs on the same field causes high competition to survive in market. A lot of literature empirically discuss the success factors of SMEs in Indonesia, but there are still few which carry out further research by structuring a model of interconnectedness between these elements to find out the key factors. This study aims to determine the key factors by classifying and structurally modelling the factors based on the level of the affecting elements of an SME's success. Interpretive Structural Modeling (ISM), which is an interactive learning process where a set of elements are structured into a comprehensive system model, is used as the approach. ISM helps in determining the sequence and purpose of complex relationships between elements in the system. This research found 24 success factors that divided into 13 levels and categorized into 4 sectors including autonomous, dependent, linkage, and independent sector. The sub-elements or key factors that influence the success of SMEs include (1) motivation and experience, (2) entrepreneurial spirit and leadership, (3) formal education level, and (4) technical guidance.

1. Introduction

Small and Medium Enterprises (SMEs) are known to be one of the most important parts of the country's economy in terms of creating new jobs for the community. Various literatures and researches have carried out many analysis on what factors influencing the success of SMEs in Indonesia, but only few have carried out further research by structuring interconnectedness between these factors as in a research [1]. This research states that the influencing factors consist of external and internal factors. If further research is carried out on the structural modeling and the linkages of factors in the SMEs' success, the main factors (key elements) that have a big role in the success of SMEs and the ways of improving work performance will be discovered.

The approach used to further reaching the objectives is Interpretive Structural Modeling. ISM itself is first introduced by J. Warfield in 1975 [2] to analyze the complexity of the system and to find solutions to complex problems or to involve many factors in it and to interact [3]. The methodology of ISM is an interactive learning process whereby a set of elements are structured into a comprehensive system model. ISM helps in determining the sequence and purpose of complex relationships between elements in the system [4].

Various literatures use the ISM approach to discuss on how the factors correlating with each other and discover the key elements in many cases. However, only few have brought the field of SMEs to the table. For instance, 'The Research on the Key Success Factors of Mobile Internet with Interpretative Structural Modeling' discusses how the linkages of factors in success / key elements of mobile internet using ISM, resulting four key elements including product and service innovation, sharp market environment wisdom, user experience, and core technology from ten previously determined factors [5]. Another research said that twelve factors that determined in Six Sigma implementation had a balanced relationship [6]. There are no insignificant factors. However, there are two factors strong enough that they cannot be affected by the others: education and training and the factors of understanding the tools and techniques in six sigma. With the aim of knowing the ranking of six factors in green-lean implementation, it results that green human resource management is the strongest factor found [7]. Then, [8] discussing how to prevent prisoners from escaping through risk factor analysis of prisoners fleeing from prison, the results show that the ISM is a powerful way to express and understand how these risk factors lead to the escape of prisoners in prison. Furthermore, ten software projects of risk management are proven with result in helping V Corp.'s e-business enhance the ability of their software projects of risk management [10].

Therefore, to describe the linkages of success factors in SMEs, this research will utilize the Interpretive Structural Modelling (ISM) approach and then classify the factors based on the level and structurally model the linkages between these factors. By knowing the success factor and its linkages, SMEs could plan the action to improve their performance. A case study of ISM approch for success factor in a SMEs in Indonesia will be conducted.

2. Literature Review

Interpretive Structural Modelling (ISM) is an interactive learning process in which a series of different elements and relationships are directly arranged into a comprehensive systemic model [2]. ISM helps in identifying linkages between variables. ISM also helps to determine the order and direction in the complexity of the relationships between elements in a system and analyze the influence of one variable with other variables. ISM has been used by researchers to understand the direct and indirect relationships between various variables in different industries.

This ISM technique can be used to conduct program analysis in accordance with the vision and mission. Broadly speaking, the ISM technique is divided into two parts, element classification and hierarchical arrangement. The first step that needs to be done in the ISM analysis is to determine the elements that correspond to the existing problems. Sub-elements are then arranged for each element selected. The selection of elements and the arrangement of sub elements is carried out from discussions with the experts. The results of the assessment are arranged in the Structural Self Interaction Matrix (SSIM) which is made in the form of a Reachability Matrix (RM) table by replacing V, A, X, O into numbers 1 and 0. The classification of elements is based on Structural Self Matrix (SSM) rooted from VAXO system as follows:

- V if $e_{ij} = 1$ and $e_{ji} = 0$;
- A if $e_{ij} = 0$ and $e_{ji} = 1$;
- X if $e_{ij} = 1$ and $e_{ji} = 1$;
- O if $e_{ij} = 0$ and $e_{ji} = 0$.

The matrix is then changed to a closed matrix. This is done to correct the matrix by fulfilling the transience rules, namely if A affects B and B affects C, then A must also affect C. Value 1 means that there is a contextual relationship between the i element and the j element, while eij = 0 means there is no contextual relationships between them. SSM is converted to reachability matrix by changing VAXO to 1 and 0, the transience testing is then done until a closed matrix occurs. The matrix that has fulfilled the transience is proceed in order to get the reachability matrix, the Driver Power (DP), and the Dependence (D).

The last step is to group the sub-elements into four sectors as follows:

IOP Conf. Series: Materials Science and Engineering **697** (2019) 012015 doi:10.1088/1757-899X/697/1/012015

a. Weak driver - weak dependent variables (AUTONOMOUS), variables in this sector are generally not related to the system that they have a slight linkages.
b. Weak driver strongly - dependent variables (DEPENDENT), variables that belong to this group are independent variables.
c. Strong driver strongly dependent variables (LINKAGE), variables in this sector must be carefully examined because the interaction can have an impact and feedback on the system.
d. Strong driver weak dependent variables (INDEPENDENT), variables in this sector have a strong influence on the system and greatly determine the success of the program.

3. Methodology

Reference used in determining the success factors of SMEs is conducted by [11]. In his research, Munizu concludes that the success factors of SMEs based on previous literature include external factors and internal factors. This study uses most of those factors combined with the ones resulted from discussions with the experts that it finally consists of the following 24 factors as presented in Table 1.

Table 1. Success factors of SMEs

NO	FACTOR	NO	FACTOR
1	Motivation	13	Market demand
2	Formal education level	14	Competitive pricing
3	Experience	15	Promotion
4	Entrepreneurial spirit and leadership	16	Distribution channels and marketing areas
5	Own capital	17	Business climate and investment
6	Borrowed capital	18	Number of same-field competitors
7	Rate of profit and capital accumulation	19	Economic growth
8	Financial governance (personal and business)	20	Community income level
9	Production capacity	21	Government policy
10	Modern technology and quality control	22	Capital assistance from related institutions
11	Material selection	23	Technical guidance
12	Technology selection (machine / equipment)	24	Monitoring and evaluation

The primary data collection in this study was conducted by interviews and questionnaires to 4 experts representing SMEs. Representatives of each SME are experts who have an important role in establishing the business such as the owner, founder, manager, or section head. Questionnaire method is done by giving a list of questions to be filled in by respondents (experts). They are asked to fill in opinions about the relationship between the factors that have been determined using the ISM scale. Secondary data is obtained from various library sources related to the research.

4. Result and Discussion

4.1. Structural Self Interaction Matrix

Based on the results of the questionnaire, the Structural Self Interaction Matrix (SSIM) is made. At this stage, the results of the questionnaire are inputted in ISM Professional V.4.0. The 4 answers of the experts are needed to be covered under 1 SSIM, therefore, the ISM software itself is used (Table 2).

IOP Conf. Series: Materials Science and Engineering **697** (2019) 012015 doi:10.1088/1757-899X/697/1/012015

Table 2. Structural Self Interaction Matrix (SSIM)

Key Success Factors for SMEs	A1	A2	A3	A4	A5	A6	A7	A8	A9	A10	A11	A12	A13	A14	A15	A16	A17	A18	A19	A20	A21	A22	A23	A24
Motivation		V	A	V	V	V	A	V	V	V	A	V	V	V	V	V	V	V	V	O	O	V	V	V
Formal education level			V	V	V	O	V	A	V	V	X	V	O	V	A	A	O	V	X	O	O	O	O	X
Experience				V	V	V	V	V	V	V	V	V	V	V	V	V	V	X	O	O	O	O	V	V
Entrepreneurial spirit and leadership					V	V	V	V	V	V	V	V	V	V	V	V	V	V	V	O	O	O	V	V
Own capital						O	O	X	V	V	V	V	V	O	O	V	V	O	A	A	O	O	O	O
Borrowed capital							O	V	V	V	V	V	V	O	V	V	V	O	A	O	O	V	O	O
Rate of profit and capital accumulation								V	V	A	A	A	V	A	V	V	A	A	A	A	O	O	A	A
Financial governance (personal and business)									V	V	V	V	O	V	V	V	O	O	O	O	O	O	O	O
Production capacity										A	A	A	A	V	V	V	A	A	A	A	O	A	A	A
Modern technology and quality control											O	V	A	V	V	V	V	O	O	O	O	O	X	X
Material selection												O	V	V	V	X	X	X	X	O	O	O	O	O
Technology selection (machine / equipment)													A	V	A	O	O	A	O	O	O	A	A	O
Market demand														V	V	X	X	V	V	A	O	O	O	V
Competitive pricing															V	V	A	V	V	A	O	O	A	O
Promotion																V	V	A	A	O	O	O	A	A
Distribution channels and marketing areas																	O	A	V	V	O	O	A	O
Business climate and investment																		V	X	A	O	A	A	A
Number of same-field competitors																			A	A	O	A	A	A
Economic growth																				A	A	V	O	V
Community income level																					A	O	A	O
Government policy																						O	A	A
Capital assistance from related institutions																							O	O
Technical guidance																								X
Monitoring and evaluation																								

IOP Conf. Series: Materials Science and Engineering **697** (2019) 012015 doi:10.1088/1757-899X/697/1/012015

Table 3. Reachability Matrix

Key Success Factors for SMEs	A1	A2	A3	A4	A5	A6	A7	A8	A9	A10	A11	A12	A13	A14	A15	A16	A17	A18	A19	A20	A21	A22	A23	A24	DP
Motivation	1	1	0	1	1	1	0	1	1	1	0	1	1	1	1	1	1	1	1	0	0	1	1	1	**19**
Formal education level	0	1	1	1	1	0	1	0	1	1	1	1	0	1	0	0	0	1	1	0	0	0	0	1	**13**
Experience	1	0	1	1	1	1	1	1	1	1	1	1	1	1	1	1	1	1	0	0	0	0	1	1	**19**
Entrepreneurial spirit and leadership	0	0	0	1	1	1	1	1	1	1	1	1	1	1	1	1	1	1	1	0	0	0	1	1	**18**
Own capital	0	0	0	0	1	0	0	1	1	1	1	1	1	0	0	1	1	0	0	0	0	0	0	0	**9**
Borrowed capital	0	0	0	0	0	1	0	1	1	1	1	1	1	0	1	1	1	0	0	0	0	1	0	0	**11**
Rate of profit and capital accumulation	1	0	0	0	0	0	1	1	1	0	0	0	1	0	1	1	0	0	0	0	0	0	0	0	**7**
Financial governance (personal&business)	0	1	0	0	1	0	0	1	1	1	1	1	0	1	1	1	0	0	0	0	0	0	0	0	**10**
Production capacity	0	0	0	0	0	0	0	0	1	0	0	0	0	1	1	1	0	0	0	0	0	0	0	0	**4**
Modern technology and quality control	0	0	0	0	0	0	1	0	1	1	0	1	0	1	1	1	1	0	0	0	0	0	1	1	**10**
Material selection	1	1*	0	0	0	0	1	0	1	0	1	0	1	1	1	1	1	1	1	0	0	0	0	0	**12**
Technology selection (machine)	0	0	0	0	0	0	1	0	1	0	0	1	0	1	0	0	0	0	0	0	0	0	0	0	**4**
Market demand	0	0	0	0	0	0	0	0	1	1	0	1	1	1	1	1	1	1	1	0	0	0	0	1	**11**
Competitive pricing	0	0	0	0	0	0	1	0	0	0	0	0	0	1	1	1	0	1	1	0	0	0	0	0	**6**
Promotion	0	1	0	0	0	0	0	0	0	0	0	1	0	0	1	1	1	0	0	0	0	0	0	0	**5**
Distribution channels and marketing areas	0	1	0	0	0	0	0	0	0	0	1	0	1	0	0	1	0	0	1	1	0	0	0	0	**6**
Business climate and investment	0	0	0	0	0	0	1	0	1	0	1	0	1	1	0	0	1	1	1	0	0	0	0	0	**8**
Number of same-field competitors	0	0	1	0	0	0	1	0	1	0	1	1	0	0	1	1	0	1	0	0	0	0	0	0	**8**
Economic growth	0	1	0	0	1	1	1	0	1	0	1	0	0	0	1	0	1*	1	1	0	0	1	0	1	**12**
Community income level	0	0	0	0	1	0	1	0	1	0	0	0	1	1	0	0	1	1	1	1	0	0	0	0	**9**
Government policy	0	0	0	0	0	0	0	0	0	0	0	0	0	0	0	0	0	0	1	1	1	0	0	0	**3**
Capital assistance from related institutions	0	0	0	0	0	0	0	0	1	0	0	1	0	0	0	0	1	1	0	0	0	1	0	0	**5**
Technical guidance	0	0	0	0	0	0	1	0	1	1	0	1	0	1	1	1	1	1	0	1	1	0	1	1	**13**
Monitoring and evaluation	0	1	0	0	0	0	1	0	1	1	0	0	0	0	1	0	1	1	0	0	1	0	1	1	**10**

4.2. Reachability Matrix and Diagram

The next step in finding out the key factors that influence the success of SMEs and their structure modeling is to make Reachability Matrix by changing the SSIM symbols, VAXO, to become a binary matrix 1 and 0 (Table 3). Then, partition is done to the RM table after making further corrections that it already a closed matrix meeting the rules of transience. Transience is the completeness of the causal loop, for example, A influences B and B affects C, so it can be said that A must influence C too. From the partition, it can be known of the ranking, level, and driver power of each sub-element. The following as shown in table 3 is the result of the RM matrix as well as the RM partition using the transience rules.

After knowing the levels order / ranks of each sub-element, level partition is applied by ranking the objects from the highest to the lowest, making it easier to read in the next step, which is making the diagraph. From the Final Reachability Matrix, structured results are found based on the levels. It is then the structural model can be made according to its ranks and reveal the key success factors of SMEs which are motivation (A1) and experience (A3), as shown in Figure 1.

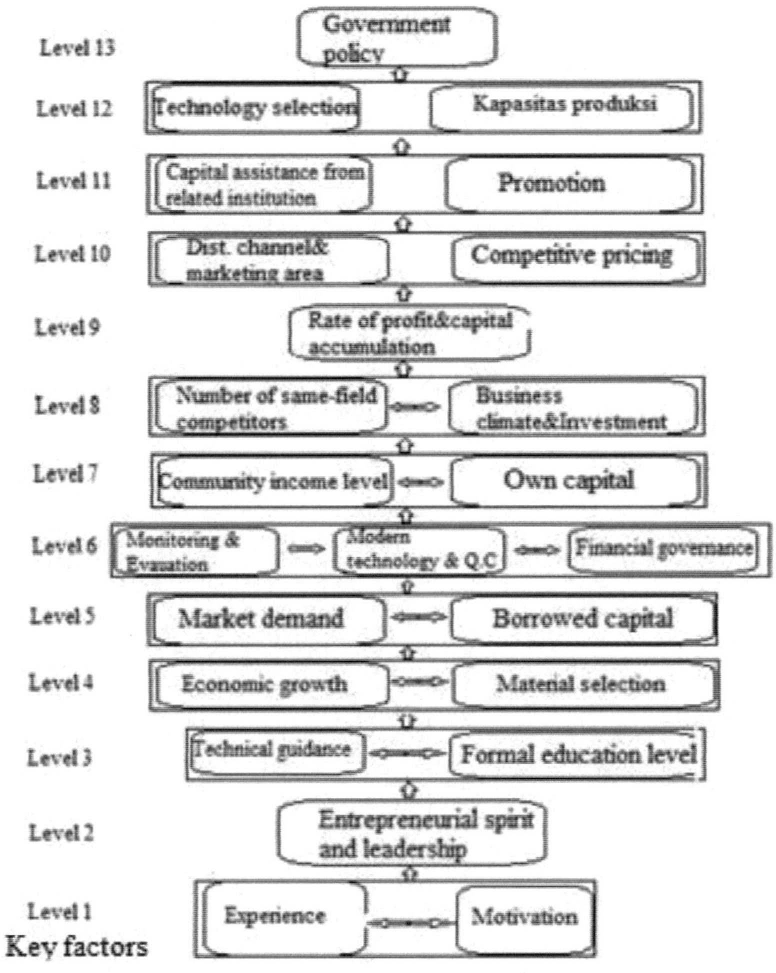

Figure 1. Diagraph (ISM)

The result of data processing that have been done previously have shown that the linkages between factors influencing the success of SMEs are divided into 13 levels. In this case, the 13 levels will be classified as sub-elements / factors based on the result of the RM that have fulfilled the rules of transience as presented on Figure 2. The result is carried out from the Power Driver (DP) and Dependence (D) values, which will determine the classification of the sub-elements as follows:

a. Sector 1 : weak driver - weak dependent variables (autonomous). Sub elements included in this sector are not related to the system, and maybe there are only a few linkages, even though the linkages can be strong. Sub-elements included in sector 1 are factors A5, A6, A8, A10, A11, A13, A19, A20, A21, A22, and A24.
b. Sector 2 : weak driver - strongly dependent variables (dependent). Generally, sub-elements included in this sector are dependent. Sub-elements included in sector 2 are factors A7, A9, A12, A14, A15, A16, A17, and A18.
c. Sector 3 : strong drivers - strongly dependent variables (linkage). Sub-elements included in this sector must be carefully examined, because the relationship between sub-elements is unstable. Every action on the sub elements will have an impact on other sub-elements and the feedback can increase the impact. There are no sub-elements that meet the requirements for this sector.
d. Sector 4 : strong driver - weak dependent variables (independent). Sub elements that enter this sector have a strong influence on the system and greatly determine the success of the program. Sub-elements included in sector 4 are factor A1, A2, A3, A4, and A23.

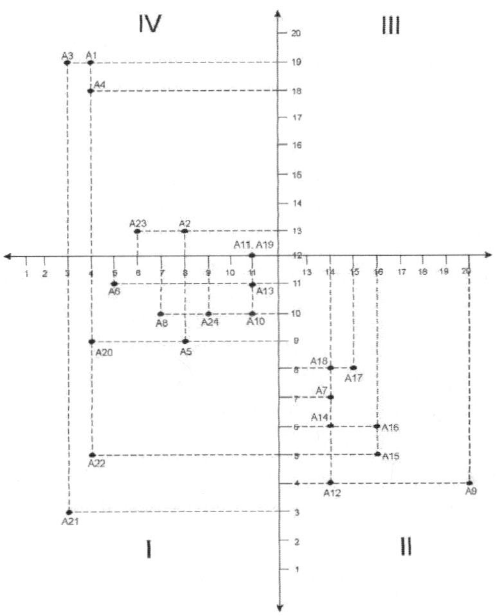

Figure 2. Sub-element Classification

5. Conclusions and Recommendations

This research found twenty-four success factors that divided into thirteen levels and categorized into four sectors including autonomous, dependent, linkage, and independent sector. The key success factors of SMEs are experience (A3) and motivation (A4). It is supported by the previous literature [10] saying that motivation is directly proportional to the success of a business and experience positively and significantly affects its development.

In addition, business motivation is also directly proportional to success. The higher the motivation, the more successful it will be. Motivation here refers to business motivation that comes from motives, hopes, incentives, profits, freedom, personal dreams, and independence.

Furthermore, based on conclusions, it is recommended to improve the performance of SMEs by actively participating in education and training for owners and employees to improve the quality of human resources. For motivation, both quality and quantity must be maintained and improved by minimizing excessive workload, improving the family atmosphere in the workplace, and praising the good work of employees and giving advice to each other for improved performance.

References

[1] Purwaningsih, R. & Kusuma, P. D., "Analisis Faktor-Faktor yang Mempengaruhi Kinerja Usaha Kecil Menengah (UKM) dengan Metode Structural Equation Modeling," *Prosiding SNST*, ISBN 978-602-99334-4-4, 2015.

[2] Warfield, J., "Developing Interconnected Matrices in Structural Modelling," *IEEE Transactions on Systems Men and Cybernetics*, Vol. IV (**1**), pp. 51-81, 1974.

[3] Mohammed, I. S. R. B. D., "Creating flex-lean-agile value bhain by," *Business Process Management Journal*, Vols. I, II, XIV, no. **3**, pp. 338-389, 2008.

[4] Pfohl, H. C., Gallus, P., & Thomas, D., "Interpretive structural modeling of supply chain," *International Journal of Physical Distribution & Logistics Management,* Vol. XLI, no. **9**, pp. 839-859, 2011.

[5] Wan, J., Zhu, Y., & Liang, L., "The Research on the Key Success Factors of Mobile Internet with Interpretative Structural Modeling," *Scientific Research*, p. 6, 2013.

[6] Alidrisi, H., "Prioritizing Critical Success Factors for Six Sigma Implementation Using Interpretive Structural Modeling," *American Journal of Industrial and Business Management*, pp. 697-708, 2014.

[7] Jadhav, J. R., Mantha, S. S., & Rane, S. B., "Interpretive Structural Modeling for Implementation of Integrated Green-Lean System," *International Journal of Computer Applications*, p. 0975 - 8887, 2013.

[8] Song, Y., "Risk Factors Analysis of Criminals Escape from Prison Based on Interpretative Structural Modeling," *Advances in Applied Sociology*, Vol. **VII**, pp. 365-377, 2017.

[9] Wan, J. & Wang Z., "Case Study on E-Business V Corp. Software Project Risk Management with Interpretive Structural Modeling," *Open Journal of Social Sciences*, Vol. **III,** pp. 1-7, 2015.

[10] Gemina, D., Silaningsih, E., & Yuningsih, E., "Pengaruh Motivasi Usaha terhadap Keberhasilan Usaha dengan Kemampuan Usaha sebagai Variabel Mediasi pada Industri Kecil Menengah Makanan Ringan Priangan Timur-Indonesia". *Jurnal Manajemen Teknologi*, pp. 297-323, 2016.

[11] Munizu, Musran, "Pengaruh Faktor-Faktor Eksternal dan Internal Terhadap Kinerja Usaha Mikro dan Kecil (UMK) di Sulawesi Selatan", *Jurnal Manajemen dan Kewirausahaan* 12, pp.33-41, 2010.

Acknowledgments

This research was supported by Industrial Engineering Department, Faculty of Technology Industry, Universitas Islam Indonesia, Yogyakarta, Indonesia.

IOP Conf. Series: Materials Science and Engineering **697** (2019) 012016 doi:10.1088/1757-899X/697/1/012016

Professional Skills Requirement of Mechanical Engineers

W Omar Ali Saifuddin Wan Ismail[1], Noraini Hamzah[2], Ireana Yusra Abdul Fatah[1], Azami Zaharim[2]

[1] Faculty of Innovative Design and Technology, Universiti Sultan Zainal Abidin, 21300 Kuala Nerus, Terengganu, Malaysia
[2] Faculty of Engineering and Built Environment, Universiti Kebangsaan Malaysia, 43600 Bangi, Malaysia.

Abstract. Professional Skills is the basic requirement that all engineers need to master in the current job market. This study was conducted throughout peninsular Malaysia using the Soft Skills Model introduced by the Ministry of Education (MoE). Among the objectives of this study is the level of requirement of the Soft Skills (SS) element to produce a competitive Mechanical Engineer and identify the most dominant Soft Skills needed in the current job market against Mechanical Engineers. The questionnaire distributed to respondents and data was analysed using XL-STAT 2014 software to obtain Discriminant Analysis (DA) to achieve the objective of the study. The overall percentage for (SS) elements is 91% respectively. This study will be a complementary set for the students of Mechanical Engineering in Malaysia. Mastering the features required by the industry makes it easy for students to put themselves in the real job world.

1. Introduction
Soft Skills
The Ministry of Education (MoE) has emphasised on standard methods and objectives to measure the students' soft skills. Among the elements emphasised are psychomotor, communication, English proficiency, leadership, ethics, spirituality and emotional intelligence. All universities need to incorporate these elements into the program's learning outcomes to evaluate student achievement [1]. Soft skills, generic skills and employability skills refer to modern engineering graduates [2] who need to be skilful in every aspect of this skill if they want to succeed at work then carry out successful work at the workplace [3] in helping to improve the capabilities of a company [4]. It is the primary asset for any workforce at the professional level, manager or business owner [5].

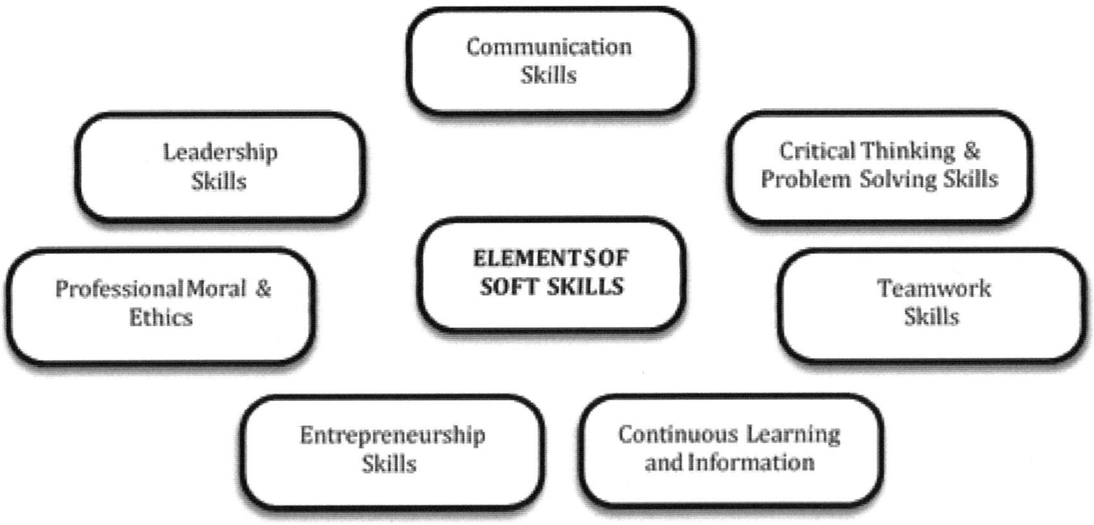

Figure 1. Soft Skills Model

Based on Figure 1, there are seven elements of Soft Skills outlined by (MoE) for reference of all universities. i. **Communication Skills**: through effective communication skills, one will try to absorb and adapt [6] with news or situation that has the potential to reduce concerns and improve performance [7] as claimed by employers to meet high market demands in the industry employment [8] especially engineering graduates [9]. ii. **Critical Thinking and Problem Solving Skills**: reform in education is now increasingly focusing on essential processes of thinking, including problem-solving items [10,11]. Critical thinking and judgment are needed for engineering at a professional level to solve workplace problems [12,13]. Critical thinking is a step to make assumptions, testing of legitimacy, seeing things from different points of view, and taking into account every action [14,15,16]. As future graduates, university students need to equip themselves with critical thinking and problem-solving skills [17,18] as this is the focus of employers in hiring a new workforce [19]. iii. **Teamwork Skills**: teamwork is a high-level need that needs to be addressed [20]. Students believe that they are studying teamwork skills through the curriculum approach [21], joining sports teams or volunteers to work in an organisation that does not involve profit [22]. iv. **Continuous Learning and Information Management**: through the development of the methods used by incorporating the adapted learning, it can address the various learning styles of students and improve the effectiveness of the use of technology and information management [23]. v. **Entrepreneurial Skills**: nowadays, entrepreneurial skills are considered the ideal strategy for expanding productivity and overall development of any organisation [24] and being recognised as one of the critical skills [25], among engineering graduates [26]. Graduates not only have business opportunities and manage their businesses but also to create jobs for others [27,28,29]. vi. **Professional Ethics and Moral**: professional ethics is the best practice for all engineering students and engineering practitioners and advisable to stick to this principle [30] for the well-being of society which is the essence of engineering profession [31]. vii. **Leadership Skills**: essential skills are personality skills that are related to interpersonal skills such as leadership skills [32,33] that can develop the quality of competent leadership skills [34].

Based on the Program Outcomes (PO), engineering students are expected to master the skills, knowledge and attitudes through programs offered. That includes the following items: i. Engineering Knowledge; ii. Problem Analysis; iii. Design/Development of Solutions; iv. Investigation; v. Modern Tool Usage; vi. The Engineer and Society; vii. Environment and Sustainability; viii. Ethics; ix. Communication; x. Individual and Team Work; xi. Life-Long Learning; and xii. Project Management and Finance [35]. The items contained in (PO) are closely related to the employability of mechanical engineering graduates also involve the professional skills required by the industry.

2. Materials and methods

The research was conducted throughout the Peninsular of Malaysia. It is also involved in Mechanical Engineering industrial activities. A set of questionnaires is distributed by researchers to obtain feedback from industry, employers and organisations involved in the process of finding and securing the workforce in this field. All surveys have distributed, updated into XL-Stat software. All of the data generated, and the results of the study included Descriptive Statistics (DS) and Discriminant Analysis (DA).

3. Results and discussions

A total of 300 respondents have responded to the question item with a different current position. Eight items for Soft Skills (Communication) - (SSA), seven items Soft Skills questions (Critical Thinking and Problem Solving Skills) - (SSB), five Soft Skills questions (Teamwork) - (SSC), three-question item of soft skills (Continuous Learning and Information Management) - (SSD), four Soft Skills questions (Entrepreneurship Element) - (SSE), three Soft Skills questions (Professional Ethics and Moral) - (SSF) and four Soft Skills questions (Leadership) - (SSG).

Table 1. Descriptive Statistics Demographics

No.	Gender	Frequency	Percentage
1	Man	240	80.00
2	Woman	60	20.00

Table 1 shows the Histogram and Descriptive Statistics Gender Demographics with the number of respondents noted, 240 were men and 60 women. It involves staff in heavy, medium and small industries throughout Peninsular Malaysia.

Table 2. Descriptive Statistics Demographics

No.	Current Position	Frequency	Percentage
1	Management & Administration	100	33.33
2	Senior Engineers	72	24.00
3	Senior Technologist	38	12.67
4	Engineers	90	30.00

Table 2 shows the Histogram and Descriptive Statistics Current Position Demographics of respondents have work experience of more than five years in the industry, especially in the mechanical engineering field and engineering department.

Table 3. Percentage of Classification of Level Needed

Formula	Percentage	Level	Justification
100 - 1 / 5 = 19.8	80.6 – 100	Five	Very High
	60.7 - 80.5	Four	High
	40.8 - 60.6	Three	Moderate
	20.9 - 40.7	Two	Low
	1.00 - 20.8	One	Very Low

Percentages used in this study were divided into five categories, as shown in Table 3 based on the Likert Scale (1932).

i. **Soft Skills (Communication) - (SSA)**

Mean of the question item i. SSA1 (4.37); ii. SSA2 (4.43); iii. SSA3 (4.47); iv. SSA4 (4.34); v. SSA5 (4.42); vi. SSA6 (4.30); vii. SSA7 (4.31); and viii. SSA8 (4.32).

Table 4. Descriptive Statistic Question Item Soft Skills (Communication)

Question Item	Frequency	Percentage	Level of Need
SSA1: Ability to communicate ideas, effectively, and with confidence, verbally and in writing	273	91.00	Very High
SSA2: Ability to practice active listening skills and provide feedback	292	97.34	Very High
SSA3: Ability to make a presentation expressly with confidence and appropriate to the audience level	276	92.00	Very High
SSA4: Ability to use technology in a presentation	260	86.67	Very High
SSA5: Ability to negotiate and reach agreement	278	92.67	Very High
SSA6: Ability to communicate with participants of communication with different cultures	260	86.67	Very High
SSA7: Ability to develop personal communication skills	275	91.66	Very High
SSA8: Ability to use non-verbal skills	258	86.00	Very High

Table 4 shows the question items SSA2 are ranked first with a percentage (97%) followed by SSA5 (93%), SSA3 (92%), SSA7 (92%), SSA1 (91%), SSA4 (87%), SSA6 (87%) finally is SSA8 (86%). The researcher collected the result shows significant as a ref [6,7,8,9].

ii. Soft Skills (Critical Thinking and Problem Solving Skills) - (SSB)

Mean of the question item i. SSB1 (4.52); ii. SSB2 (4.47); iii. SSB3 (4.57); iv. SSB4 (4.28); v. SSB5 (4.43); vi. SSB6 (4.56); and vii. SSB7 (4.48) as shown.

Table 5. Descriptive Statistics Question Item Soft Skills (CTPSS)

Question Item	Frequency	Percentage	Level of Need
SSB1: Ability to identify and analyse problems in complex and blurring situations, and make a justified evaluation	282	94.00	Very High
SSB2: Ability to expand and improve thinking skills such as describing, interpreting and evaluating discussions	282	94.00	Very High
SSB3: Ability to find ideas and find alternative solutions	281	93.66	Very High
SSB4: Ability to think beyond the reach	259	86.33	Very High
SSB5: Ability to make decisions based on substantial evidence	287	95.66	Very High

SSB6: Ability to survive and to give full responsibility	300	100.00	Very High
SSB7: Ability to understand and adapt to a new community and working culture	274	91.33	Very High

Table 5 shows the question item SSB6 lead the highest value with percentage value (100%), followed by item SSB5 with percentage (96%), SSB1 (94%), SSB2 (94%), SSB3 (94%), SSB7 (91%) and SSB4 (86%). Percentage gives the impression that the element of the Critical Thinking and Problem-Solving Skills needed by engineering students as a ref [10,11,12,13,14,15,16,17,18,19] mentioned in the research before.

iii. Soft Skills (Teamwork)

The Mean values recorded are as follows: i. SSC1 (4.46); ii. SSC2 (4.38); iii. SSC3 (4.41); iv. SSC4 (4.43); and SSC5 (4.46).

Table 6. Descriptive Statistic Question Item Soft Skills (Teamwork)

Question Item	Frequency	Percentage	Level of Need
SSC1: Ability to build relationships, interact with others and work effectively with them to achieve the same objectives	279	93.00	Very High
SSC2: Ability to understand and take alternate roles between the group leader and group members	267	89.00	Very High
SSC3: Ability to recognise and respect, behaviours and beliefs of others.	263	87.66	Very High
SSC4: Ability to contribute to planning and aligning the outcome of the group's efforts	280	93.33	Very High
SSC5: Responsible for group decisions	287	95.66	Very High

Table 6 shows the SSC5 question items got the highest percentage requirement reading (96%), followed by SSC4 (93%), SSC1 (93.00%), SSC2 (89%) and SSC3 (88%). As mention before in ref [20,21,22], all the item gives their contribution to producing the competitive of Mechanical Engineers in the future.

iv. Soft Skills (Continuous Learning and Information Management)

The Mean values recorded by the researcher are as follows: i. SSD1 (4.39); ii. SSD2 (4.44); and iii. SSD3, (4.56).

Table 7. Descriptive Statistic Question Item Soft Skills (CLIM)

Question Item	Frequency	Percentage	Level of Need
SSD1: Ability to locate and manage relevant Information from multiple sources	279	93.00	Very High
SSD2: Ability to accept new ideas and capable of autonomous learning	267	89.00	Very High

SSD3: Ability to develop curiosity and a thirst for knowledge	287	95.66	Very High

Table 7 shows the question item SSD3 gets the highest percentage requirement reading (96%), followed by SSD1 (93%) and SSD2 (89%) as a ref [23] mention before, the effectiveness of the use of technology and information management should be improved and up to date.

v. Soft Skills (Entrepreneurship)

Mean values recorded are as follows: i. SSE1 (4.29); ii. SSE2 (4.17); iii. SSE3 (4.21) and iv. SSE4 (4.12).

Table 8. Descriptive Statistic Question Item Soft Skills (Entrepreneurship)

Question Item	Frequency	Percentage	Level of Need
SSE1: Ability to identify business opportunities	258	86.00	Very High
SSE2: Ability to design business planning	239	79.67	High
SSE3: Ability to build, explore and seize business and employment opportunities	251	83.67	Very High
SSE4: Ability to work alone	252	84.00	Very High

Table 8 shows the SSE1 question items get the highest percentage requirement reading (86%), followed by SSE4 (84%), and SSE3 (84%). The only question in a high group is SSE2 (80%). According to the ref [24,27,28,29], entrepreneurship is one of the strategies to manage properly by the graduate to fulfil the needed of the job and task on the engineering fields.

vi. Soft Skills (Professional Moral and Ethics)

The Mean values recorded are as follows: i. SSF1 (4.41); ii. SSF2 (4.38); and iii. SSF3 (4.47).

Table 9. Descriptive Statistic Question Item Soft Skills (Professional Ethics and Moral)

Question Item	Frequency	Percentage	Level of Need
SSF1: Ability to understand economic, environmental and socioeconomic effects in professional practice	283	94.83	Very High
SSF2: Ability to analyse and decide on ethical-related issues	270	90.00	Very High
SSF3: Ability to practise ethical behaviour, also to have a sense of responsibility towards society	281	93.66	Very High

Table 9 shows the question item SSF1 got the highest percentage (95%), followed by SSF3 (94%), and SSF2 (90%). To produce the competitive of the mechanical engineer, an institution of higher education must show good manners and principle, as mention before in ref [30,31].

vii. Soft Skills (Leadership)

The Mean values recorded are as follows: i. SSG1 (4.22); ii. SSG2 (4.35); iii. SSG3 (4.30); and iv. SSG4 (4.36).

Table 10. Descriptive Statistic Question Item Soft Skills (Leadership)

Question Item	Frequency	Percentage	Level of Need
SSG1: Knowledge of basic theories of leadership	247	82.34	Very High
SSG2: Ability to lead the project	275	91.66	Very High
SSG3: Ability to understand and take alternate roles between team captains and team members	271	90.34	Very High
SSG4: Ability to oversee team members	269	89.67	Very High

Table 10 shows the SSG2 question items got the highest percentage requirement reading (92%), followed by SSG3 (90%), SSG4 (90%), finally is SSG1 (82%). According to ref [32,33,34], the quality of competent leadership skills can be developed and learn time to time with the best attitude of the leader.

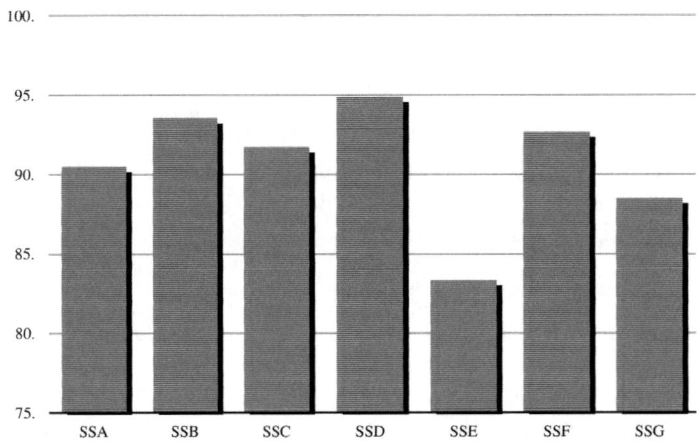

Figure 2. Histogram Percentage of Level of Needs

Fig. 2 shows a histogram ratio of the percentage of the requirement that involves seven Soft Skills: i. SSA - Communication (91%); ii. SSB - Critical Thinking and Problem Solving Skills (94%); iii. SSC - Teamwork (92%); iv. SSD - Continuous Learning and Information Management (95%); v. SSE - Entrepreneurship Elements (84%); vi. SSF - Professional Ethics and Morals (93%); and vii. SSG - Leadership (89%). The results of the study have shown that all the elements strongly required by the employers and the industry to the graduates of Mechanical Engineering. It is complementary to each other to produce the best result on an engineering organisation. It must be mastered completely before facing the real world, as mentioned in a ref [35].

Discriminant Analysis

Discriminant Analysis (DA) is used in systematic scientific research [36,37,38] to identify clusters [39], discriminating against two or more groups [40,41,42]. Four types of clusters are involved in Management and Administration, Senior Engineers, Senior Technicians and Engineers. DA is also used to validate the classification of careful observations [43] classification of identified groups [44], and it is a classification technique [45] on observations performed [46] by researchers at the early stage of study [47].

Table 11. Wilks' Lambda Test (Rao Estimate - Position)

i: Management & Administration		**ii: Senior Engineers**	
Criteria	**Value**	Criteria	Value
Lambda	0.069	Lambda	0.069
F (Observed value)	11.145	F (Observed value)	11.095
F (Critical value)	1.261	F (Critical value)	1.262
DF1	102	DF1	102
DF2	788	DF2	785
p-value	<0.0001	p-value	<0.0001
alpha	0.05	alpha	0.05
iii: Senior Technologist		**iv: Engineers**	
Criteria	Value	Criteria	Value
Lambda	0.070	Lambda	0.071
F (Observed value)	10.988	F (Observed value)	10.804
F (Critical value)	1.262	F (Critical value)	1.262
DF1	102	DF1	102
DF2	782	DF2	779
p-value	<0.0001	p-value	<0.0001
alpha	0.05	alpha	0.05

Table 11 (i, ii, iii, and iv) shows the Discriminant Analysis of the results of the Wilks' Lambda Test (Rao Estimate) obtained for the Soft Skills - (position) clusters for i. Management and Administration; ii. Senior Engineers; iii. Senior Technologist and; iv. Engineers. The test shows that H0: The vector mean of the four clusters is the same and Ha: At least one mean vector is different from the others. Based on the analysis, the resulting p-value is lower than the significant level of alpha = 0.05. Hence, the three hypotheses H0 Zero rejection, and alternative Ha hypotheses are accepted. The risk of rejecting the H0 zero hypothesis is when the accuracy is lower than 0.01%. The four clusters are different from each other [48].

Table 12. Error Matrix for Buget Samples

From / To	1	2	3	4	Total	% Correct
1	75	21	0	4	100	75.00
2	8	61	0	3	72	84.72
3	0	10	28	0	38	73.68
4	7	20	0	63	90	70.00

Table 12 shows the formulation of the matrix of errors in the sample estimates involving four Cluster Type Positions. Based on the table: i. One hundred respondents from Management and Administration with 75% correct percentage; ii. Seventy-two respondents from Senior Engineer with 85% correct; iii. Thirty-eight respondents from Senior Technician with percentage right 74%; and iv.

Ninety respondents among engineers with a percentage of 70%. That means Senior Engineer answered questions item of the study with the highest level of accuracy.

4. Conclusions

Overall, it can be stated by the researcher that industry in Malaysia needs the seven elements of Soft Skills in obtaining competitive Mechanical Engineers. It can be proved by the evidenced available in the results of the study. The most dominant Soft Skills is Continuous Learning and Information Management. Employers and engineering organisations today expect every engineer always strive to find and get the latest inputs as well as efficient in technologically advanced. These soft skills are combined with the knowledge and attitudes to produce future high-quality Mechanical Engineering workforce following the expectations expected by the (MoE) and the Malaysian industry.

References

[1] Mentransformasi Sistem Pendidikan. Rancangan Malaysia Kesebelas, 2016-2020: Hala Tuju. Unit Perancang Ekonomi, Jabatan Perdana Menteri.
[2] Kilam I. K., Jyoti Sharma (2013). Analysis and evaluation of various soft skills training models employed by private technical educational institutions in india. Vol. 2, Issue 2.
[3] Inayatullah Kakepoto, Noor Abidah Mohd Omar, Yusuf Boon, S M Zafar Iqbal (2012). Perspectives on oral communication skills for engineers in engineering profession of Pakistan. International Journal of Applied Linguistics & English Literature. Vol. 1 No. 5; September 2012 [Special Issue on General Linguistics] Perspectives.
[4] Cushing Anderson, John F. Gantz. (2013). Skills requirement for tomorrow's best jobs helping educators provide students with skills and tools they need. IDC Analyse the future.
[5] Dipawalee Santosh Mishra, (2016). Engineering employability skills required by employers in india. International Research Journal of Engineering and Technology (IRJET). Vol. 3, Issue: 02 Feb-2016.
[6] Najib Ahmad Marzuki, Che Su Mustaffa & Zarina Mat Saad. (2015). Emotional intelligence: it's relationship with communication and information technology skills. Asian Social Science. Retrieved.
[7] Bramhall, E (2014). Effective communication skills in nursing practice. Nursing Standard. 29, 14, 53-59.
[8] Rosdiadee Nordin (2013). Technical communication skills among recent electrical and electronics engineering graduates in job industries. 164 (2013). Universiti Kebangsaan Malaysia, Bangi, Selangor, Malaysia.
[9] Vijaya Lakshmi D. (2016). Significance of language labs in engineering education. International Journal of Humanities in Technical Education. Institute of Technology, Bhimavaram, (Andhra Pradesh).
[10] Pusat Perkembangan Kurikulum (2002). Kemahiran Berfikir dalam Pengajaran dan Pembelajaran. Pusat Perkembangan Kurikulum, Kementerian Pendidikan Malaysia.
[11] Bahagian Jaminan Kualiti (2004). Panduan Kriteria dan Standard bagi Program Bidang Pendidikan. Jabatan Pendidikan Tinggi, Kementerian Pendidikan Malaysia.
[12] Stevens R., A. Johri, and K. O'Connor, Professional Engineering Work, Chapter 7 in Cambridge Handbook of Engineering Education Research, edited by A. Johri and B.M. Olds, Cambridge University Press, New York, 2014.
[13] Hector Montiel Campos, Alenjandro Magos Rubio, Gerardo Haces Atondo & Yarissa Marcela Palma Chorres (2015). Relationship between Creativity, Personality and Entrepreneurship: An Exploratory Student, Mexico.
[14] Brookfield S.D., Teaching for Critical Thinking, Jossey-Bass, San Francisco, 2012.
[15] Mansoor Fahim., Maryam Pezeshki (2012). Manipulating Critical Thinking Skills in Test Taking. International Journal of Education. Vol. 4, No. 1.
[16] Forrest Thompson, H. Lee Washington (2015). Critical Thinking Skills and Teaching

Accounting: A Comparative Study. Journal of Finance and Accountancy. Vol. 19, March, 2015.
[17] Colley, B. M., Bilics, A. R., & Lerch, C. M. (2012). Reflection: A Key Component to Thinking Critically. The Canadian Journal for the Scholarship of Teaching and Learning, 3(1), 1-19.
[18] Charles E. Baukal (2015). Promoting Critical Thinking during Problem Solving: Assessing Solution Credibility. Proceedings of the 2015 Zone III Conference of the American Society for Engineering Education.
[19] Shazaitul Azreen Rodzalana, Maisarah Mohamed Saatb (2015). The Perception of Critical Thinking and Problem Solving Skill among Malaysian Undergraduate Students. Procedia-Social and Behavioral Sciences 172 (2015) 725 – 732 Global.
[20] David Burkus ((2013). 10 Practices from the Most Innovative Organizations.
[21] Cathie McClellan (2016). Assessment Update • January–February 2016 • Vol. 28, Number 1
[22] Salpeter, M. (2012). "7 Soft Skills You Need to Get Hired in 2013." Aol Jobs. Retrieved.
[23] Roselainy Abdul Rahman, Yudariah Mohammad Yusof, Hamidreza Kashefi, Sabariah Baharun (2012). Developing Mathematical Communication Skills of Engineering Students. Procedia-Social and Behavioral Sciences 46 (2012) 5541 – 5547.
[24] Amineh Aali, Nader Naderi, Bijan Rezaie, Habib Jafari, Mohammad Bagher Aali (2016). Examine the Relationship between Organizational Justice and Organizational Commitment with Eentrepreneurial Skills (Case Study: Urmia Industrial Estate). International Journal of Contemporary Applied Sciences. Vol. 3, No. 2, February.
[25] Salwah Che Mat, Siti Mistima Maat, Norhatta Mohd (2015). Identifying Factors that Affecting the Entrepreneurial Intention among Engineering Technology Students. 2nd Global Conference on Business and Social Science-2015, GCBSS-2015, 17-18 September 2015, Bali, Indonesia. Procedia-Social and Behavioral Sciences 211 (2015) 1016 – 1022.
[26] Daniela Cristina Momete. Joining Economic and Engineering Perspectives – A Tool for Successful Entrepreneurs. The 6th International Conference Edu World 2014 "Education Facing Contemporary World Issues", 7th - 9th November 2014. Procedia-Social and Behavioral Sciences 180 (2015) 395 – 400.
[27] Hasniyati Hamzah, Zahiriah Yahya, Abdul Ghani Sarip & Yasmin Mohd Adnan (2016). Impact of Entrepreneurship Education Programme (EEP) on Entrepreneurial Intention of Real Estate Graduates.
[28] Aleksander Kucel, Peter Robert, Marian Buil & Nuria Masferrer (2016). Entrepreneurial Skills and Education-Job Matching of Higher Education Graduates. European Journal of Education, Vol. 51, No. 1, 2016.
[29] European Council (2012). Towards Growth Friendly Consolidation and Job Friendly Growth, In: Council, M.O. T. E. (Ed) (Brussels, European Council Publishing Department).
[30] Krishnamurthy. N. Forensic Civil Engineering and Professional Ethics. Proceedings of the International Conference on Forensic Civil Engineering, Nagpur, India 21,22,23 January 2016.
[31] Jaghannath. K. Ethics and values: The Need for Student Awareness of Workplace Value Systems. International Journal of Engineering Research and General Science Vol. 4, Issue 3, May-June, 2016. Ethics.
[32] David J. Demiang (2015). The Growing Importance of Social Skills in the Labor Market, Harvard University and NBER.
[33] Aseel Berglund and Fredrik Heintz (2014). Integrating Soft Skills into Engineering Education for Increased Student Through put and more Professional Engineers, IDA, Linkoping University, Sweden.
[34] Manshoor Hussain Abbasi, Attiya Siddiqi, Rahatul Ain Azim (2011). Role of Effective Communications for Enhancing Leadership and Entrepreneurial Skills in University Students. International Journal of Business and Social Science Vol. 2 No. 10; June 2011.
[35] Engineering Programme Accreditation Manual (2012). Malaysia.
[36] Mutalib, S. N. S. A., Juahir, H., Azid, A., Sharif, S. M., Latif, M. T., Aris, A. Z., Zain, S.M.,

& Dominick, D. (2013). Spatial and temporal air quality pattern recognition using environmetric techniques: a case study in Malaysia. Environmental Science, Processes & Impacts 1-12.

[37] Singh, K. P., Malik, A., Mohan, D., & Sinha, S. (2004). Multivariate statistical techniques for the evaluation of spatial and temporal variations in water quality of Gomti River (India) a case study. Water Research, 38, 3980–3992.

[38] Singh, K. P., Malik, A., & Sinha, S. (2005). Water quality assess- ment and apportionment of pollution sources of Gomti River (India) using multivariate statistical techniques-a case study. Analytica Chimica Acta, 538, 355–374.

[39] Aminu Ibrahim, Hafizan Juahir, Mohd Ekhwan Toriman, Adamu Mustapha, Azman Azid, Hamza A Isiyaka. Assessment of Surface Water Quality Using Multivariate Statistical Techniques in the Terengganu River Basin.Malaysian Journal of Analytical Sciences, Vol. 19 No 2 (2015): 338-348

[40] Juahir, H., Zain, S.M., Yusoff, M.K., Hanidza, T.I.T., Armi, A.S.M., Toriman, M.E., Mokhtar, M., 2011. Spatial water quality assessment of Langat River Basin (Malaysia) using environmetric techniques. Environ. Monit. Assess. 173, 625–641.

[41] Fermo, P., Beretta, G., Facino, R.M., Gelmini, F., Piazzalunga, A., (2013). Ionic profile of honey as a potential indicator of botanical origin and global environmental pollution. Environ. Pollut. 178, 173–181.

[42] Retnam, A., Zakaria, M.P., Juahir, H., Aris, A.Z., Zali, M.A., Kasim, M.F., (2013). Chemometric techniques in distribution, characterisation and source apportionment of polycyclic aromatic hydrocarbons (PAHS) in aquaculture sediments in Malaysia. Mar. Pollut. Bull. 69 (1–2), 55–66.

[43] Azimah Ismail a, Mohd Ekhwan Toriman, Hafizan Juahir, Sharifuddin Md Zain, Nur Liyana Abdul Habir, Ananthy Retnam, Mohd Khairul Amri Kamaruddin, Roslan Umar, Azman Azid (2015) Spatial assessment and source identification of heavy metals pollution in surface water using several chemometric techniques. Marine Pollution Bulletin.

[44] Manjunath, B.G., Frick, M. and Reiss, R.D. (2012). Some Notes on Extremal Discriminant Analysis. Journal of Multivariate Analysis. 103: 107–115.

[45] Mutalib, S. N. S. A., Juahir, H., Azid, A., Sharif, S. M., Latif, M. T., Aris, A. Z., Zain, S. M. and Dominick, D. (2013). Spatial and temporal air quality pattern recognition using chemometric techniques: a case study in Malaysia. Environmental Science: Processes and Impact. 15(9):1717-1728.

[46] Azid, A., Juahir, H., Aris, A.Z., Toriman, M.E., Latif, M.T., Zain, S.M., Yusof, K.M.K.K. and Saudi, A.S.M. (2014). Spatial analysis of the air pollutant index in the southern region of peninsular Malaysia using environmetric techniques. From Sources to Solution. 307-312.

[47] Azid, A., Juahir, H., Ezani, E., Toriman, M.E., Endut, A., Rahman, M.N.A., Yunus, K., Kamarudin, M.K.A., Hasnam, C.N.C., Saudi, A.S.M., and Umar, R. (2015). Identification source of variation on regional impact of air quality pattern using chemometric. Aerosol and Air Quality Research. 1-14

[48] Mohamad Romizan Osman, Azman Azid, Hafizan Juahir, Kamaruzzaman Yunus, Mohammad Azizi Amran, Ahmad Dasuki Mustafa, Fazureen Azaman, Syahrir Farihan Mohamed Zainuddin (2015). Indoor air quality at higher institution's laboratory: a study on pre-symptoms, awareness and understanding among occupants. Jurnal Teknologi.

Acknowledgments

The research fully funded under the Faculty of Innovative Design and Technology, Universiti Sultan Zainal Abidin (UniSZA), Centre for Engineering Education Research, Faculty of Engineering and Built Environment, The National University of Malaysia (UKM), and Ministry of Higher Education (MoE).

IOP Conf. Series: Materials Science and Engineering **697** (2019) 012017 doi:10.1088/1757-899X/697/1/012017

The essential of engineering education involving critical thinking and problems solving skills among mechanical engineer employees

W Omar Ali Saifuddin Wan Ismail[1*], Noraini Hamzah[2], Ireana Yusra Abdul Fatah[1], and Azry Khoiry Muhammad[2]

[1] Faculty of Innovative Design and Technology, Universiti Sultan Zainal Abidin, Gong Badak Campus, 21300 Kuala Terengganu, Malaysia

[2] Faculty of Engineering and Built Environment, Universiti Kebangsaan Malaysia, 43600 Bangi, Malaysia.
[*]Email: woasaifuddin@unisza.edu.my

Abstract. Engineering Education is a course that emphasis on soft skills that are related to critical thinking and the ability in solving problems. Creative and Critical Thinking Skills (CCTS) is a type of skill that had introduced and strongly emphasized in the education system, starting from primary school. Recently, all the exam questions in primary, secondary schools and public universities oriented to the principle of Higher Order Thinking Skills (HOTS) in (RMK 11 - Transforming Education System). The objective of this education program is to produce students that are more competent in solving problems by applying effectively various forms of knowledge that had studied theoretically. This study aims to examine the extent to which the needs of elements of critical thinking and problem-solving skills to produce competitive engineers from industries point of view. Feedback from respondents was taken through quantitative methods using a set of questionnaires. A total of 300 respondents have participated in this study. The study is expected to be a guide for future mechanical engineers who would be involved by the requirements of the mainstream industry.

1. Introduction

In the 19th century, critical thinking referred to the ability to analyse information, determine the relevance of the data collected and interpreted in solving the problem [1-5]. By the 20th century, critical thinking consists of things that show our thoughts by thinking of the activities we do [6] in a higher-level (High Level Thinking skills) which involves the process of analysis, valuation, fairness, and reflection [7,8]. Critical thinking also helps in the cognitive development of students when they try to collect as much information as possible to solve problems, organise information, and to look at the issue from a different perspective [9,10]. Critical thinking involves the purposeful, goal-directed thinking during the process of making decisions based on the evidence in the scientific problem-solving process and not just a mere guesses [11,12]. Through the development of this ideology, critical thinking and judgment method are required for engineers at a professional level to solve the big part of the problems at work [13], and it is part of the requirements for getting a good job in the engineering field in particular. [14-16].

The process of problem-solving is an essential requirement for any engineer [17], and it is significant for all other professional position [18]. Therefore, the need to foster a tendency to think critically viewed as an essential component in the development of the thinking process and the approach to problem-solving that forms into expertise [19]. Focusing on problem-solving skills have been disclosed at all levels of education particularly at the Higher Learning Institute, but the results of the study showed that the ideology of engineering students to solve problems are at low levels despite the various efforts that had been carried out by the trainers in the four-year bachelor's degree program [20]. Engineering students are thinkers who need more critical thinking than students in the social sciences, basic sciences and humanitarian courses [21-23].

The need for critical thinking and problem-solving skills will continue to increase in years to come, given the urgent need for new insights about the ways of people handling complexity and uncertainty [24]. Implementation of the i-think program began in 2014 as a thinking tool to catalyse high-level thinking to enhance and cultivate thinking skills among students [25]. All employers and the organisation in the industry is trying to employ and train engineers of all disciplines based on the problem-solving and critical thinking skills with the combination of communication skills [26, 27].

2. Experimental

Discriminant Analysis Method was used to process the data. A total of 300 respondents were involved in answering these questions. All the respondents from the industry consist of management and administration groups, senior engineers, senior technologist, and engineers. Items of the issues show the extent to which the requirements for Critical Thinking and Problem-Solving Skills element according to current industry needs. Through this sum then comes the reading percentage and mean for Critical Thinking and Problem-Solving Skills element with values in the table.

2.1 The Model Soft Skills Higher Education Institute

According to figure 1, Model Elements of Soft Skills has been introduced by the Ministry of Education (MoE) Malaysia and has become the primary source of success and outcomes for the students at a certain level.

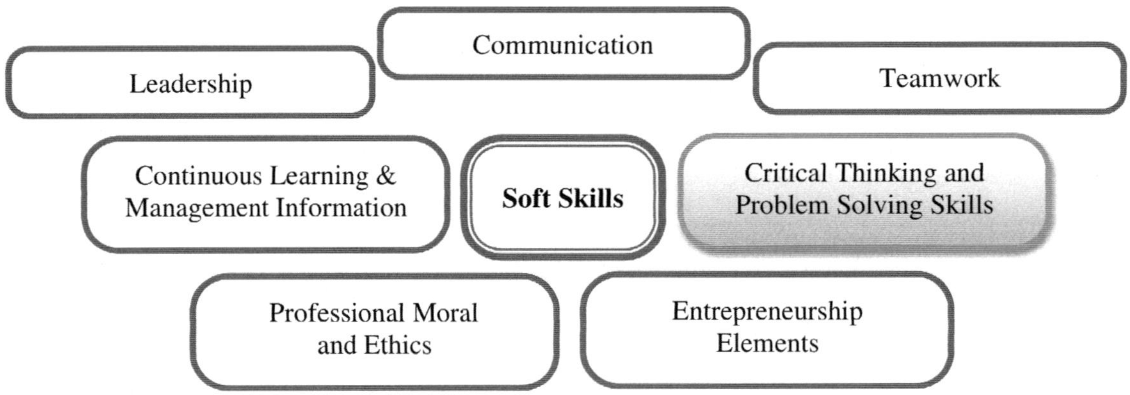

Figure 1. Soft skills model (MoHE).

Throughout this study, the researchers have used this model, which consist of seven elements where one of the important highlighted factors listed was the Critical Thinking and Problem-Solving Skills. Later will represent the importance of applying thinking skills among students in both public and private universities apart from their earlier education. Since reforms in education are focusing on critical thinking process including the item solving problem [28-30] as a cognitive starting point [31], the development progress continues for a positive impact towards the employment environment.

Model of Formation of First Class Mentality has been wrought by the Ministry of Education (MOE) Malaysia as shown in figure 2.

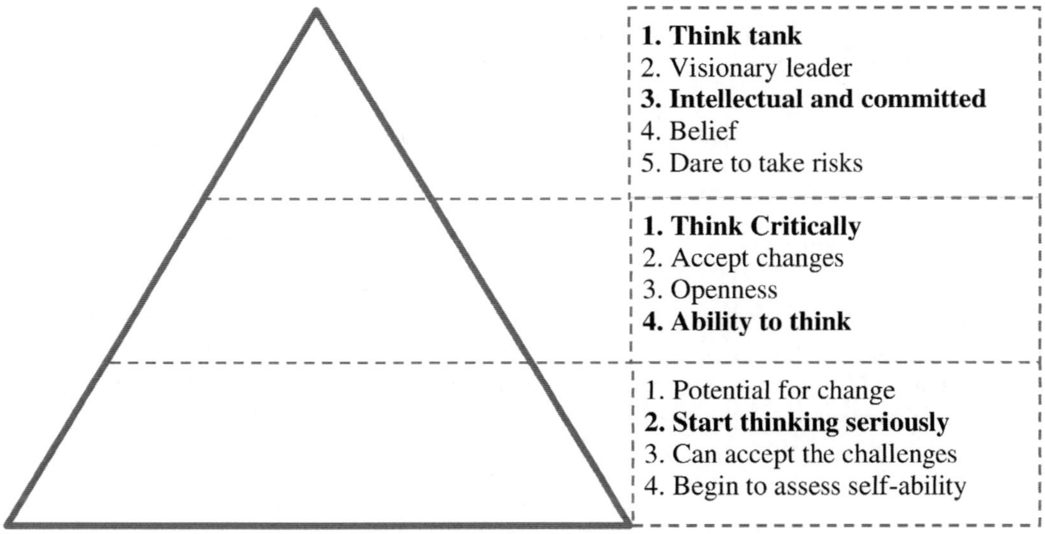

Figure 2. Model of formation of the first class mentality.

The composition of the importance and the need has been clearly in the thinking skills competency model. The Think Element was thought to be a fundamental part of every stage of the process known as creative and critical thinking in achieving the ultimate goal of engineering education at universities. Criteria contained in this model detailed in the construction of the questionnaire. Each level emphasises the importance of thinking skills variously in problem-solving and decision making.

Currently, the ability and availability to identify and problem-solving among engineering students in Malaysia are highly required. This essential element in the practice of critical thinking is the practice of reflection [32] which will assist the individual vital and analytical thinking in solving any problem and methods to develop a positive attitude towards addressing all of the issues [33]. In applying critical thinking, students need to explore creativity by defining the problems using one of the common creative tools (5w+1h - what, who, where, when, why and how) to identify the problem and solve them [34] with distinction variety of solutions.

Today's scenario, graduates had face problems in getting jobs due to their lack of critical thinking ability other than knowledge and soft skills performance [35]. Measures for the better idea is to provide a model for teaching and assessment following the criteria in Higher Order Thinking Skills (HOTS) openly to encourage students to use creative and critical thinking skills in problem-solving [36, 37]. Engineering students not only need to have the qualifications and technical skills to manage their role in carrying out daily work but, without the soft skills, employers are less likely to make these graduates as workers in their organization at a professional level, managers or even business partners [38]. Focusing through a modern engineering student, soft or generic skills is an element that must be mastered and blended in each skill aspect if they want to carry out the work successfully at the same time, help improve the ability and development of the company [39, 40].

3. Results and discussions

3.1 Descriptive Statistics Demographics Analysis

Table 1 shows the histogram of the current position of respondents and descriptive statistics demographics have work experience of more than five years in the industry, especially in the mechanical engineering field and department of engineering. Five years of working experience was suppose known to be the starting point for the maturity period in career path.

Table 1. Descriptive statistics demographics.

No.	Current Position	Frequency	Percentage
1	Management & Administration	100	33.33
2	Senior Engineers	72	24.00
3	Senior Technologist	38	12.67
4	Engineers	90	30.00

Table 2 shows the histogram gender of respondents and descriptive statistics demographics with the number of respondents noted, 240 were men and 60 women. It involves staff in heavy (HICOM), medium and small (SME) industries throughout the peninsular of Malaysia. Generally, the majority staff in the mechanical engineering field led by male.

Table 2. Descriptive statistics demographics.

No.	Gender	Frequency	Percentage
1	Man	240	80.00
2	Woman	60	20.00

3.2 The Findings of Seven Items in Questionnaire
The questionnaires were distributed to the respondents for the constructive answer and respond. The sets of selected questions that have risen in the questionnaire are as follows table 3.

Table 3. Items of the questionnaire.

Questions	Item Questions
Q1	Ability to identify and problem analysis in a complex and ambiguous situation and make justified evaluations.
Q2	Ability to expand and improve thinking skills such as explaining, analysing and evaluating the discussion.
Q3	Ability to search for ideas and find alternative solutions.
Q4	Ability to think outside the box.
Q5	Ability to make decisions based on strong evidence.
Q6	Ability to persevere and give its full responsibilities assigned.
Q7	Ability to understand and adapt to the culture of the community and a new work environment.

Table 4 shows the findings of the seven-item questionnaire involved in this study. Based on the results, all questions are at a very high level of need. Surprisingly, for Q6, 100% of the respondents agreed that mechanical engineering students must have high resilience to face a variety of possibilities where they also need to be responsible for the trust that has an entrusted to them. Results for the Q5 found that 96% of respondents agreed that students in this field should be able to create an effect based on the authentic evidence. Meanwhile, 94% believe that mechanical engineering students should have the ability to identify and analyse any problem in any situation and develop their thinking skills for both

Q1 and Q2 respectively. They have to be smart in interpreting the assessment and elaborate a clear explanation or idea. This ability can be practised and applied during assessment and group discussions.

Table 4. The findings of the seven items questionnaire.

Questions	Need Level	Percentage
Q1	Very High	94.00
Q2	Very High	94.00
Q3	Very High	93.66
Q4	Very High	86.33
Q5	Very High	95.66
Q6	Very High	100.00
Q7	Very High	91.33

The findings for Q3 also showed that almost 94% of these mechanical engineers have to be smart in generating new ideas. They need to be more creative and aware with current issue/technology in generating new ideas as an alternative to complete the task. The resulting of 91% for Q7 shows that the students in the mechanical engineering stream have to be smart to face and adapt to any situation so that they understand the work culture despite being in a new place. Lastly, for Q4, the results showed that 86% of respondents agreed that thinking outside the box is important for every mechanical engineering student to ensure that any action taken is more matured and beyond limit expectation.

3.3 Evaluation of Response from Questionnaires

Table 5 shows the justification of percentage and the mean score used in this study. The researcher used the score mean: i. 3.68 - 5.00 (High Needs); ii. 2.34 - 3.67 (Medium Needs); and iii. 1.00 - 2.33 (Low Needs). This method is used to simplify the process of classifying the results.

Table 5. Classification mean requirement levels.

Formula	Mean	Level	Justification
5 - 1 / 3 = 1.33	3.68 - 5.00	Three	High
	2.34 - 3.67	Two	Medium
	1.00 - 2.33	One	Low

(Source: Modified from Likert Scale, 1932)

Figure 3 shows box plots of mean Critical Thinking and Problem-Solving Skills generated based on the questions that had been answered by the respondents. From figure 3, the information transferred into Table 6 and all the results clearly shown.

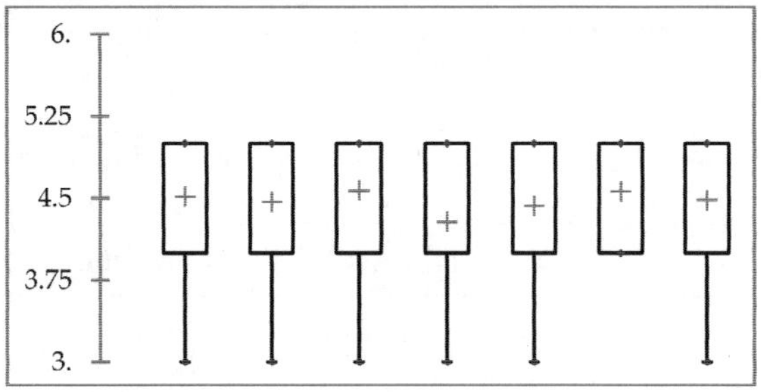

Figure 3. Box plots mean Critical Thinking and Problem Solving Skills.

Table 6 shows the mean and summary statistics of Critical Thinking and Problem-Solving Skills. From the results obtained, each respondent answered all the questionnaire with a minimum score at 3.00, and a maximum score at 5.00. Likewise, there is no respondent prefer to tick at the rating 1.00 and 2.00 for the entire questionnaire item. Items that had the highest mean value is Q3 (4.57), followed by Q6 (4.56), Q1 (4.52), Q7 (4.48), Q2 (4.47), Q5 (4.43) and the lowest mean value is item Q4 (4.28). It fulfilled the requirement and in line with Table 4, which accumulate the mean value of 3.68 - 5.00 at a high level. Therefore, the mean results interpreted those seven questions items are at a high level of need, and it also corroborated by Table 6 and references [28-31].

Table 6. Statistical summary Critical Thinking and Problem Solving Skills.

Question Item	Respondent	Minimum	Maximum	Mean Item
Q1	300	3.00	5.00	4.52
Q2	300	3.00	5.00	4.47
Q3	300	3.00	5.00	4.57
Q4	300	3.00	5.00	4.28
Q5	300	3.00	5.00	4.43
Q6	300	4.00	5.00	4.56
Q7	300	3.00	5.00	4.48

Based on the Descriptive Analysis (DA), the percentage of the overall reading with a view to the needed and very needed. The only two items show the need for any findings either high or low requirements of the industry in producing competitive mechanical engineers the percentage shown in table 7.

Table 7. Classification percentage of requirement level.

Formula	Percentage	Level	Justification
100 - 1 / 5 = 19.8	80.6 - 100	Five	Very High
	60.7 - 80.5	Four	High
	40.8 - 60.6	Three	Moderate
	20.9 - 40.7	Two	Low

1.00 - 20.8	One	Very Low

(Source: Modified from Likert Scale, 1932)

Figure 4 represents the item questions that are at high-level requirements. The horizontal axis represented a question answered by the respondent, and the vertical axis indicates the portion of the results of each item based on the percentage. The seven issues are at the high-level requirements, and it is also fulfilling the condition in table 7, which is the percentage value of 80.6 - 100.00 is the very high level. It gives the impression that the element of the Critical Thinking and Problem-Solving Skills are needed by engineering students, as mentioned in reference [18] and [33].

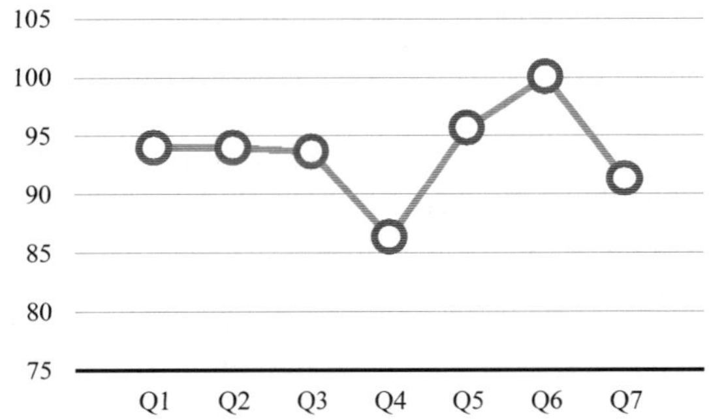

Figure 4. Graph plot item question (Critical Thinking and Problem Solving Skills).

Table 8 shows the percentage of the whole item question formulation of Critical Thinking and Problem Solving Skills. The percentage of 94% showed a very high level of requirements of demand for competitive Mechanical Engineers according to the current needs and industries. It must entirely have mastered before facing the real world according to a ref [39].

Table 8. Summary of the overall percentage.

Item	Need Level	Value
Percentage	Very High	93.57

Based on ref [38,40] generic skills are a guide for each engineer to assess the extent of their level of knowledge, skills and attitude towards engineering, but more than that, today the industry needs additional skills and one of them is critical thinking and problems solving abilities among engineers in Malaysia to compete with other countries around the globe.

3.4 Reinforcement of critical thinking component

In designing a firm professional engineer, many approaches and methodology be practiced in the engineering module at the university level that related to critical thinking components. There have five basic critical thinking components that are necessary can be implemented in the curricular university course for the Mechanical Engineering student. These components include specific knowledge, experience, competencies, attitude and professional standards. Blends of these five components will produce a competent engineering output as suggested in [41,42]. The particular content of an engineering education syllabus that involved written documentation as well as scoring rubric should revise according to critical thinking and problem-solving scope. During degree programme, engineering student must be able to demonstrate a representable, creative and rational presentation with a high critical thinking effect. While, in solving any kinds of problem, students must be able to explore several concepts simultaneously to reach a practical solution.

Under the Eleventh Malaysia Plan 2016 - 2020: A Roadmap (10.77), all the higher education institute are required to review the program regularly offered to drop the program that overlaps and is no longer relevant to the industry and provided no benefits. Continuous efforts to improve the quality and the introduction of the new program will be through review and rigorous screening to ensure that this program has a better curriculum, the added value that appropriate and meet the needs of the industry. Based on a study done by [42], critical thinking and problem-solving approached should be taught to various academic centres including both institutes, not only in public university but also private university that offers engineering courses.

An overview of the important element of thinking skills and problem-solving skills applied in the curriculum of engineering education had stressed in this particular study. Lacking skills of critical thinking and problem-solving among job applicants have commonly reported the essential skills deficit to 45% among job applicants based on the JobStreet.com website, and it is a worrying situation for the next-generation engineer.

4. Conclusions

In producing high potential mechanical engineering students to fulfil the gap of the current job market, the emphasis should be givens on the elements of critical thinking and problem-solving skills. The item has become an interest and is no longer a requirement. Students in this area are advised not to only rely on their academic textbooks alone as this will limit their level of creative thinking. So, instructors and academician must also act as facilitators in guiding and practising self-reflection processes which are essential elements in critical thinking and abandon chalk and talk teaching techniques. Research also proves this element of soft skills needed by the industry towards a positive and high qualified engineer. Mechanical Engineering environment is circulating and moving forward according to the rapid technological advancement, so it is not too much to say, graduates and engineers of Mechanical Engineering need to pay more attention in applying and exploring critical thinking in their future career endeavours. Therefore, competence in soft-skill is main criteria required by the industry in hiring skilled proficiency employees and holistic engineers.

References

[1] Gagne R 1988 *Instructional Science.* **17(4)** 387-390

[2] Halpern DF 1989 *Thought and knowledge: An introduction to critical thinking* (Hillsdale, NJ: Lawrence Erlbaum)

[3] Ennis RH 1991 *Goals for a critical thinking curriculum* ed AL Costa

[4] Snyder LG and Snyder MJ 2008 *Delta Pi Epsilon Journal* **L(2)** 90-99

[5] Halpern DF 1998 *American Psychologist* **53(4)** 449-455

[6] Dantas-Whitney M 2002 *System* **30(4)** 543-555

[7] Jeevanantham LS 2005 Why teach critical thinking? *Africa Education Review* **2(1)** 118-129

[8] William D, Hendricson MA, Sandra MS, Andrieu C, Chadwick DG, Jacqueline E, Chmar BA, James R, Cole, Mary C, George RDH, Gerald N, Glickman, Joel FJD, Glover, Jerold S, Goldberg, Haden NK, Cyril Meyerowitz, Neumann L, Pyle M, Lisa A, Tedesco, Richard W, Valachovic MPH, Richard G, Weaver, Ronald L, Winder, Young SK, Kenneth L and Kalkwarf 2006 *J. Dental Education* **70(9)** 925-936

[9] Kathpalia SS and Heah C 2008 *RELC J.* **39** 300-317

[10] Reinstein-Staton R 2008 The rights and wrongs of strategic planning', security: for buyers of products, systems & services. **45(7)** 34-36

[11] Nugent PM and Vitale BA 2008 *Fundamental success: a course review applying critical thinking to test taking*, USA: F. A. Davis Company

[12] Lau J 2012 A Mini Guide to Critical Thinking

[13] Stevens R, Johri A and O'Connor K 2014 *Professional engineering work* ed Johri A and Olds BM (Chapter 7: Cambridge handbook of engineering education research, Cambridge University Press, New York)

[14] Mohamad S, Rasul R, Amnah AR and Mohd YH 2014 *Kemahiran kebolehdapatan kerja suatu keperluan pekerjaan* (Penerbit Universiti Kebangsaan Malaysia, Bangi)

[15] Kamarudin MT, Ruhizan MY and Ramlee M 2014 *Pemangkin kebolehpasaran graduan, status pekerjaan graduan, 30* (Penerbit Universiti Kebangsaan Malaysia, Bangi)
[16] Mohd Huzairi J, Yuzainee MY, Mohd Zaidi O and Azami Z 2014 *Kemahiran kebolehgajian graduan kejuruteraan* (Penerbit Universiti Kebangsaan Malaysia, Bangi)
[17] Aldridge MD 1994 *J. Eng. Education* **83(3)** 231-236
[18] Eraut M 1994 *Developing professional knowledge and competence* (Falmer Press, London)
[19] Walker SE 2003 *J. Athletic Training* **38(3)** 263-267
[20] Nickerson RS 1994 *The teaching of thinking and problem-solving* ed Sternberg RJ *Thinking and problem-solving* (2nd ed. San Diego: Academic Press)
[21] Leach BT and Good DW 2011 *Int. J. Humanities and Social Science* **1(21)** 100-106
[22] Aliakbari M and Sadeghdaghighi A 2011 Investigation of the relationship between gender, field of study, and critical thinking skill: the case of Iranian students. *Paper Presented at The 16th Conference of Pan-Pacific Association of Applied Linguistics*, The Chinese University of Hong Kong 301-310
[23] Mahdyeh N and Arefi MA 2014 *Indian J. Fundamental and Applied Life Sciences* **4(1)** 153-162
[24] Funke J 2013 *J. Problem Solving* **6(1)** 2-19
[25] Hasnah I and Jamaludin B 2017 *Malay Language Education J.* **7(1)** 56-65
[26] Vanderbilt University School of Engineering
[27] Rosdiadee N 2013 *Technical communication skills among recent electrical and electronics engineering graduates in job industries 164* (Universiti Kebangsaan Malaysia, Bangi, Selangor, Malaysia)
[28] Pusat Perkembangan Kurikulum 2002 *Kemahiran berfikir dalam pengajaran dan pembelajaran* (Pusat Perkembangan Kurikulum, Kementerian Pendidikan Malaysia)
[29] Bahagian Jaminan Kualiti 2004 *Panduan kriteria dan standard bagi program bidang pendidikan* (Jabatan Pendidikan Tinggi, Kementerian Pendidikan Malaysia)
[30] Roselainy AR, Yudariah MY, Hamidreza K and Sabariah B 2012 *Developing mathematical communication skills of engineering students* (Procedia-Social and Behavioral Science)
[31] Collins JP 2008 *Modern approaches to teaching and learning anatomy.* Student BMJ 337:a1753
[32[Colley BM, Bilics AR and Lerch CM 2012 *The Canadian J. for the Scholarship of Teaching and Learning* **3(1)** 1-19
[33] Jake M and Laguador 2013 *Developing students' attitude leading towards a life-changing Career* (Educational research international)
[34] Infor Resources System Sdn. Bhd. 2001 *Creativity and creative problem-solving* (IRSSB Selangor)
[35] Ili Atiqah Md D and Ruslin A 2016 *J. Psikologi Malaysia* **30(1)** 16, http://www.wolcottlynch.com
[36] Campos HM, Rubio AM, Atondo GH, Palma YM and Chorres 2015 *Relationship between creativity, personality, and entrepreneurship: An exploratory study* (Mexico)
[37] Mishra DS 2016 *Int. Research J. Eng. and Tech.* **3(2)**
[38] Inayatullah K, Noor Abidah MO, Yusuf B and Zafar Iqbal SM 2012 *Int. J. Applied Linguistics & English Literature* **1(5)** 176-183
[39] Anderson C and John FG 2013 *Skills requirement for tomorrow's best jobs helping educators provide students with skills and tools they need* (IDC Analyse the Future)
[40] Desmond A and Martin J 2016 *Int. J. Higher Education* **5(2)** 23-59
[41] Hairuzila I, Hazadiah MD and Normah A 2010 *2nd Int. Congress on Engineering Education*, Kuala Lumpur, Malaysia 258-263

Acknowledgments
The research was fully funded by the Faculty of Innovative Design and Technology, Universiti Sultan Zainal Abidin (UniSZA), Centre for Engineering Education Research, Faculty of Engineering and Built Environment, The National University of Malaysia (UKM), and Ministry of Higher Education (MoHE).

IOP Conf. Series: Materials Science and Engineering **697** (2019) 012018 doi:10.1088/1757-899X/697/1/012018

Exploring Quality Influencing Factors for Frozen Food Industry

Aishah Abdul Aziz [1], Sarbani Daud [1], Shahryar Sorooshian[2*]

[1]Universiti Malaysia Pahang, Gambang campus, 26300, Pahang, Malaysia
[2]University of Gothenburg, 405 30, Gothenburg, Sweden
*E-mail: sorooshian@gmail.com

Abstract. Quality is the grade of excellence or level of acceptability by the buyers. For the frozen food industry, it can be defined as the composite of those features which have significance in determining the level of acceptability by the consumer. This study was to analyse the relationship between storage, transportation, stock control, packaging, and raw material on the quality of frozen food. The research methodology of this work was a combination of observation and interview, and survey from 55 Malaysian frozen-food small-enterprise. From the findings, only two factors, Packaging, and raw material came to significantly and positively affect the quality of the frozen food. The findings of this study could be an asset for food industries to have a better understanding of the quality control of their products.

1. Introduction

From year to year, frozen food becomes more popular among consumers in Malaysia and it has been rapidly growing. Generally, frozen food is preserved by freezing the food using various freezing techniques. Frozen food ease the consumers because consumers do not have to prepare all the ingredient of the specific food, and they just heat the food before being served. According to Adeyeye [1], Fikiin [2], George and Gormley [3], and Jessen et al [4] the global food industry has invested a lot of money in new freezing technologies in order to differentiate the quality of frozen food to compete for each other. The quality of frozen food is defined when the food is maintaining in its color, flavor, texture and nutritional value from starting ingredient that the food is not preserved yet [5-7]. The quality of a product will impact the business [8,9]. This research focuses on listing the operational factor which influenced the quality of products of the frozen food industry.

2. Research methodology

For this study, data are being related to observation, interview, and self-developed questionnaire. As shown in Figure 1, for the first phase of this study, some data was collected from an interview with the managers in 55 frozen food small enterprise. The Researcher has to visit the sites where the frozen food is produced to do an interview. The language used for the interviews session was the Malay language. At the first interview, the researcher asked the company's manager about the background of the company, also ask about what is common problem occur in the frozen food that the owner produce. From the interview sessions, five factors identified which might influence the quality of frozen food. The factors are storage, transportation, stock control, packaging, and raw material. Then after the initial understanding of the influencing factors, a survey conducted to test the factors with a quantitative approach. Data collected from a survey with a questioner as the data collection instrument and the respondents were top-manager of the enterprises.

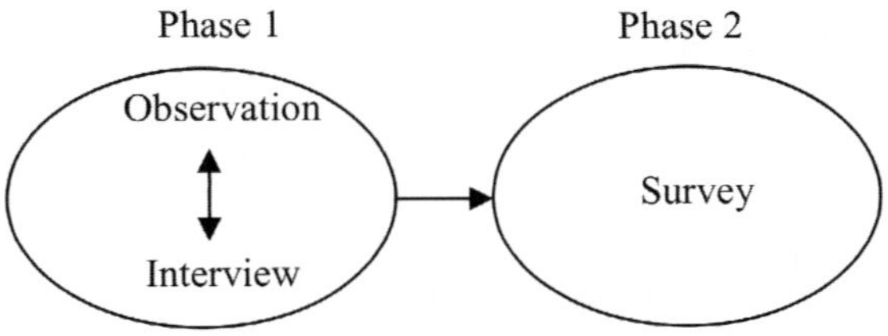

Figure 1. Research phases

3. Results

Total of 55 questionnaires regarding on factor influence the quality of frozen food was created and distributed to respondents. From the total number of the distributed, only 50 questions that have been fully answered. The data collected are obtained from 32% from male and 68% is obtained from the female. This show the majority of the respondents at frozen food are female. The collected data shown 48% of work experience is less than 5 years, follows by 5-10 years which shows 28%. Besides, 14 % is gain experience in 11-20 years and the lowest percentage is more than 20 years which only 10%. Also, from the data, it shows most of the frozen food SME are established less than 5 years which the percentage is 46, follows by 5-10 years which 26%. Next, 11-20 years shows 16% and lastly is more than 20 years with the percentage of 12%. The first step before further data analyzing, the reliability of the dependent data is tested. One of the ways to test the reliability of the data is using Cronbach's Alpha. The Cronbach Alpha of this study was acceptable, 0.79.To test relationship between storage and the quality of the frozen food, the Pearson Correlation is 0.316. It shows the weak correlation between storage and quality of frozen food. Besides, the Significant level is <0.05 thus, it has enough evidence to reject the Hypothesis of "There is relationship between storage and quality of frozen food". The graph, in Figure2, proves that there is only weak correlation between quality of the frozen food and the storage

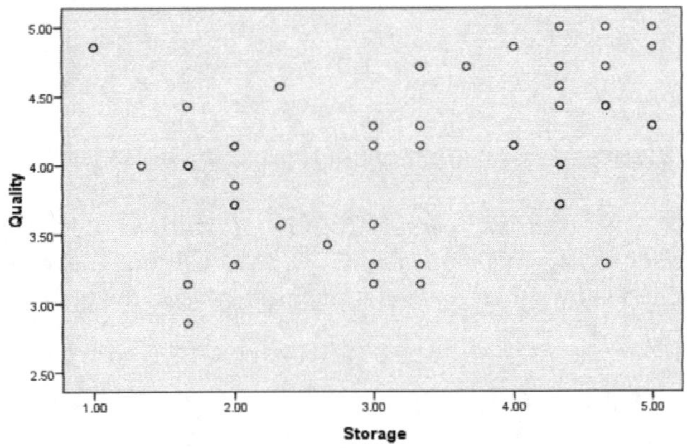

Figure 2. Correlation graph for quality and storage

The data prove that there is a very weak relationship between transportation and the quality of frozen foods. The Confidence level of transportation is 0.476 which exceeds 0.05. The graph, in Figure 3, proves that there is only a weak correlation between the quality of frozen food and transportation standards.

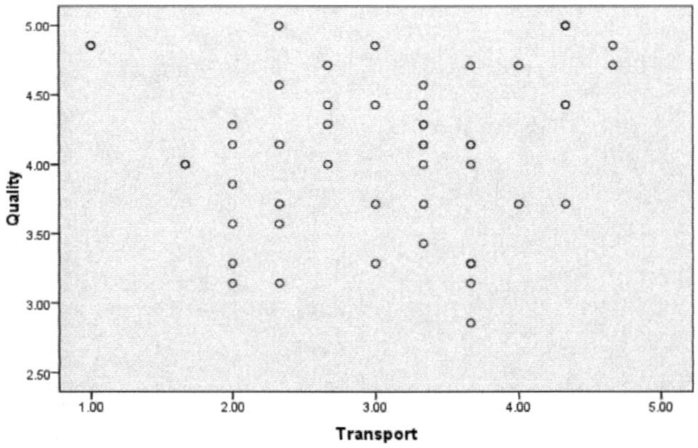

Figure 3. Correlation graph for quality and transportation

It also shows the result of the correlation between the quality of frozen food and stock control cannot be an acceptable hypothesis. From the data obtained, the Pearson Correlation is 0.071. Figure 4 shows the data collected are not fit in a line.

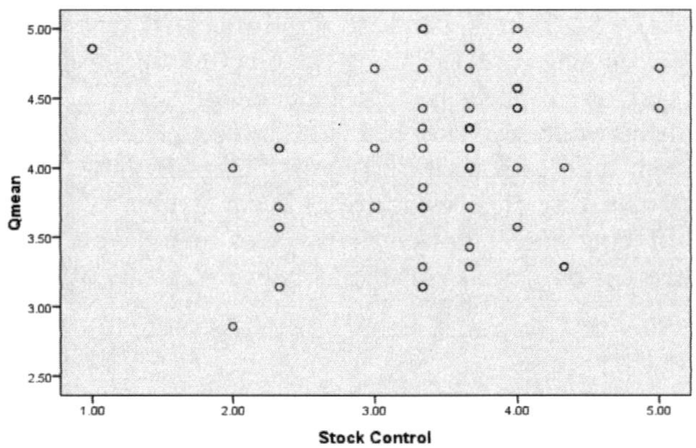

Figure 4. Correlation graph for quality and stock control

From Table 1 and Figure 5, it shows Pearson Correlation value is 0.35 which is a weak correlation for the relationship between packaging and quality of frozen foods. However, the Significant level $p<0.05$ which is 0.012 which shows there is a relationship between the packaging and quality of frozen food.

Table 1. Correlation table for quality and packaging

		Qmean	PGmean
Qmean	Pearson Correlation	1	.354*
	Sig. (2-tailed)		.012
	N	50	50
PGmean	Pearson Correlation	.354*	1
	Sig. (2-tailed)	.012	
	N	50	50

*. Correlation is significant at the 0.05 level (2-tailed).

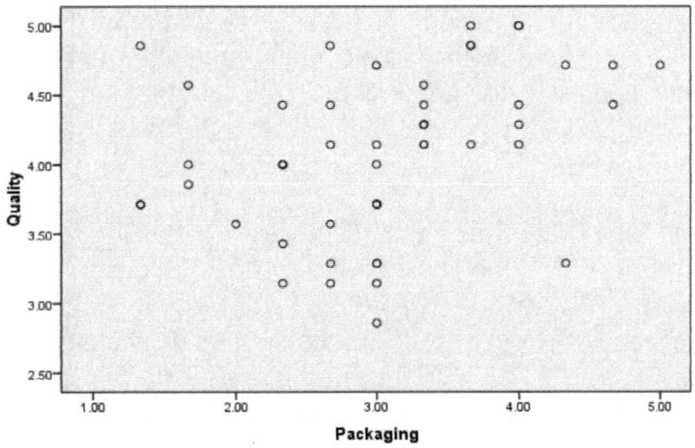

Figure 5. Correlation graph for quality and packaging

Table 2 and figure 6 show the correlation between the raw material and quality. This proves by Pearson Correlation which shows 0.5 and the significant level also accepted the hypothesis for "There is a relationship between quality and raw material", which is 0.041.

Table 2. Correlation table for quality and raw material

		Qmean	RMmean
Qmean	Pearson Correlation	1	.511*
	Sig. (2-tailed)		.041
	N	50	50
RMmean	Pearson Correlation	.291*	1
	Sig. (2-tailed)	.041	
	N	50	50

*. Correlation is significant at the 0.05 level (2-tailed).

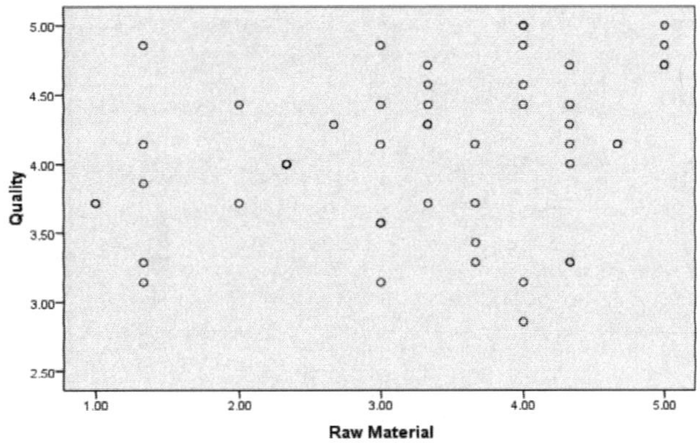

Figure 6. Correlation graph for quality and raw material

4. Conclusion

This study has explored influencing factors for Quality of frozen food as it was recommended by past studies [10]. It provides valuable information to several groups. First, this research provides valuable information for owners of the Companies to understand what the factor influence is the quality of frozen

IOP Conf. Series: Materials Science and Engineering **697** (2019) 012018 doi:10.1088/1757-899X/697/1/012018

food. A consistent pattern was found where emphasized on similar of the factor which is raw material, storage and packaging indicate these factors are the important factor that should be a focus on. Next, the finding of this paper may also assist the government in developing new strategies, such as subsidizing quality raw material as well as quality packaging, to further the Malaysian industries.

References

[1] Adeyeye, S. A. O. 2017 The role of food processing and appropriate storage technologies in ensuring food security and food availability in Africa. Nutrition & Food Science, **47** 122–139. https://doi.org/10.1108/NFS-03-2016-0037

[2] Fikiin, K. Handbook of frozen food processing and packaging. International Journal of Refrigeration, **30** (2007), 745–746. https://doi.org/10.1016/j.ijrefrig.2006.08.002

[3] George, M., & Gormley, R. 2000 Managing the Cold Chain for Quality and Safety. Europe, 5–33. Retrieved from http://www.teagasc.ie/ashtown/research/preparedfoods/managing_the_cold_chain.pdf

[4] Jessen, F., Nielsen, J., & Larsen, E. Chilling and Freezing of Fish. Seafood Processing: Technology, Quality and Safety, (2014), 33–59. https://doi.org/10.1002/9781118346174.ch3

[5] Pannirselvan, M. D., Bin Rahamaddulla, S. R., Muuhamad, P. F., Maarof, M. G., & Sorooshian, S. 2016 Innovative solution for barriers of green logistics in food manufacturing industries. International Journal of Applied Engineering Research, **11(18)** 9478-9487.

[6] Sun, D. W. Handbook of frozen food processing and packaging. CRC Press, 2016.

[7] Creed, P. G. Quality and safety of frozen ready meals. Handbook of frozen food processing and packaging. CRC Press, 2016.

[8] WE How, S Sorooshian, 2013 Service quality and its relationship with customer satisfaction, Journal of Management and Science, **3** 109-114

[9] Salimi, M., Hadjali, H. R., & Sorooshian, S. 2012 A moderator's perspective from service quality and purchase intentions relationship: A case study of AirAsia airlines. Research Journal of Applied Sciences, Engineering and Technology, **4(23)** 5074-5078.

[10] Xie, A., Sun, D. W., Xu, Z., & Zhu, Z. 2015 Rapid detection of frozen pork quality without thawing by Vis–NIR hyperspectral imaging technique. Talanta, **139** 208-215.

IOP Conf. Series: Materials Science and Engineering **697** (2019) 012019 doi:10.1088/1757-899X/697/1/012019

Mastery of leadership skills among mechanical engineers from the industrial Perspective

W Omar Ali Saifuddin Wan Ismail[1], Noraini Hamzah[2], Ireana Yusra Abdul Fatah[1]

[1] Faculty of Innovative Design and Technology, Universiti Sultan Zainal Abidin, 21300 Kuala Nerus, Terengganu, Malaysia
[2] Faculty of Engineering and Built Environment, Universiti Kebangsaan Malaysia, 43600 Bangi, Malaysia.
*E-mail: woasaifuddin@unisza.edu.my

Abstract. Leadership is an essential element of all types of work. This skill is also a focus in the field of Mechanical Engineering. Mechanical Engineers are said to be less familiar with leadership skills, and this will negatively affect the reputation of an industry. They aimed of this study to look at the extent of the need for leadership skills among Mechanical Engineers. The quantitative method used to obtain the study data involves 300 respondents who need to answer the questionnaire. The results show that the elements of leadership skills are at a very high level of need of 88.50%. Therefore, graduates in Mechanical Engineering should be more prepared to prepare for the workplace.

1. Introduction

Another essential skill is related to personality skills with interpersonal skills involving leadership skills [1,2]. Leadership is one of the processes that take place over time several years on an ongoing basis and requires attention to strength and the unique leader context [3]. Leadership skills look to the ability of a person to become a leader in whatever situation and theirs can control the situation wisely [4]. Besides, skills leadership plays a significant role in the event of anxiety and on time. That is all the skills, such as communication and management, will play a role.

Leadership Skill

Communication skills, inspiration and cognitive skills were performance management and leadership. It shows that the senior manager it is necessary to have better oral communication skills and less roar when communication goes on. The proactive responsibility of the leader more experienced need to be mobilised by taking time to introduce inclusive leadership development [5]. Some of the right manager lack of concern about public management, interpersonal, and non-verbal worries. Executive coaching is a combination of organisational development and leadership. Operational training is conducted through learning method through experience because leaders need to reach the target and report on success; It provides a way for leaders to understand its learning priorities and adapt to various behaviours Leadership includes building relationships with followers, peers, and stakeholders [6]. Executive guidance important part for teaching them about conversation skills and provide them with variety the aspect of managing others [7] and this is a significant benefit coaching through fast learning for leadership development for leaders who are behind science education.

The University of Kent has conducted a study on leadership guided. The findings show that different individuals also need to style a different direction. You may be able to manage a talented team member, efficient and reliable by using transformation techniques, whereas a lazy and unreliable person may need a style autocratic. Leadership skills and skills required in making an assessment and further action. The concept of mutual respect should be practised in engineering to make engineers more humane for carrying out their responsibilities as leaders in their careers. Research from Australia and European universities found educators reported lack of engineering leadership progress program if compared to the American engineering faculty in the United States [8] organising programs in the formation of led leadership.

One study requires respondents to answer the question of the type of skills that can develop the most effective leadership qualities among students universities involve listening, speaking, reading and writing skills. The study found that out of 100 respondents, 59 agreed that skill communication could develop the most effective leadership qualities, 72 respondents agree leadership is a natural quality. But there is which denies that it can polish through training and other practices. The analysis found to develop leadership qualities, communication and education are the most important. It is one of the most effective communication skills and can create a variety of competent leadership skills [9].

Many authors in their studies presented in Professional Writing Seminar for Engineering expresses empathy as the primary skill for effective leadership and management. Managers and leaders practising empathy skills can communicate honestly and proactively even they have excellent listening skills. It makes the process achieving success to navigate their organisation easier, including while handling the uncertain transition times.

Confident leaders' behaviour in this leadership practice includes building trusting relationships with followers, listening and responding actively against various views, support the decisions made by followers, and give followers the opportunity as a leader [10,11,12]. The Leadership Practice Inventory has proven the authenticity of the group in the field of nursing leadership, teaching, education, and law enforcement has a high level of leadership though there has been a significant change in the environment over the last decade [13]. Research in leadership development has found in the fields of nursing education, education and psychology, and the development of female leadership [14,15].

Leadership is a practice of development for managers, especially in an organisations team structure [16]. Guidance for leadership development contributes to value financial business [17]. Leadership theory was more interesting than doctoral writing research [18]. Progress in the application of learning and improving the curriculum combines the way to learn leadership to accelerate leadership development [19]. Engineering students record the lowest score in practice leadership at the beginning of the year of study compared to students in the field of art, business and education. But they showed the highest increase in five leadership practices by the end of the year of study [20]. It is the hope of engineering firm, technology employer, faculty engineering at universities, licensing agencies, engineering unions and societies Science, Technology, Engineering and Mathematics (STEM) [4] in ensuring that the resulting Mechanical Engineers are competitive and ready with various possibilities to lead any governmental organisation later.

2. Materials and methods

By using the Discriminant Analysis method, the processing data involved, 300 respondents from the industry consist of management and administration groups, senior engineers, senior technologist and engineers. Items of the issues show the extent to which the requirements for Leadership Skills element according to current industry needs. Through this sum then comes the reading percentage and mean for Leadership Skills element with values in the table.

The seven elements contained in the model but only the aspects of leadership skills are using as a study element. The soft skills model introduced by the Malaysian of Education (MoE) was used throughout the study emphasis placed on aspects of leadership skills that centred on the mechanical engineers' workforce. The model shows in figure 1.

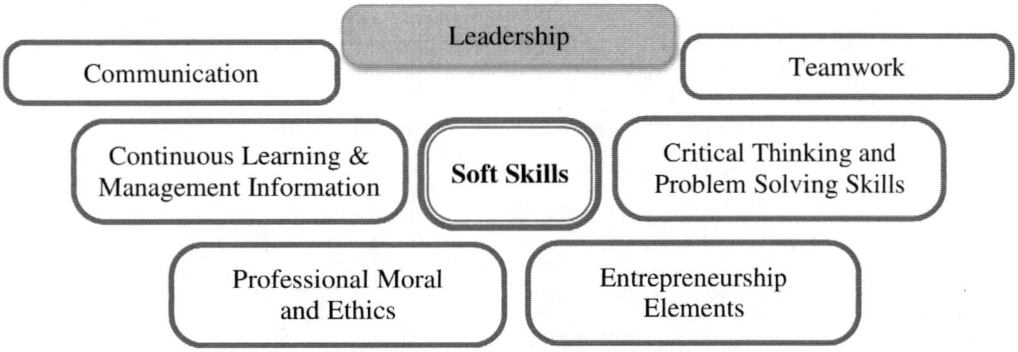

Figure 1. Soft Skills (MoE)

Modern engineering graduates need to be proficient in every aspect of the soft and generic skills if they want to succeed doing work at the workplace [21] to the best possible. This soft skill is essential to help to enhance the capabilities of a company [22], and this has proven by a study among students and faculty members of the engineering that, soft skills are especially crucial for engineering students with the added value of mastery English [23].

3. Results and discussions

The number of respondents involved in the process of answering this question item is 300 respondents. Among the questions items included are as follows (L1) Knowledge of essential theoretical leadership, (L2) Ability to lead the project, (L3) Ability to understand and take other roles between team captains and team members and (L4) Ability to oversee team members. Figure 2 shows the Box plot mean Soft Skills for elements (Leadership).

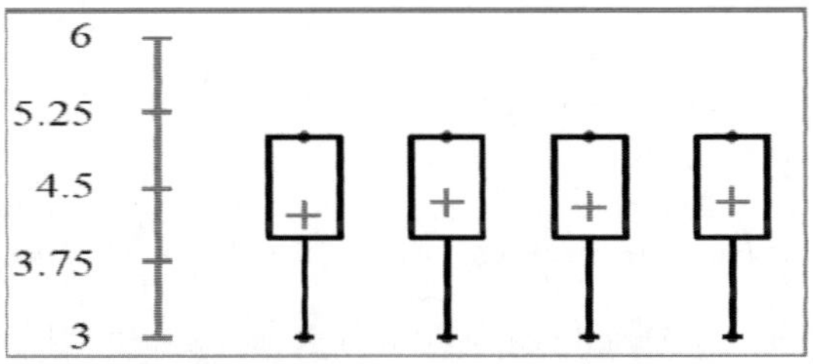

Figure 2. Box plot mean (Leadership)

Table 1 shows the leadership statistics formula. The table shows detailed information on mean readings that have produced in the study. Clearly shows the highest mean of the question item (L4), followed by (L2), (L3) and (L1).

Table 1. Mean (Leadership)

Question Item	Respondent	Minimum	Maximum	Mean Item
L1	300	3.00	5.00	4.22
L2	300	3.00	5.00	4.35
L3	300	3.00	5.00	4.30
4. L4	5. 300	6. 3.00	7. 5.00	8. 4.36

IOP Conf. Series: Materials Science and Engineering **697** (2019) 012019 doi:10.1088/1757-899X/697/1/012019

Table 2, 3, 4 and 5 show the descriptive statistics of items questions for (L1), (L2), (L3) and (L4). Table 2 shows the five levels of requirements studied involving items questions (Knowledge of essential theoretical leadership).

Table 2. Descriptive statistics of question items (L1)

Item	Level of Needs	Frequency	Percentage
1	very unnecessary	0	0
2	no need	0	0
3	simple need	53	17.67
4	need	128	42.67
5	very necessary	119	39.67

Table 3 shows the five levels of requirements studied involving items questions (Ability to lead the project).

Table 3. Descriptive statistics of question items (L2)

Item	Level of Needs	Frequency	Percentage
1	very unnecessary	0	0
2	no need	9. 0	0
3	simple need	25	8.33
4	need	145	48.33
5	very necessary	130	43.33

Table 4 shows the five levels of requirements studied involving items questions (Ability to understand and take intermediate roles between the captain and team member.

Table 4. Descriptive statistics of question items (L3)

Item	Level of Needs	Frequency	Percentage
1	very unnecessary	0	0
2	no need	0	0
3	simple need	29	9.67
4	need	152	50.67
5	very necessary	119	39.67

Table 5 shows the five levels of requirements studied involving items questions (Ability to supervise team members).

Table 5. Descriptive statistics of question items (L4)

Item	Level of Needs	Frequency	Percentage
1	very unnecessary	0	0
2	no need	0	0
3	simple need	31	10.33
4	need	131	43.67
5	very necessary	138	46.00

Figure 3 shows Scree plot question items (Leadership) on an overall involving four elements of the questions reviewed. Question items (L2) the highest percentage was 92%, followed by (L3) 90%, (L4) 90% and the lowest rate (L1) with a percentage reading of 82%. The four items of questions indicate a very high level of need.

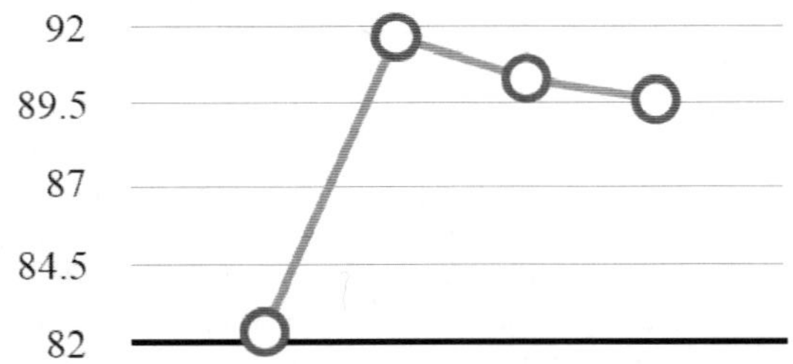

Figure 3. Scree plot question items (Leadership)

Table 6 shows the overall formulation of Soft Skills (Leadership) produced through research. The reading of the percentage is at a very high level of need. A very high rate has given an overview of the requirements of the Leadership elements in producing Mechanical Engineers such as which is expected by all parties.

Table 6. Question item summary (Leadership)

Item	Level of Needs	Value
Percentage	very necessary	88.50

The mean is at a high level and the overall percentage shown in the findings of the study conducted by the researcher at a very high standard for this Leadership Element. Explanation by [13], there exists Inventory Leadership Practice in a particular professional group over the last decade, although there are still significant changes in the environment. These shows this element has its interests and needs in the field of Mechanical Engineering in particular to produce Mechanical Engineers competitive. The rapid changes in the environment will not affect the interest in these leadership skills.

In evaluating based on these elements, the items used knowledge of the basic theory of leadership. There are many leadership theories which have been around for so long. However, the application in everyday work needs to be wise. The use of these theories depends on the situation and current problems encountered according to the efficiency of the Mechanical Engineer itself and careful discretion. It is a necessary fundamental thing known and controlled. This finding supported by [3], leadership is a continuous process by giving attention to strength and the uniqueness of a command. Effective communication is also one the process of competent leadership as stated by [9], and It is great to hope for organisations and employers in the field Mechanical Engineering as indicated by [4] in its findings.

The ability to lead the project is also the item studied in this element. In line with what [4] has stated, the ability to drive is seen in the handling of any circumstances and wise to control the situation. Through a study conducted by [16], leadership is a developmental practice for managers to handle something projects either lead horizontally or vertically within an organisation. The wisdom of choosing this kind of leadership at the same time will bear fruit positive and encouraging. Plans are being made easier for completed under a leader of the authoritative Mechanical Engineer.

Researchers also look at the ability to understand and take alternating roles between a team captain and team member. The findings of this study parallel and supported by [7] states, the principle of coaching with providing training and guidance is the best guidance for applying the value of leadership in each. This exercise will be creating mutual trust and exchanging roles in a state of affairs, which requires someone to act to make a decision. Refer to [6], also in conformity with the outcome of the study, all the necessary members and leaders adapt to various forms of behaviour, including in building a good relationship with each other. Mechanical Engineers should be prepared for leadership and at the same time lower their power on under his command to decide if the current situation requires such action for the common good.

The last thing reviewed in this element involves the ability to supervising troop members. The experience will differentiate the skills of the supervisor. The findings show it is consistent with the statements made by [5] older leaders need to be more proactive in driving and introducing inclusive leadership development amongst those who led. Supervisors or leaders also need to provide opportunities people led to expressing opinions as stated by [10,12] include maintaining good relations with each other, and this is consistent with the findings of the study conducted by the researcher. A mechanical engineer should be able to be a good supervisor and be fair to all workers under supervision. The way supervisors will manifest through pattern leadership, which is practised by a Mechanical Engineer itself and manner how the leadership qualities had been before.

4. Conclusions

Based on the elements of Leadership Skills, the researcher has concluded the following. Program in progress leadership should design as did engineering schools in America. They prioritise the concept of leadership in the effort to apply values of empathy in leadership and effective management. Study at the level the university also needs to make improvements in learning and curriculum by incorporating organised leadership learning. Leadership led will contribute to the financial value of an organisation and industry. Apart from providing services, one of the other features of a company or organisations needs to be a continuous element of revenue generation. It can realise through a sustainable leadership structure and expertise among experts which has entrusted.

References

[1] Berglund, A., & Heintz, F. 2014. *Integrating soft skills into engineering education for increased student throughput and more professional engineers*. IDA, Linkoping University, Sweden.
[2] Plumblee, J., Cattano, C., Bell, L., & Klotz, L. 2012. *Fulfilling engineering program objectives through service learning campaigns in developing countries. Leadership and management in engineering* **12(2)**: 46- 52.
[3] Rowe, W.E., Heykoop, C., & Etmanski, C. 2015. *A master's degree in global leadership.* In Ngunjiri F.W. & Madsen, S.R. (Eds.), women as Global leaders, 187-208. Charlotte, NC.
[4] MacIntyre, P. 2016. *Advancing leaders in engineering: ways of learning leadership.* Fairleigh Dickinson University.

[5] Dinpolfo, S., Silva, C. & Carter, N.M. 2012. *High potentials in the pipeline: leaders pay it forward.* New York, NY: Catalyst.
[6] Turesky, E.F. & Gallagher, D. 2011. Know thyself: coaching for leadership using Kolb's experiential learning theory. *The Coaching Psychologist* **7(1):** 5-14.
[7] Adams, R., Evangelou, D., Dia de Figuerrdo, A., Mousoulides, N., Pawley, A., Schifellite, Stevens, R., Svincki, M., Trenor, J., & Wilson, D. 2011. *Ten faces of engaging future engineers.* Journal of Engineering Education **100 (1):** 1-41.
[8] Khattak, H., Ku, H. & Goh, S. 2012. Courses for teaching leadership capacity in professional engineering degrees in Australia and Europe. *European Journal of Engineering Education* **37(3):** 279-296.
[9] Manshoor Hussain Abbasi, Attiya Siddiqi, Rahatul Ain Azim. 2011. Role of effective communications for enhancing leadership and entrepreneurial skills in university students. *International Journal of Business and Social Science*, **2(10).**
[10] Kouzes, J.M., & Posner, B.Z. 2011. Leadership practices inventory (LPI) New norms November, 2012.
[11] Kouzes, J.M., & Posner, B.Z. 2011. *The leadership challenge (5th ed.).* San Francisco, CA: Jossey-Bass.
[12] Kouzes, J.M., & Posner, B.Z. 2013. Leadership practices inventory (LPI) 4/e Facilitator's Guide. San Francisco, CA: Pfeiffer.
[13] Clavelle, J.T., Drenkard, K., Tullai-McGuinness, S., & Fitzpatrick, J.J. 2012. Transformational leadership practices of chief nursing officers in magnet organizations. *Journal of Nursing Administration* **42(4):** 195-201.
[14] De Haan, E. & Duckworth, A. 2012. Signalling a new trend in executive coaching outcome research. *International Coaching Psychology Review* **8(1)**: 6-19.
[15] Ely, R.J., Ibarra, H. & Kolb, D. 2011. Taking gender into account: theory and design for women's leadership development programs. INSEAD Faculty & Research Working Paper. Retrieved April 15, 2015.
[16] Joo, B.K., Sushko, J.S., & McLean, G.N. 2012. Multiple faces of coaching: manager as coach, executive coaching, and formal mentoring. *Organization Development Journal* **30(1):** 19-38.
[17] Bower, K.M. 2012. Leadership coaching: does it really provide value? *Journal of Practical Coaching* **4(1):** 1-5.
[18] MacIntyre, P.L. 2014. A quantitative correlational research study of leadership development for women engineers. PROQUEST dissertations.
[19] Kassotakis, M.E. & Rizk, J.B. 2015. Advancing women's leadership development: effective practices for the design and delivery of women's leadership programs. In Ngunjiri F.W., & Madsen, S.R. (Eds.), Women as global leaders 163-185. Charlotte, NC: Information.
[20] Posner, B.Z., Crawford, B., & Stewart, D.R. 2015. A longitudinal study of Canadian student leadership practices. *Journal of Leadership Education Spring* 161-181.
[21] Inayatullah Kakepoto, Noor Abidah Mohd Omar, Yusuf Boon, S M. & Zafar Iqbal. 2012. Perspectives on oral communication skills for engineers in engineering profession of Pakistan. *International Journal of Applied Linguistics & English Literature* ISSN 2200-3592 (Print), ISSN 2200-3452 (Online), **1(5),** September 2012 [Special Issue on General Linguistics] Perspectives.
[22] Anderson, C., John, F., & Gantz. 2013. Skills requirement for tomorrow's best jobs helping educators provide students with skills and tools they need. IDC analyse the future.
[23] Kilam, I.K., Jyoti Sharma. 2013. Analysis and evaluation of various soft skills training models employed by provate technical educational institutions in India **2(2).**

Acknowledgments
The research fully funded under the Faculty of Innovative Design and Technology, Universiti Sultan Zainal Abidin (UniSZA), Centre for Engineering Education Research, Faculty of Engineering and Built Environment, The National University of Malaysia (UKM), and Ministry of Higher Education (MoHE).

Local search algorithm for solving periodic location routing problem

N Prasanti[1*], A Andriansyah [2], and F Fitriadi[1]

[1]Industrial Engineering Department, Universitas Teuku Umar, Meulaboh, Meulaboh, 23681, Indonesia
[2]Industrial ngineering, Universitas Syiah Kuala, Jl. Tgk. Syech Abdul Rauf No.7 Banda Aceh 23111 Aceh, Indonesia
[*] E-mail: nissaprasanti@gmail.com

Abstract. This study examines the problem of facility location and vehicle routes determination in supply chain system periodically. The problem is determining the location and route to meet the supply of commodity to the power plant that taken from the case of an electricity company in Indonesia. In this study, the problem is called Periodic Location Routing Problem (PLRP). Periodic Location Routing Problem is a non-polynomial hard problem (NP-hard), if locations considered is increase, the computation time used is increasing exponentially. Therefore, to solve a large-scale of this problem, a local search algorithm is proposed. Local search is a very simple metaheuristic algorithm. To explore the solution space for the problem, this algorithm will use the relocation operator. The results of computational experiments from 10 instances indicate that the results of the solution have a small standard deviation with short computing time in less than 2 hours.

Keyword: PLRP, NP-hard, Metaheuristic, Local search, Relocation

1. Introduction

Distribution system is a sub-system in supply chain management. The distribution system has the role of ensuring the delivery of commodities from suppliers to users in a supply chain. To ensure the arrival of this commodity, the costs incurred will be very large if the existing supply chain network is ineffective. To create an effective supply chain network, resource points and routes must be determined in advance in the commodity delivery process.

Determining the resource points of a supply chain is usually called the facility location problem [1]. The route that must be traversed in a supply chain network is called vehicle routing problem [2]. If these two problems are integrated, it is called the Location Routing Problem (LRP). [3], [4] and [5] divide LRP models into several structures and hierarchies based on the characteristics of the problem, the solution search method and the number of objective functions. There are three decisions made in LRP, namely the location of the facility, the route of the vehicle, and the allocation of requests for each facility [4].

In this study, the problem was more like LRP with two echelon (2ELRP). This type of LRP was first discussed by [6] in the newspaper delivery system. The problems studied are commodity delivery systems in a company in Indonesia. Commodity are taken from several permanent resources. These commodities are collected in the warehouse before being delivered to user points. In its implementation, the delivery of commodity must be based on a particular delivery schedule. The problems in this study

IOP Conf. Series: Materials Science and Engineering **697** (2019) 012020 doi:10.1088/1757-899X/697/1/012020

are called Two Echelon Periodic Location Routing Problems (2EPLRP). The objective function of this problem is to minimize shipping costs and operations within the planning horizon.

2EPLRP is an NP-Hard problem [3]. This problem requires very long computational time to find the optimal solution if using analytical methods. Therefore, to obtain near-optimal solutions, one algorithm is needed to explore the solution space of the problem, for example the heuristic algorithm. The Tabu Search (TS) algorithm is used by [7] for 2ELRP problems. This algorithm is able to provide a good solution to the problems studied with efficient computing. The Local Search algorithm (LS) is used by [8]. This algorithm is then compared to TS. This algorithm looks for solutions faster than TS with similar quality solutions.

This paper proposes the LS algorithm at 2EPLRP. This algorithm is used because the implementation is quite easy, it also provides a promising quality solution. The difference between this paper and previous studies is the characteristic problem that considers delivery schedules. The objective of this study is minimizing the total cost, including fixed cost and operational cost both in echelon-1 and echelon-2.

Theater distribution was studied by [9]. That is the problem which is the model of 2ELRP in multicultural shuttle routes, time windows, limited heterogeneous vehicles, and capacity constraints. Field of logistics for the city was developed by [10]. [11] presents the Mixed Programming Integer (MIP) based on arc-variable formulation for one echelon-1 facility and echelon-2 facility of 2ELRP. These facilities were capitalized except a fixed cost of the opening facility. This study [11] was the first to introduce the term multi-echelon LRP. [12] developed another of 2ELRP variant with no-capacity echelon-1 facilities and capacitated echelon-2 facilities that consider window time constraints. Research [13] examined a special case of 2E-LRP where the customer set was divided into a number of clusters and customer requests consisted of two parts that had to be fulfilled from inside and outside the cluster. [14] study 2E-LRP with the time window as soft-constraint. [15], [7] and [16] present the MIP formulation for problems with several echelon-1 facilities which are then resolved using the TS algorithm. [7] and [16] are technical reports that explain the results obtained in [15]. [17] developed a mathematical model of three indices and Genetic Algorithm for 2E-LRP problems with several no-capacity facilities of echelon-1, capacitated facilities of echelon-2, and route decisions with heterogeneous fleets. [18] uses a multi-start heuristic algorithm consisting of two components, namely Greedy randomized adaptive search procedure (GRASP) and Embedded Evolutionary Local Search (ELS) / Iterated LS (ILS) to obtain solutions to 2E-LRP problems. [11] added limitation of quality and developed two metaheuristic methods for the model [11]. [19] developed an algorithm [18] using multi-start ILS combined with taboo lists and Relinking Path (PR). [20] solved the variants studied in two previous studies ([18] and [19]) and developed GRASP which was reinforced by learning process and PR. [21] present branch-and-cut and Adaptive Large Neighborhood Search (ALNS) algorithms. The most recent development that can be found include research [22] which develops research [14] to be multi-objective in the context of perishable food supply chains. For this bi-objective case, [22] developed the MHPV algorithm which is a combination of Multi Objective Particle Swarm Optimization (MOPSO) and Adapted Multi Objective Variable Neighborhood Search (AMOVNS). In the same year, [23] developed bi-objective 2E-LRP which was solved using the Scatter TS procedure for non-linear Multi-Objective Optimization (SSPMO) and Non-dominated Sorting GA II (NSGA-II) methods.

PLRP combines standard LRP with Periodic VRP, where the route of visit is planned along the horizon that have a multi period. A visit patterns set that have been predetermined chose the visit period for each customer. PLRP aims to determine (i) configuration of facilities to be used in all planning periods, (ii) allocation of visit patterns to customers, (iii) allocation of customers to facilities for each period during the planning horizon where one customer does not have to be served by the same facility in each period, and (iv) vehicle routes from each facility for all periods. The objective function of PLRP is generally to minimize the total cost of opening and operating the facility, the fixed cost of the vehicle, and the cost of administering the route. Therefore, the fixed cost of the vehicle for each facility is based on the maximum number of vehicle routes carried out from each facility in each period along the planning horizon.

2. Methods

2.1. Local Search Approach

The local search algorithm is a heuristic method that uses several optimization techniques in finding solutions [2]. In generating solutions, the operators used are usually insertions or exchanges a point in a solution. In this study, the objective function of the problem is divided into two: product delivery costs which consist of two echelon and operating costs of selected resource points. The constraints on the problem is:

1. Vehicle capacity in echelon 1
2. Vehicle capacity in echelon 2
3. Allocation of delivery period
4. Capacity of resource in echelon 1
5. Capacity of resource in echelon 2 (warehouse)
6. Customer allocation
7. Constraint related vehicle routing problem

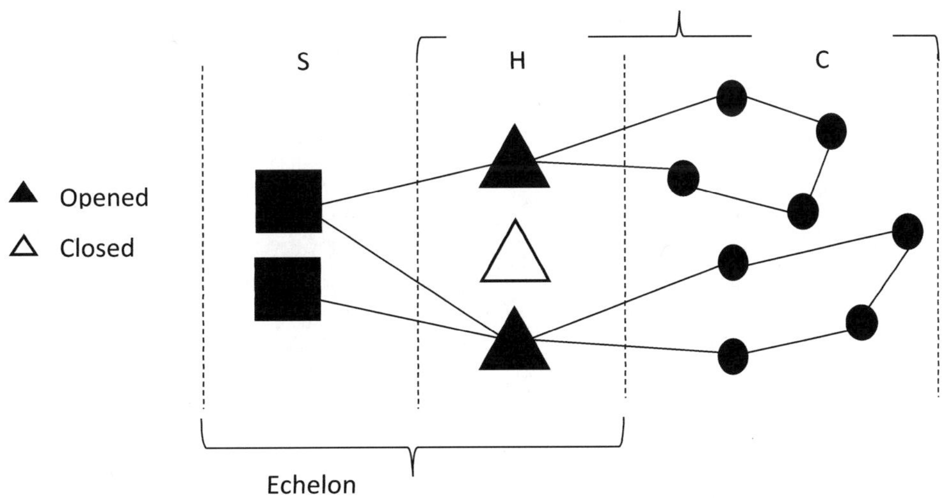

Figure 1. Solution example of 2E-LRP

Step of proposed LS in this paper are given as follows:

0. Initialize the number of generation parameters I and number of relocation G.
1. Generate initial solution σ_o using insertion heuristic.
2. Set the current solution $\sigma_c = \sigma_o$.
3. Repair current solution σ_o and generate new solution using relocation operator based on echelon.
4. Take one best solution σ_b based on objective function.
5. Calculate $\Delta = f(\sigma_b) - f(\sigma_c)$.
6. If $\Delta < 0$, set $\sigma_c = \sigma_b$, Otherwise, reject all new solutions.
7. Investigate I. If it is the same as the number of generations, set $\sigma_a = \sigma_c$ Stop, otherwise, back to step 3.

3. Result

The data used in the experiment is hypothetical data. The hypothetical data used consists of 2 resource points (S), 4 potential hub nodes (H) and 4 customer nodes (C). Customer demand and source capacity are randomly varied for each data set while taking into account the limitations of the algorithm. In this performance analysis 10 hypothetical data sets are used. Naming the test data set, for example 8N_01, means that 8N is a data set with eight nodes and 01 means the first set of data. The processing for each data set is done ten times replication and then the best value is taken, the data set with the best objective function. The LS algorithm used is implemented in the programming language MATLAB™.

Experimental results can be seen in table 1. Based on these results, the resulting objective has a small deviation. This can be seen from the average. The computing time needed to get the results is also short based on the table. This applies to all data tested. Based on the examination of all constraints, the solution obtained shows that there are no constraints that is violated. This can be seen in table 2. The resulting solution is the optimal local solution. This is because the operators used are still very random.

Overall, the LS can solve all kinds of instances with short computation. After being examined, there are no violating constraints which can conclude that the LS can solve the problem with large scale of data.

Table 1. Results of Experiment

Data	Best Objective	Average	Time (seconds)
8N_01	37364	37455,6	101,2
8N_02	25366	25466,7	150,3
8N_03	19733	19677,5	133,4
8N_04	30167	30166,8	148,2
8N_05	23345	23446,6	114,1
8N_06	18721	18733,7	129,3
8N_07	28932	28995,3	133,5
8N_08	24566	24666,7	178,2
8N_09	26615	26744,2	144,3
8N_10	27954	28123,9	126,5

Table 2. Results of constraints violation investigation

Constraints	State
Echelon 1 vehicle capacity	Unviolated
Echelon 2 vehicle capacity	Unviolated
Delivery period allocation	Unviolated
Echelon 1 resource capacity	Unviolated
Echelon 2 resource capacity	Unviolated
Customer allocation	Unviolated
Vehicle routing problem constraint	Unviolated

4. Conclusion

This study proposed the 2E-LRP to address the problems on a company in Indonesia. The objective of this problem is to minimize product delivery costs which consist of two echelon and operating costs of selected resource points. The contribution of this paper is considered delivery schedules at the customer. The problem is solved using the local search algorithm. The results shown by the algorithm are a good solution, but are still local optimal. This is caused by the selection of operators that are still random.

Future work can improve this algorithm, or can use other algorithms to improve the quality of the solution. In addition, future work can develop mathematical models as a comparison of the quality of the solutions produced by the algorithm.

References

[1] Syam S S 1997 A model for the capacitated p-facility location problem in global environments *Computers & Operations Research* **24** 1005–1016
[2] Toth P and Vigo D 2014 *Vehicle Routing: Problems, Methods, and Applications* (SIAM)
[3] Nagy G and Salhi S 2007 Location-routing: Issues, models and methods *European Journal of Operational Research* **177** 649–672
[4] Lopes R B, Ferreira C, Santos B S and Barreto S 2013 A taxonomical analysis, current methods and objectives on location-routing problems *International Transactions in Operational Research* **20** 795–822
[5] Drexl M and Schneider M 2015 A survey of variants and extensions of the location-routing problem *European Journal of Operational Research* **241** 283–308
[6] Jacobsen S K and Madsen O B G 1980 A comparative study of heuristics for a two-level routing-location problem *European Journal of Operational Research* **5** 378–387
[7] Boccia M, Crainic T G, Sforza A and Sterle C 2010 A Metaheuristic for a Two Echelon Location-Routing Problem *Experimental Algorithms* Lecture Notes in Computer Science ed P Festa (Springer Berlin Heidelberg) pp 288–301
[8] Derbel H, Jarboui B, Hanafi S and Chabchoub H 2010 An Iterated Local Search for Solving A Location-Routing Problem *Electronic Notes in Discrete Mathematics* **36** 875–882
[9] Burks R 2006 An Adaptive Tabu Search Heuristic for the Location Routing Pickup and Delivery Problem with Time Windows with a Theater Distribution Application
[10] Crainic T G, Ricciardi N and Storchi G 2009 Models for Evaluating and Planning City Logistics Systems *Transportation Science* **43** 432–454
[11] Perboli G, Tadei R and Vigo D 2011 The Two-Echelon Capacitated Vehicle Routing Problem: Models and Math-Based Heuristics *Transportation Science* **45** 364–380
[12] Aksen D and Altinkemer K 2008 A location-routing problem for the conversion to the "click-and-mortar" retailing: The static case *European Journal of Operational Research* **186** 554–575
[13] Ambrosino D, Sciomachen A and Scutellà M G 2009 A heuristic based on multi-exchange techniques for a regional fleet assignment location-routing problem *Computers & Operations Research* **36** 442–460
[14] Nikbakhsh E and S.H.A.D Z 2010 A Heuristic Algorithm and A Lower Bound For The Two-Echelon Location-Routing Problem with Soft Time Window Constraints
[15] Sterle C 2009 Location-Routing models and methods for Freight Distribution and Infomobility in City Logistics
[16] Crainic T G, Sforza A and Sterle C 2011 Tabu Search Heuristic for a Two-Echelon Location-Routing Problem
[17] Li Jin, Yunlong Zhu, Hai Shen and Tao Ku 2010 A hybrid genetic algorithm for two-layer location-routing problem *4th International Conference on New Trends in Information Science and Service Science* 4th International Conference on New Trends in Information Science and Service Science pp 642–645
[18] Nguyen V-P, Prins C and Prodhon C 2010 A Multi-Start Evolutionary Local Search for the Two-Echelon Location Routing Problem *Hybrid Metaheuristics* Lecture Notes in Computer Science ed M J Blesa, C Blum, G Raidl, A Roli and M Sampels (Springer Berlin Heidelberg) pp 88–102
[19] Nguyen V-P, Prins C and Prodhon C 2012 A multi-start iterated local search with tabu list and path relinking for the two-echelon location-routing problem *Engineering Applications of Artificial Intelligence* **25** 56–71

[20] Nguyen V-P, Prins C and Prodhon C 2012 Solving the two-echelon location routing problem by a GRASP reinforced by a learning process and path relinking *European Journal of Operational Research* **216** 113–126
[21] Contardo C, Hemmelmayr V and Crainic T G 2012 Lower and upper bounds for the two-echelon capacitated location-routing problem *Computers & Operations Research* **39** 3185–3199
[22] Govindan K, Jafarian A, Khodaverdi R and Devika K 2014 Two-echelon multiple-vehicle location–routing problem with time windows for optimization of sustainable supply chain network of perishable food *International Journal of Production Economics* **152** 9–28
[23] Martínez-Salazar I A, Molina J, Ángel-Bello F, Gómez T and Caballero R 2014 Solving a bi-objective Transportation Location Routing Problem by metaheuristic algorithms *European Journal of Operational Research* **234** 25–36

Simulated annealing algorithm for solving pickup and delivery problem with LIFO, time duration and limited vehicle number

A Andriansyah[1*], N Prasanti [2], and H Y Sastra[3]

[1]Laboratory of Industrial Computation and Optimization, Industrial Engineering, Universitas Syiah Kuala, Jl. Tgk. Syech Abdul Rauf No.7 Banda Aceh 23111 Aceh, Indonesia
[2]Industrial Engineering Department, Universitas Teuku Umar, Meulaboh, Indonesia
[3]Industrial ngineering, Universitas Syiah Kuala, Jl. Tgk. Syech Abdul Rauf No.7 Banda Aceh 23111 Aceh, Indonesia
[*]E-mail:andriansyah@unsyiah.ac.id

Abstract. Nowadays, the process of delivering goods or documents from one place to another can be done faster by utilizing a courier service. This courier must be able to create a travel route to pick goods and deliver them with a short time. Route creation will be more difficult if there is a time window for each customer and the goods being transported are heavy or dangerous items. In this case, the courier company must implement a last-in-first-out rules. In addition, with the limitation of vehicles number, companies have to minimize the number of underserved customers. This problem called pickup and delivery problem with last-in-first-out, time duration, and limited vehicle number. In this study, to solve the problem, simulated annealing algorithm was used for some dataset. To explore the solution space for the problem, this algorithm will use pair relocation operator. In several experiments, the results show a good solution and short computation time.

1. Introduction

Vehicle Routing Problem (VRP) is the problem determining best routes with the aim of obtaining minimal transportation costs. Many VRP variants that have been developed in recent decades, for example VRP with time windows, VRP pickup and delivery and several other variants [1].

This study will discuss one of the VRP variants that are often encountered in life, namely the Pickup and Delivery Problem (PDP). This variant can usually be found in companies engaged in pickup and delivery services (door-to-door). This problem can be defined as follows: there is a set of customer locations that have the same pickup location and delivery location. These locations have information about the number of items to be transported and delivered. The location of the depot which is the location of the initial departure of the vehicle will assign several vehicles to service these locations. Each location will be visited right once by one vehicle. Each transport vehicle assigned has a certain maximum capacity that must not be violated. Many studies discuss VRP with the characteristics described. For surveys on PDP, it is recommended to read [2], [3] and [4] who classify these problems in several types.

PDP has its variant. Technically, if the assigned vehicle only has a single door for loading-unloading access, it can be another problem. The problem is about the rules of goods that will be sent first. If the last transported item is a hazardous material, fragile, too heavy, it is prioritized to deliver the item first [5]. If there are no delivery rules, it might be done, but if you consume too much time because you have to do the demolition, this should be avoided [6]. This rule is called Last-In-First-Out (LIFO). An example of a route that shows this rule can be seen in figure 1(a), Figure 1(b) does not comply with the rules because the last item taken is goods from location P2, then the vehicle must prioritize location D2, not location D1. In this paper, the problem is limited by the maximum duration for each vehicle traveling because usually the companies that deal with ensure the maximum time to arrive at the destination. It is called Pickup and Delivery Problem with LIFO and Time Duration (PDPLT).

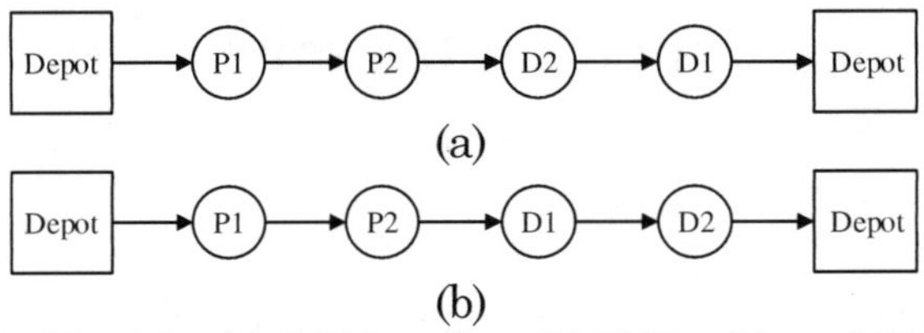

Figure 1. (a) PDP with LIFO policy, (b) PDP without LIFO policy

The assumption of PDP is unlimited number of vehicles and all customers can be served by the vehicles [6]-[8]. In fact, there is no company that have unlimited number of vehicles. This issue affected that not all customers can be served by the vehicles. Therefore, there should be best planning to minimize the number of unserved customer. [9] has discussed about this. The problem is called Pickup and Delivery Problem with LIFO, Time Duration and Limited Vehicle Number (PDPLTV). Mathematical formulation has been developed in that study.

Simulated annealing (SA) algorithm has been created for solving PDPLTV. This is a contribution of this study. For literature review, many algorithm has been developed for solving PDP. [10] developed Variable Neighborhood Search (VNS) algorithms for these problems and successfully implemented on several data samples with good performance. [5] developed the Branch and Bound algorithm for the same problem, [11] developed branch and cut algorithms and successfully completed sample data with up to 50 customer locations. The VNS algorithm for this problem with a successful tree representation resolves the TSPPDL problem developed by [12]. The same problem but the existence of distance constraints and solved by the two-stage Simulated Annealing (SA) algorithm with the VNS-Tabu Search (TS) developed by [13]. If the problem allows the LIFO rule to be violated, it is called the Traveling with Problem Pickup, Delivery and Handling Cost (TSPPDH), but is subject to a penalty cost for the violation that has been committed. [14] developed branch and cut exact algorithms for these problems. The metaheuristic approach was developed by [15] uses the TS algorithm and successfully solves the problem.

The problems most similar to the research in this paper were examined by [6], but did not consider the limited vehicle number and resolved using the TS algorithm. [7] discusses the problems considering time windows, but does not limit the time duration for each vehicle and solve by branch and cut and price algorithm. In the following year [16] developed the Genetic Algorithm (GA) -Local Search algorithm (LS) to solve large-scale cases. The last paper that discussed about PDPLTV is [9], but solved by mathematical model that implemented in optimization software.

2. Methods

2.1. Model Formulation

A set of nodes N and a set of arcs A that incorporated in a network G. $N = O \cup P \cup D$ which $O = \{0, 2n+1\}$is the depot, $P = \{1, 2, \dots, n\}$ is pickup nodes and $D = \{n+1, \dots, 2n\}$ is delivery node. Each node $i \in P$ has a request $q_i > 0$ that should be sent to $(i+n) \in D$. That is represented by request $q_{i+n} = -q_i$. The opening time e_i and closing time l_i are conducted with each point $i \in O$. They are also used as the maximum duration that the vehicle must depart from the depot and return to the depot. There is travel time t_{ij} that is associated with each link $(i,j) \in A$. The set $K = \{1, 2, \dots, v$ is also used in this problem for describing vehicles. Each $k \in K$ has a vehicle capacity Q. The problem is formulated into integer linear programming (ILP) model and use several decision variables. The variables that are related to vehicle load and vehicle travel time is L_i^k and T_i^k as a non-negative continuous decision variable. x_{ij}^k is a decision variable. 1 if the vehicle travels and 0 to vice versa. PDPLTV is represented in a three-

IOP Conf. Series: Materials Science and Engineering **697** (2019) 012021 doi:10.1088/1757-899X/697/1/012021

index integer linear programming. The formulations are based on [17] and [18]. The completed ILP model can be seen in [9].

2.2. Simulated Anneling Algorithm

The construction algorithm used in this study is the Insertion Heuristic (IH) algorithm. This algorithm forms a solution by inserting one by one the customer's location between the arcs formed and then the route with the minimum insertion cost is selected.

SA algorithm is an algorithm that belongs to a heuristic improvement where the resulting solution is near optimal. This algorithm uses random techniques to explore the solution space. The SA algorithm is found based on the analogy of the annealing process of a material. The material is heated at a certain temperature to its melting point, then cooled to the speed of cooling slowly to a certain temperature to find the best atomic structure. This cooling speed is commonly called a cooling schedule [19]. The implementation of the SA algorithm is always to accept a better solution and allow it to accept a worse solution with a certain probability. The probability function can be seen in equation 1.

$$P = e^{-\Delta/T_c} \tag{1}$$

With P is the probability of acceptance, Δ is the difference between the evaluation of the new solution $f(\sigma_b)$ and the evaluation of the current solution $f(\sigma_c)$ and T_c are the current temperature. The solution produced by the SA algorithm is very dependent on the parameters used such as the initial temperature T_0, the final temperature T_a and the cooling schedule α. To find the new temperature T_a in the cooling process can use equation 2.

$$T_b = \alpha \cdot T_c \tag{2}$$

Step of proposed LS in this paper are given as follows:

0. Control parameters initialization T_0, T_a, α, l, E .
1. Generate initial solution (σ_0) using insertion.
2. Set current temperature (T_c) equal to initial temperature (T_0), $e = 1$, dan current solution (σ_c) equal to initial solution (σ_0).
3. Using pair relocation operator, generate current solution
4. Collect the set of new solutions (σ_b).
5. Calculate $f(\sigma_b)$.
6. Take $f(\sigma_b)$ minimum objective.
7. Calculate difference $f(\sigma_b)$ and $f(\sigma_c)$ or Δ. If $\Delta < 0$, continue to 12, otherwise, 9.
8. Generate random number r between 0-1.
9. Calculate $\exp(-\Delta/T_c)$. If $r < \exp(-\Delta/T_c)$, continue to 12, otherwise, 11.
10. Set $e = e + 1$. Back to 4.
11. Set $\sigma_c = \sigma_b$.
12. $e = E$?. If yes, continue to 14. otherwise, 11.
13. ($T_c \leq T_a$)?. If yes, continue to 17, otherwise L15.
14. $(T_b) = \alpha\ (T_c)$.
15. $(T_c) = (T_b)$, back to 4.
16. $\sigma_a = \sigma_c$.
17. Stop.

3. Result

Intel® Core™ i5-3470 3.20 GHz (4CPUs) 4 GB RAM, under Windows 10 operating system are used to the computational experiment. The SA algorithm used is implemented in the programming language MATLAB™.

[6] generates the instance that the passenger and package request randomly generated the following uniform distributions. These instances are used in this study for the computational experiments. Each request has pickup and delivery nodes. Parameter $\alpha = 40$ and $\beta = 60$, $v = 2$, $n = 7$ are used in this experiment, same with [9].

Table 2 shows that SA algorithm is able to complete several instances. The solutions offered are similar to solutions in previous studies. But at some data, only shows optimal local solutions. This is because the SA algorithm is probabilistic. Each time (when running) will show different results. In addition, the selection of the appropriate SA parameters will also affect the quality of the solution produced. However, the SA algorithm is able to provide a solution with a shorter time. This is because the SA algorithm does not need to explore all the solutions from problem space.

Table 1. Computational results from [9]

No	Instance	Objective	Unserved Customer	Solution	Time (s)
1	lc101_20_0	251,33	2	Optimal global	121
2	lc102_20_0	363,33	4	Optimal global	249
3	lc103_20_0	353,33	4	Optimal global	18
4	lc104_20_0	379,33	5	Optimal global	1
5	lc105_20_0	349,33	4	Optimal global	5
6	lc106_20_0	243,33	2	Optimal global	182
7	lc107_20_0	368,67	5	Optimal global	6
8	lc108_20_0	371,33	4	Optimal global	815

Table 2. Computational results from SA

No	Instance	Objective	Unserved Customer	Solution	Time (s)
1	lc101_20_0	265,33	2	Optimal local	4,2
2	lc102_20_0	383,21	4	Optimal local	6,3
3	lc103_20_0	359,33	4	Optimal local	1,2
4	lc104_20_0	379,33	5	Optimal global	0,3
5	lc105_20_0	349,33	4	Optimal global	0.8
6	lc106_20_0	255,25	2	Optimal local	5,3
7	lc107_20_0	388,63	5	Optimal local	2,2
8	lc108_20_0	400,33	4	Optimal local	10,2

4. Conclusion

This study proposed SA algorithm for solving the PDPLTV. Minimize total travel time and number of unserved customers are the objective of this study. The contribution of this paper is considering the limited vehicle number. The results shown by the algorithm are a good solution, but are still local optimal. This is caused the probabilistic of algorithm. Future work can improve this algorithm, or can use other algorithms to improve the quality of the solution. To improve this algorithm, other operator can be chosen. The operator is very important, because this tool work to explore space of solution. Using other parameter value is another option. For example, number of iterations, bigger number can produce good quality solution, but increase running time. Future work, this algorithm can be implemented with large scale of data.

Acknowledgments

The authors would like to thank to the Universitas Syiah Kuala for laboratory facilities of Industrial Engineering and partial financial assistance under Laboratory Grant of Fiscal Year 2019 Number: 305/UN11.2/PP/PNBP/SP3/2019 Date May 3, 2019. Thanks, and high appreciation to Rector, Head of Integrated Laboratory and Head of LPPM Universitas Syiah Kuala.

References

[1] Toth P and Vigo D 2014 *Vehicle Routing: Problems, Methods, and Applications* (SIAM)
[2] Berbeglia G, Cordeau J-F, Gribkovskaia I and Laporte G 2007 Static pickup and delivery problems: a classification scheme and survey *TOP* **15** 1–31
[3] Parragh S N, Doerner K F and Hartl R F A survey on pickup and delivery problems: Part I: Transportation between customers and depot *Journal für Betriebswirtschaft* 2008
[4] Parragh S, Doerner K and Hartl R 2008 A survey on pickup and delivery problems: Part II: Transportation between pickup and delivery locations *Journal für Betriebswirtschaft* **58** 81–117
[5] Carrabs F, Cerulli R and Cordeau J-F 2007 An Additive Branch-and-Bound Algorithm for the Pickup and Delivery Traveling Salesman Problem with LIFO or FIFO Loading *INFOR: Information Systems and Operational Research* **45** 223–238
[6] Benavent E, Landete M, Mota E and Tirado G 2015 The Multiple Vehicle Pickup and Delivery Problem with LIFO Constraints *European Journal of Operational Research* **243** 752–762
[7] Cherkesly M, Desaulniers G and Laporte G 2014 Branch-Price-and-Cut Algorithms for the Pickup and Delivery Problem with Time Windows and Last-in-First-Out Loading *Transportation Science* **49** 752–766
[8] Furtado M G S, Munari P and Morabito R 2017 Pickup and Delivery Problem with Time Windows: A New Compact Two-Index Formulation *Operations Research Letters* **45** 334–341
[9] Andriansyah, Prasanti N and Sentia P D 2018 Pickup and delivery problem with LIFO, time duration, and limited vehicle number *MATEC Web Conf.* **204** 07003
[10] Carrabs F, Cordeau J-F and Laporte G 2007 Variable Neighborhood Search for The Pickup and Delivery Traveling Salesman Problem with LIFO Loading *INFORMS J. on Computing* **19** 618–632
[11] Cordeau J-F, Iori M, Laporte G and González J J S A branch-and-cut algorithm for the pickup and delivery traveling salesman problem with LIFO loading *Networks* **55** 46–59
[12] Li Y, Lim A, Oon W-C, Qin H and Tu D 2011 The tree representation for the pickup and delivery traveling salesman problem with LIFO loading *European Journal of Operational Research* **212** 482–496
[13] Cheang B, Gao X, Lim A, Qin H and Zhu W 2012 Multiple pickup and delivery traveling salesman problem with last-in-first-out loading and distance constraints *European Journal of Operational Research* **223** 60–75
[14] Battarra M, Erdoğan G, Laporte G and Vigo D 2010 The Traveling Salesman Problem with Pickups, Deliveries, and Handling Costs *Transportation Science* **44** 383–399
[15] Erdogˇan G, Battarra M, Laporte G and Vigo D 2012 Metaheuristics for the traveling salesman problem with pickups, deliveries and handling costs *Computers & Operations Research* **39** 1074–1086
[16] Cherkesly M, Desaulniers G and Laporte G 2015 A Population-based Metaheuristic for the Pickup and Delivery Problem with Time Windows and LIFO Loading *Comput. Oper. Res.* **62** 23–35
[17] Furtado M G S, Munari P and Morabito R 2017 Pickup and delivery problem with time windows: A new compact two-index formulation *Operations Research Letters* **45** 334–341
[18] Cherkesly M, Desaulniers G and Laporte G 2014 Branch-Price-and-Cut Algorithms for the Pickup and Delivery Problem with Time Windows and Last-in-First- Out Loading Branch-Price-and-Cut Algorithms for the Pickup and Delivery Problem with Time Windows and Last-in-First-Out Loading *Transportation Science, Article in Advance* 1–15

[19] Yang X S 2010 *Engineering Optimization An Introduction with Metheuristic Application* (New Jersey: John Wiley & Sons, Inc)

IOP Conf. Series: Materials Science and Engineering **697** (2019) 012022 doi:10.1088/1757-899X/697/1/012022

An Integrated approach for Path Planning for Mobile Robot Using Bi-RRT

Faiza Gul[1], Wan Rahiman[2]
[1]School of Electrical and Electronic Engineering, Universiti Sains Malaysia 14300 Nibong Tebal, Pulau Penang, Malaysia,
[2]Sustainable Platform Air, Land and Naval (SPALAN) Research Group, Universiti Sains Malaysia
*E-mail: faizagul@student.usm.my

Abstract. In this paper, an integrated effort has been exploited for path planning along with path smoothness for a mobile robot. An extension of randomly exploring tree is used to plan the path for robot along with path length smoothness. Our proposed system influences the concept of Bi-directional Rapidly Exploring Random Trees (Bi-RRTs) for path planning together with path smoothness for minimizing the processing time to search for the path length. We compared our system with standard method in trials with the autonomous robotic platform. Produced results show that the presented approach as high robustness and use less computational effort.

1. Introduction

Path planning and obstacle avoidance for mobile robot navigation has gain a lot of importance in recent years. It is considered as complex task yet very important. Autonomous robots are well known for their intelligent behaviour and capabilities. They have numerous applications in the field of industry, medical, rescue operations and transportation [1]. Due to their intelligent capabilities' researchers are working to further increase their tendency to behave autonomous. The expectations are to develop real time monitoring system to avoid the collision, interpret information collected from environment in right way and performs smooth and uninterpreted navigation. The path planning involves the collision free path for vehicle/robot to navigate in environment freely. It involves the safe delivery from one point to another point decreasing the path cost. Depending on this nature path planning is mainly divided into static (obstacles do not change position w.r.t time) and dynamic (obstacles do change position w.r.t time) environment [2]. Up till now numerous techniques have been proposed and applied to solve the path planning and obstacle avoidance of mobile robot. For example, road map approach [3], cell decomposition [4] and APF [5]. These methods are considered as inefficient because of high computational complexity and ability to easily stuck in local minima problem. To ensure the stability and good tracking trajectory, path planning utilize control system strategy to steer the autonomous mobile robot. A robot must exploit all the forces to make a turn. Like, for example if a robot is turning at an angle of 90 degree, all external forces will act while taking a turn. Robot needs to position the tires with robotic trajectory direction. This act is contrast with human driver who will exploit all forces on vehicle as it is heading towards the end of a turn. The main idea of this work is based on the utilization of conventional Randomly Exploring tree to Bi-directional RRT. To overcome the nonlinear mapping, an integrated path planning and control strategy is incorporated that leverage's the Bi-RRT path planner together with pure pursuit controller. The utilization of Bi-RRT works efficiently for path planning and to keep a track of mobile robot, pure pursuit controller is used.

The paper is arranged as follows: In section II describes the proposed method of path planning and control method. Section III presents the modelling of mobile robot wheels. Section IV proves the effectiveness of proposed method through experimental results. In Section V, performance analysis is done with APF and PRM methods. Section VI contains the conclusion.

2. Methodology

Our integrated control and path planning methodology depends on two major factors i.e. a path planner using Bi-RRT and steering controller to keep track of desired trajectory. The method is used to help the robot to navigate towards goal position following the optimal path via avoiding obstacles in its way.

2.1 Bi-RRT based Path Planning Navigation

Firstly, RRT was first proposed by Steven M. LaValle and James J. Kuffner Jr. in [6] to handle problems with obstacles and differential constraints. The algorithm was designed to help in efficient data structure to search for high dimensional spaces by randomly building space filling tree. The tree grows in space by joining random points in free space. If the selected node/tree found colliding with obstacles it will not be selected as newly selected node of expanding tree. Since it is a stochastic method.

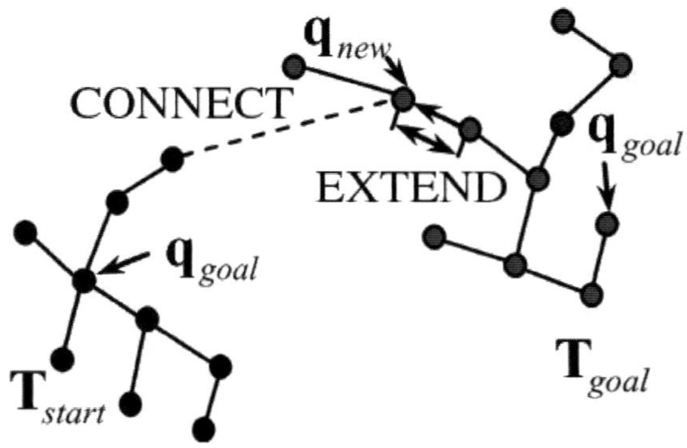

Figure 1. Bi-Directional Random Exploring Tree

The main principle of Bi-RRT is shown in Fig.1. The summary can be made in few steps. The working principle of Bi-RRT is same as RRT the only difference comes in the space configuration is decomposed into two trees, as Ta *and* Tb. Define the start point $Xinit$ and end point $Xtarget$. Set $Xinit$ as root of the tree. Select some random points in the space configuration $Xrand$. Expansion is done by expanding the nearest node in tree towards the random state by small segment of length e, where $Xnear, Xnew$ is a segment on the line joining the nearest point on tree. If no obstacles are found add the $Xnew$ as a new point on tree. The algorithm will work until it reaches to target. The brief algorithm of Bi-RRT is described in Fig 1 and 2:

```
RRT_BIDIRECTIONAL(x_init, x_goal)
  T_a.init(x_init); T_b.init(x_goal);
  for k = 1 to K do
      x_rand ← RANDOM_STATE();
      if not (EXTEND(T_a, x_rand) = Trapped) then
          if (EXTEND(T_b, x_new) = Reached) then
              Return PATH(T_a, T_b);
      SWAP(T_a, T_b);
  Return Failure
```

Figure 2. Algorithm for Bi-RRT

Several advantages and disadvantages of RRTs and Bi-RRT are discussed below:

[1.] A tree rapidly explores the whole space configuration, instead of circling around the start node.

IOP Conf. Series: Materials Science and Engineering **697** (2019) 012022 doi:10.1088/1757-899X/697/1/012022

[2.] It easy to implement so lead to faster analysis to search for a path.
[3.] It can be used for solving different optimization problems.
[4.] It can be applied to Kino dynamic and nonholonomic planning.
[5.] It is considered as consistent due to its distribution of vertices approaches sampling distribution.
[6.] It is incorporated with different path planning systems, so it is considered as important module of path planning.

The major drawback of RRT and Bi-RRT algorithm are:
[1.] It is not deterministic and optimal.
[2.] It depends on the probability of picking up a point from the space which makes it harder to explore smaller areas and find path through them.
[3.] With numerous numbers of obstacles in space and choosing a point that does not lie on obstacle and then connect it to the tree is very small.
[4.] With sharp edged shaped obstacles path finding is not easy.

3. Path Tracking

After planning a path for robot path tracking is equally important for its motion control process. As the path planning helps in generating collision free path in space occupied by different obstacles, path process schedules the motion of mobile robot along a desired trajectory. Trajectory following is essential in motion control of mobile robot system. Therefore, a robust control law is needed to provide stabilization of an autonomous mobile robot.

Pure pursuit is a trajectory tracking algorithm for path planning purpose [7]. The pure pursuit algorithm enables a smoother steering control and enhance the robot ability to handle curved paths with lesser halts and short cuts. In this controller the vehicle moves along the circular trajectory when taking turns rather than following the straight lines, hence following up a circular arc is more natural than tracking a straight line. This could be analogizing with a car driver who follows a point further ahead and then steers smoothly toward that point therefore driving in a circular arc on a road.

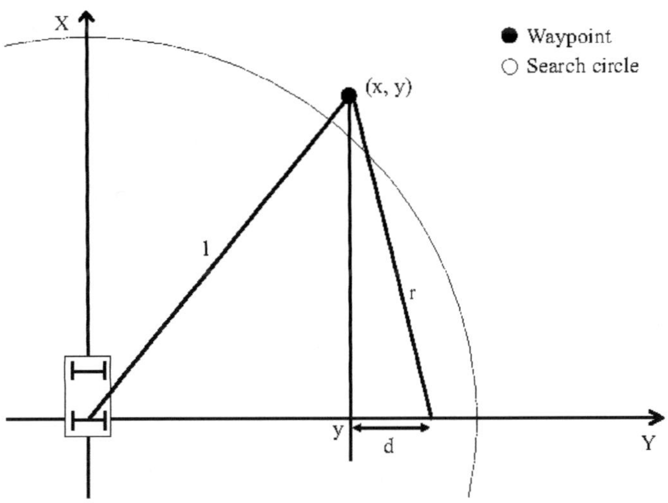

Figure 3. Pure Pursuit

Pure pursuit algorithm is demonstrated in Fig 3, described in vehicle coordinate system, where z and y axis depict downward, and forward direction of vehicle and x axis shows a right handed coordinate system. The point (x, y) is a look ahead distance from origin. the goal is to calculate and measure the curvature of the arc the joins the origin to point(x, y) of $length$ "l". In the figure, "D" is the distance between robot current position and target. γ is the curvature of the circular arc and $\frac{1}{\gamma r}$ is defined as the radius of the circle that goes from the robot start position to goal position. The derivation of curvature and radius of the arc connecting to end point and start point of robot/vehicle can be made as follows:

If the point (xc, yc) is on a look ahead $distance\ D$ from the start position of robot/vehicle.

$$x_c + y_c = D^2 \quad (1)$$

$$\Delta x + d = r \quad (2)$$

whereas, Δx is change in x of the goal to origin. The radius and curvature are derived as follows:

$$d = r - x \quad (3)$$

$$(r - x)^2 + y^2 = r^2 \quad (4)$$

and by re-arrange Equation (4) as

$$r^2 + 2rx + x^2 + y^2 = r^2 \quad (5)$$

$$2rx = D^2 \quad (6)$$

and again, re-arrange Equation (6) as

$$r = \frac{D^2}{2x} \quad (7)$$

$$\gamma_r = \frac{2x}{D^2} \quad (8)$$

Equation (1) represents the circle of radius D about robot start point and Equation represents the relationship between radius and vehicle/robot current point.
Pure pursuit algorithm work as follows:

1. Firstly, calculate the current location of the vehicle.
2. Calculating the point closets to the vehicle.
3. Find the look ahead distance (goal points) from the robot start position.
4. Convert goal points into vehicle coordinates system.
5. Calculate the curvature of the circular arc.
6. Determine the steering angle and vehicle position.

The look ahead distance is used as the main attribute of this controller because it can show how far along the pathway of a robot/vehicle can look from the start position of robot for calculating the angular velocity.

.

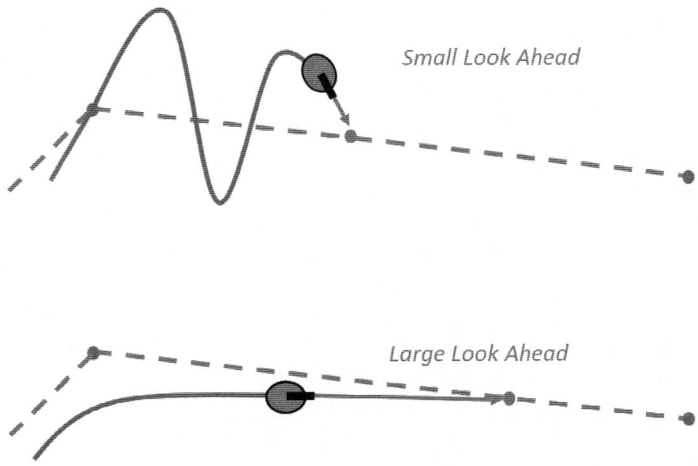

Figure 4. Changing Lookahead distance

The effect of altering the lookahead parameter results in how a robot/vehicle tracks the trajectory. These parameters help in achieving two goals i.e. maintaining the path and retrieving the path. As shown in figure 4, smaller lookahead distance makes robot to follow the path quickly but it results in oscillation due to overshoot. But these oscillations can be decreased by applying large lookahead distance, but it can also cause large curvatures neat to end corners.

4. Kinematic Modelling of Differential Drive Robot

We have considered a differential drive robot with 2 wheels, and it kinematic model can be described with global coordinate system and body frame of mobile robot. The physical configuration of mobile robot is depicted in Fig 5. The theta represents the rotation about axis, the velocity of robot can be represented in form of vectors for forward and backward motion, and angular velocity represents the rotation of mobile robot. Linear and angular velocity of mobile robot can be defined as following:

$$v = \frac{r_r}{2} w_r + \frac{r_l}{2} w_l \tag{9}$$

$$\text{w} = \frac{r_r}{d} w_r - \frac{r_l}{d} w_l \tag{10}$$

where r_r and r_l represents the wheel radii of right and left wheels and d is the width. Kinematics model of mobile robot can be transferred into global coordinate system. The system is represented in Fig 6 with the motion of equation is given as follows:

$$\dot{x} = v \cos\theta \tag{11}$$

$$\dot{y} = v \sin\theta \tag{12}$$

$$\theta = w \tag{13}$$

the position of robot can be found by integrating these equations in time differential from anytime time T1 to T. An offset can be added to give a kick start to robot if it does not start from origin.

$$x(t) = \int_{T_1}^{T} v(t) \cos(\theta(t))dt + C_0 \tag{14}$$

$$y(t) = \int_{T_1}^{T} v(t) \sin(\theta(t))dt + B_0 \tag{15}$$

$$\theta(t) = \int_{T_1}^{T} w(t)\, dt + \theta \tag{16}$$

5. Experimental Result

We used the path planner Bi-directional RRT to mobile robot in an unspecified area crowded with different shape obstacles. Five test runs were taken in completely different environment with unknown locations of obstacles. For testing the performance limits of proposed methods, in all navigation the environment and positions of obstacles were kept completely unknown. The start and target position were already set in advance for the robot. Dimensions of environment are 510*510 in all cases. Because it is Bi directional RRT the algorithm starts its search from both the ends of map (say it there are two trees which will start exploring the configuration space from both the ends of a map to reduce the processing time for search of shortest path as early as possible), the blue and red lines indicates the two trees and green line shows the intersection of two trees and indicates the end of search for algorithm. The increasing blue line indicates the Bi-RRT and obstacles are shown by black objects. The experimental results of simulation are shown in Fig 5,6,7:

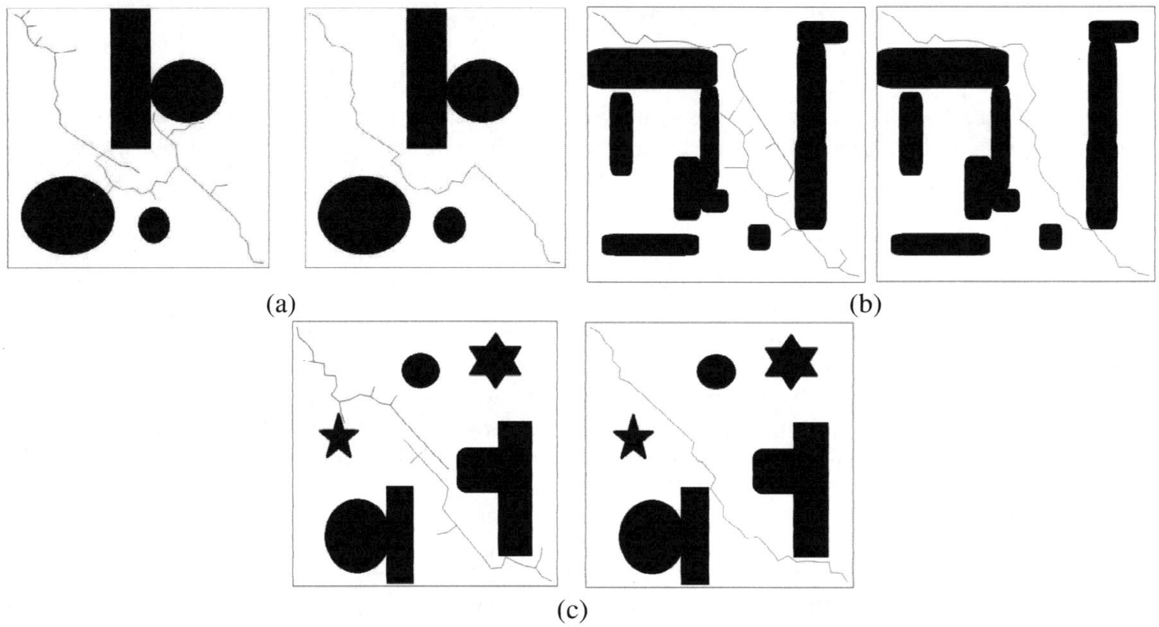

Figure 5. Different setup of obstacle shapes (a) Test 1 with rounded obstacles, (b) Test 2 with rectangular block obstacles, (c) Test 3 with various shapes of obstacles

Figures on the left corners shows the projection of Bi-RRT into the environment along with the shortest path for the robot. The Bi-directional RRT expands from the opposite sides to explore the corners of a square. Since it is Bi-directional RRT, it joins to the goal node in few iterations. Different environmental maps were checked on algorithm. After running the same algorithm on different maps, it was observed that produced solutions were optimal, and near optimal. During the testing of the algorithm, it was observed Bi-RRT can find solutions in all kind of environment in a specified time. After the path is found, pure pursuit controller is applied to track the motion of robot [8]. For smooth and plane navigation of mobile robot it would be easier to tune the linear velocity of the robot/vehicle as the path curvature. During this time period, linear velocity is held constant and angular velocity changes. The spacing between the peaks in the depicts that the rotating wheels with larger spacing results in slower speed as compared to smaller spacing which results in faster speed. This means that the proposed method guides the robot/vehicle comfortably and smoothly through difficult turns to target. During conducting experiment, it was found that the robot reaches the target in a safest way. In real time we would want that robot reaches target as early as possible.

6. Performance Analysis and Conclusion

The proposed method has been tested on three different environments occupied with distinctively sized obstacles together with other algorithms such as APF, RRT and PRM. The analysis has been made based on Path length (total distance covered from start to end end) and Processing Time (time taking by the algorithm for execution). MATLAB 2017b software is used to test the performance of algorithm. The obtained results are summarized as follows

Table 1. Experimental Results from Test 1, Test 2 and Test 3 according to Fig. 5

	Test 1		**Test 2**		**Test 3**	
	Path length	Processing Time	Path length	Processing Time	Path length	Processing Time
Bi-RRT	7.775438e+02	1.172861e+01	7.452761e+02	1.021773e+01	7.885617e+02	1.382988e+01
APF	7.784429e+02	2.322934e+00	7.926919e+02	1.018852e+01	8.144402e+02	2.835975e+00
RRT	7.83738e+02	5.101757e+00	8.398337e+02	3.514516e+00	8.565363e+02	1.723281e+01
PRM	7.951940e02	7.685013e+00	7.573221e+02	3.543044e+00	7.936524e+02	4.308038e+00

Table 1 shows that Bi-RRT is the fastest of all the other algorithms. Though APF requires less computational time but it suffers from local minima whereas PRM may uses optimal path but uses a lot of time and though RRT is quite similar to Bi-RRT (in terms of algorithm) but it is not as fastest as bidirectional one is.

References

[1] Gul, F., Rahiman, W., & Nazli Alhady, S. S. (2019). A comprehensive study for robot navigation techniques. *Cogent Engineering*, ***6*(1),** 1632046.

[2] P. K. Mohanty and D. R. Parhi, "Controlling the motion of an autonomous mobile robot using various techniques: a review," Journal of Advance Mechanical Engineering, vol. **1**, no. 1, pp. 24–39, 2013.

[3] J. M. Keil, "Decomposing a polygon into simpler components," SIAM Journal on Computing, vol. **14**, no. 4, pp. 799– 817, 1985.

[4] O. Khatib, "Real-time obstacle avoidance for manipulators and mobile robots," in Autonomous robot vehicles. Springer, 1986, pp. 396–404.

[5] D. Gallardo, O. Colomina, F. Florez, and R. Rizo, "A genetic algorithm for robust motion planning," in Tasks and Methods in Applied Artificial Intelligence, pp. 115–121, Springer, 1998.

[6] S. M. LaValle, "Rapidly-exploring random trees: A new tool for path planning," 1998.

[7] R. C. Coulter, "Implementation of the pure pursuit path tracking algorithm," Carnegie-Mellon UNIV Pittsburgh PA Robotics INST, Tech. Rep., 1992.

[8] M. Samuel, M. Hussein, and M. B. Mohamad, "A review of some pure-pursuit based path tracking techniques for control of autonomous vehicle," International Journal of Computer Applications, vol. **135**, no. 1, pp. 35–38, 2016.

Acknowledgement

The authors would like to thank to the Universiti Sains Malaysia for financial assistance under Research University Individual grant (1001\PELECT\8014097).

IOP Conf. Series: Materials Science and Engineering **697** (2019) 012023 doi:10.1088/1757-899X/697/1/012023

Supply Chain Performance Measurement Usng Supply Chain Operation Reference (SCOR) 12.0 Model : A Case Study in A A Leather SME in Indonesia

E Kusrini[1], V I Caneca[1], V N Helia[1*], S Miranda[1]

[1]Industrial Engineering Department, Faculty of Industrial Technology, Universitas Islam Indonesia, Yogyakarta, Indonesia
*Email : vembri@uii.ac.id

Abstract. The leather industry is one of the basic ingredients in fashion and textile/apparel that has good development potential. This can be used to take advantage of the opportunity for the leather industry which is still importing 60-70%. Supply chain measurement is needed to determine production capability and moreover to benchmark for its company, government, and academics. Measurement of supply chain performance is carried out using the newest-SCOR 12.0 method with a performance and processes approach. The study was conducted on XYZ SME and limited to bag product which is one of the largest leather industries in the Bantul area, Special Region of Yogyakarta. The value of supply chain performance is 54.29, which based on performance indicators. This value included in the average category. It can be inferred as the benchmark for the leather industry around areas to improve supply chain performance.

1. Introduction

The leather industry is one of the basic ingredients in fashion and textile/apparel that has good development potential. Indonesia still imports the fashion and textile apparel between 60-70%, which is also an opportunity for the local leather industry and challenges to fulfill the market potential. Then, the leather industry has challenges to keep growing and developing because of the scale of production and the high-rate of import. Other challenges are customer behavior by consuming branded, international products, and also the production level of local leather industries.

Supply Chain defined as entire activities, materials or processes to make products from raw material (by supplier's supplier) to the end product (by customer). Another definition from supply chain means from the upstream to the downstream. The supply chain also means entire processes from the purchasing, source, production, and distribution with the material and information flow [1]. Based on the definition and implementation, supply chain management are important activities for a company that can increase the value of production level in order to deliver products to the customer by managing and controlling entire the activities.

Supply chain can be measured by many methods. One of the most famous and general method is Supply Chain Operation Reference (SCOR) Model. SCOR Model measurement established by The Association for Operations Management (APICS) which deliver SCOR to measure by business process and performance. Both process and performance itself have their own metrices to defined by, such as process is defined from plan, source, make, deliver, return, and enable activities and performance by company reliability, responsiveness, agility, cost, and asset management level [2].

The purpose of this research is to measure the performance of supply chain from one of the Small-Medium Enterprise (SME) XYZ as the leather manufacture industry in Bantul, Special Region of Yogyakarta Province by using newest-SCOR Model; SCOR 12.0 version. The measurement is done by the SME process and performance including metrics. The XYZ SME is one of the largest leather industry in Bantul. Hopefully, the research not only measures the process and performance for its own company but also benchmarking with other companies, so the government can concern with the development and its impact to economics.

It has been never done before to measure leather industry supply chain and by using SCOR in previous research. The previous research with the SME in Indonesia is the research agro-industry [3], and with the mushrooms products [4]. Although, it is an opportunity to measure the supply chain management for an SME to help the production level, manage the material and information flow [5]. There are literature review about how to implement SCOR in manufactur industry [6]. Last, it is an opportunity to use the newest version of SCOR as the SCOR 12.0 version by APICS especially how it implemented in SME to gain information and development of that method for future research. By this research, the purpose is also to define the supply chain score or level in Leather Industry SME. So, the government, the owner, and the academics can give benefits to local and country economics, develop strategies, increase production level, and take the export opportunity.

2. Literature Review

2.1. Supply Chain Management

Supply Chain is related to all activities, tools and materials, or steps needed by a product from raw materials to finished products or semi-finished products to costumers, known as upstream to downstream. Related to the procurement of raw materials, suppliers, the distribution of raw materials, purchasing, production, production targets, to distribution is a discussion of the supply chain. Therefore, supply chain management is a critical activity that needs to be carried out by a company. Supply chain management is an activity that includes managing and managing the entire process of the flow of goods and business processes of a company from upstream to downstream [7]. It can be measured based on the merging of such a qualitative and quantitative approach to an *expert* or a specialist in a particular field [8].

Supply Chain Management has a role to determine the quality of performance owned by the company, and also has other roles in reducing cost, effectiveness and efficiency of resources. Specifically, it understands the quality of business process performance that applies in a related company [1].

2.2. Supply Chain Performance Measurement

Supply chain in a company can be measured by a tool or method. Activities in measuring supply chain performance is called supply chain performance measurement or assessment. Quite number of practitioners and academics are also focused on developing supply chain performance measurement theories. This is due to the complexity of the process and material contained in the supply chain, starting from procurement to distribution to customers [9]. There are several things that are expected in a supply chain measurement [10], including: (1)Identify goals or companies key performance indicator; (2) Identify the customer or target market needs; (3) Help an organization or company to understand the business process, and analyze the elements within; (4) Finding critical points in business process carried, to identify waste, bottlenecks, and the other; (5) Improve production level or business process based on data , and (6) Find out the impact of the improvement.

IOP Conf. Series: Materials Science and Engineering **697** (2019) 012023 doi:10.1088/1757-899X/697/1/012023

2.3. SCOR Model

There are several methods or approaches that can be used to measure the performance of a supply chain in an industry. SCOR, which is an acronym for the Supply Chain Operations Reference, is a model or method of approach begun in 1996 which was initiated by the Supply Chain Council called APICS (The Association for Operations Management). This SCOR Model is a framework that is used to help measure the performance of supply chains in an industry, help in terms of management, business processes, and how a company performs with market demand. SCOR Model also changes in the variable and attributes it had before, this is due to adjustments to the industrial reality or conditions. The latest SCOR Model is SCOR 12.0 which has several changes compared to the previous version [2]. The SCOR Model includes the business processes of a company that consists of (Fig.1):

a. Plan, which is the main activity in a supply chain. Regarding production planning, material needed, finance, scheduling, distribution plan, along with planning to provide value to the customer.
b. Source, related to the activity of procuring raw materials and materials needed for business processes. Therefore it will be very related to suppliers.
c. Make, is the core stage in providing added value processes to products that will be offered to customers. This stage includes the production process, work-in-process, until becomes semi-finished or finished goods.
d. Deliver, related to the stages of distributing both products and services to customers. This stage also plays an important role in supply chain performance measurement due to its relationship with the customer which is the core or main point of the product made or offered.
e. Return, is the process of returning the product, either in a condition rejected by the customer or in an effort to improve the product. This condition occurs in a certain moment, for example dismatch with market demand or with other conditions.
f. Enable, is a process related to the establishment, maintenance and monitoring of information, relationships, resources, assets, business rules, suitability, and contracts needed to carry out from processes in supply chain. This process related to High-Level Management, Finance, HR, IT, Facility Management, Product Management, Sales Processes and Support.

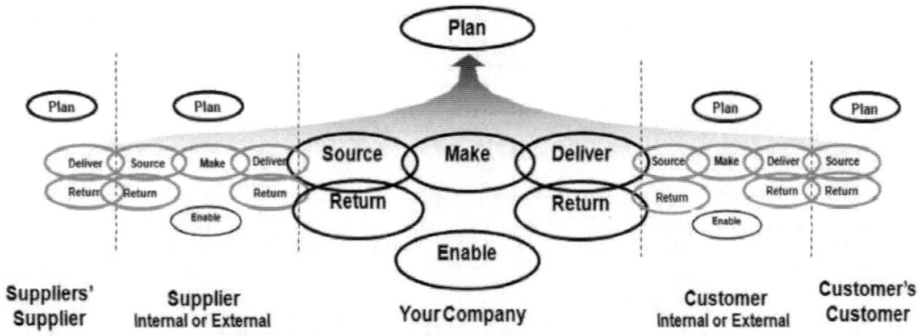

Figure 1. SCOR Model Framework
Source : The Association for Operations Management (APICS)

2.4. *SCOR Model 12.0*

In measuring supply chain performance using the SCOR 12.0 model has metrics that are used as the basis for performance assessment. These metrics are arranged at several levels which have an impact on the assessment of the supply chain performance. The metric level in question are level 1 (one), level 2 (two), and level 3 (three). Level 1 or also commonly called the Key Performance Indicator (KPI), describes the key activities contained in the company's business processes. Level 2 is the assessment criteria used for level 1 metrics. Level 2 can be used to diagnose performance from the performance found at level 1. Level 3 includes metrics that are also used to diagnose the performance of metrics at level 2.

Based on the levels above, the SCOR model evaluates based on several attributes, including reliability, responsiveness, agility, assets, and costs. Reliability is the ability to complete or work to implement in accordance with expectations and criteria that have been previously set. Responsiveness is the speed in carrying out work that is and distributes in accordance with the customer's expectations. Agility is the ability to adapt to changes in the external environment that occur, such as fluctuations in demand that occur in the market. Assets or commonly called assets management efficiency which is the ability to manage assets or manage expenses with income. Costs are components of costs owned by the company, such as material costs, transportation, and others.

Normalized Score (Snorm). Having measurement attributes in both performance and process, the data obtained in the study has unit variants that can be normalized. Data that has been obtained based on each attribute measurement according to each metric used is normalized using the Snorm de Boer method. This is done so that all data with various units owned can be used for further calculations. The use of the Snorm de Boer method can be done using the following formula [11]:

$$Snorm = \frac{(Si - Smin)}{(Smax - Smin)} \; x \; 100 \qquad (1)$$

Snorm = Normalized Score
Si = Valued Actual Indicator
Smin = Lowest Valued Actual Indicator
Smax = Highest Valued Actual Indicator

2.5. Key Performance Indicator

Key Performance Indicators is indicators that determine the value or quality of an industrial, business, and performance process of an organization. KPI values can be obtained through various discussions, some complete, qualitative and quantitative, subjective information that provides an assessment based on individual experience or expertise. In this study the calculations carried out using SCOR 12.0 will be adjusted to the standard performance indicators, based on the following table:

Table 1. Performance Indicator

System Indicator	Performance Indicator
<40	Poor
40-50	Marginal
50-70	Average
70-90	Good
>90	Excellent

3. Methodology

K-Chart is designed to understand the framework in implementing supply chain performance measurement for the research. It is the stage of the method used in the study. Started based on the type of product industry as an object of research, namely the textile and fashion apparel industry. Then, the production system used is MTO (Make To Order). Classification based on economic measurements, to the method used along with the approach. The method used is SCOR (Supply Chain Operation Reference) with an approach to Process and Performance. Supply Chain Assessment using SCOR Model in this study was carried out with several steps as follows :

a. Design a business scope diagram, as an initial step for mapping business processes run by a company
b. Mapping business processes based on processes, starting from the plan to enable
c. Design a business thread diagram
d. Set up of metrices along with performance measurement attributes
e. Performance calculation of supply chain performance
f. Weighting attributes and process

IOP Conf. Series: Materials Science and Engineering **697** (2019) 012023 doi:10.1088/1757-899X/697/1/012023

Figure K-Chart is the stage of the method used in the study. Started based on the type of product industry as an object of research, namely the textile and fashion apparel industry. Then, the production system used is MTO (Make To Order). Classification based on economic measurements, to the method used along with the approach. The method used is SCOR (Supply Chain Operation Reference) with an approach to Process and Performance.

4. Result and Discussion

4.1. Business Scope Diagram

The Business Scope Diagram describes in general the business processes that take place at the XYZ SME. The intended business process starts from the procurement of raw materials from suppliers, the production process, to arrive at customers who are divided into private and retail, as distributors (Fig.2).

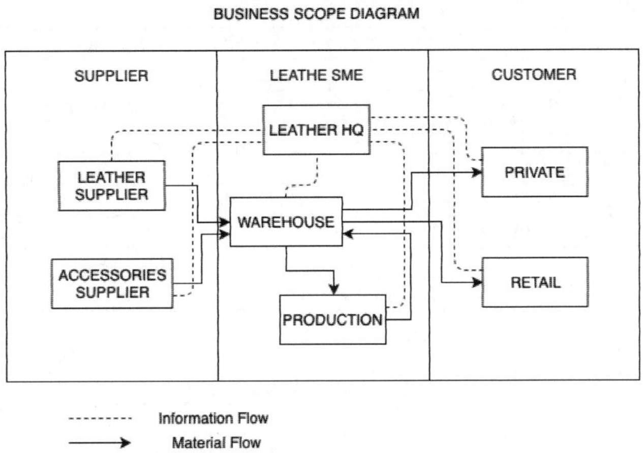

Figure 2. Business Scope Diagram for XYZ SME

4.2. Business Thread Diagram

Business Thread Diagram is a development of business scope diagrams and every stage of business processes, starting from plan, source, make, deliver, return, and enable. Figure 3 shows the following activities from second level :

1. P2 (Plan to Source), P3 (Plan to Make), P4 (Plan to Deliver), and P5 (Plan to Return)
2. S2 (Source Make-to-Order)
3. M2 (Make-to-Order)
4. D2 (Deliver Make-to-Order)
5. SR1 (Source Return) and DR1 (Product Return)

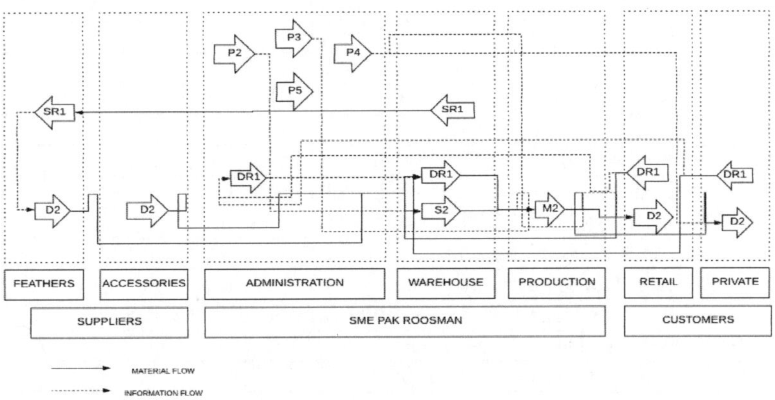

Figure 3. Business Thread Diagram for XYZ SME

Table 2. Metrics for Supply Chain Assessment

NO	SCOR LEVEL 1	METRICS	ATRIBUTES
1	PLAN	RL. 3.37 Forecast Accuracy	Reliability
2		RS. 3.29 Establish Sourcing Plans Cycle Time	Responsiveness
3		RS. 3.28 Establish Production Plans Cycle Time	
4		RS.3.27 Establish Delivery Plans Cycle Time	
5		RS. 3.26 Establish and Communicate Return Plans Cycle Time	
6	SOURCE	RL. 3.27 % Schedules Changed within Supplier's Lead Time	Reliability
7		RL. 3.18 % Orders/ Lines Processed Complete	
8		RL. 3.20 % Order/ Lines Received On-Time to Demand Requirements	
9		RL 3.23 % Orders/ Lines Received with Correct Shipping Documents	
10		RL. 3.19 % Orders/ Lines Received Defect Free	
11		RL. 3.21 % Orders/ Lines received with correct content	
12		RL. 3.24 % Orders/ Lines received damage free	
13		RL. 3.25 % Product Transferred On-Time to Demand Requirement	
14		RL. 3.26 % Product Transferred without Transaction Errors	
15		RS. 3.10 Average Days per Schedule Change	Responsiveness
16		RS 3.113 Receiving Product Cycle Time	
17		RS. 3.8 Authorize Supplier Payment Cycle Time	
18		AM. 3.28 Percentage Defective Inventory	Asset Management
19		AM. 3.37 Percentage Excess Inventory	
20	MAKE	RL.3.49 Schedule Achievement	Reliability
21		RL.3.58 Yield	
22		RL.3.31 Compliance Documentation Accuracy	
23		RL. 3.56 Warranty Cost	
24		RS.3.123 Schedule Production Activities Cycle Time	Responsiveness
25		RS.3.101 Produce and Test Cycle Time	
26		RS.3.142 Package Cycle Time	
27		CO. 3.11 Direct Material Cost	Cost
28		CO. 3.12 Indirect Cost Related to Production	
29		CO. 3.13 Direct Labor Cost	
30	Deliver	RL. 3.33 Delivery Item Accuracy	Reliability
31		RL. 3.34 Delivery Location Accuracy	
32		RL. 3.35 Delivery Quantity Accuracy	
33		RL. 3.32 Customer Commit Date Achievement Time Customer	
34		RL. 3.50 Shipping Documentation Accuracy	
35		RL. 3.41 Orders Delivered Damage Free Conformance	
36		RL. 3.42 Orders Delivered Defect Free	
37		RS. 3.117 Route Shipments Cycle Time	Responsiveness
38		RS. 3.126 Ship Product Cycle Time	
39		AG.3.1 % of labor used in logistics, not used in direct activity	Agility
40		CO. 2.4 Cost to Deliver	Cost
41	RETURN	RS.3.5 Authorized Defective Return Cycle Time	Responsiveness
42		RS.3.104 Receive Defective Product Cycle Time	
43		RS.3.136 Transfer Defective Product Cycle Time	
44		CO.2.5 Cost to Return	Cost
45		AM. 3.26 Return Rate	Asset Management
46	ENABLE	RS. 3.78 Manage Production Performance Cycle Time	Responsiveness
47		CO.3.13 Direct Labor Cost	Cost
48		AM.3.9 Capacity Utilization	Asset Management

4.3. Metrics Set Up According to SCOR 12.0

Based on business scope and business thread diagram measurement attributes can be arranged along with supply chain metrics in accordance with SCOR 12.0 guidelines. Determination of metrics and attributes of measuring supply chain performance will be carried out based on the process and performance in table 2.

So that measurements are made based on process and performance approaches. There are 48 metrics from the results of business processes in the leather industry for bag product. The metrics, in general, are found in the leather industry, so not all metrics will have a value on the measurement results in XYZ SME.

4.4. Supply Chain Performance Measurement

The calculation of supply chain performance based on historical data of XYZ SME that has normalized data using SNORM. From the data, the total score of each process and the measurement metric of SCOR 12.0 have been determined. The next step is to determine the total score of each process by giving weight to the performance attributes. The weighting of performance is given equally towards all attributes, therefore each attribute, namely reliability, responsiveness, agility, cost, and asset management has the same weight, which is equal to 0.2. Likewise for weighting the process is given the same value, which is equal to 0.17 for plan, source, make, deliver, return, and enable. So that the calculation of the total score of XYZ SME can be seen in table 3.

Table 3. Final Score XYZ SME based on Process Approach

NO	SCOR LEVEL 1	ATTRIBUTES	Score	Bobot	Score x Bobot	Scor Level 1	Bobot Level1	Final Score
1	PLAN	Reliability	0,00	0,50	0,00	50,00	0,17	8,33
2		Responsiveness	100,00	0,50	50,00			
3	SOURCE	Reliability	86,67	0,33	28,89	71,08	0,17	11,85
4		Responsiveness	26,56	0,33	8,85			
5		Asset Management	100,00	0,33	33,33			
6	MAKE	Reliability	62,25	0,33	20,75	57,48	0,17	9,58
7		Responsiveness	56,67	0,33	18,89			
8		Cost	53,54	0,33	17,85			
9	DELIVER	Reliability	100,00	0,25	25,00	55,05	0,17	9,18
10		Responsiveness	83,33	0,25	20,83			
11		Agility	9,09	0,25	2,27			
12		Cost	27,78	0,25	6,94			
13	RETURN	Responsiveness	33,33	0,33	11,11	64,44	0,17	10,74
14		Cost	80,00	0,33	26,67			
15		Asset Management	80,00	0,33	26,67			
16	ENABLE	Responsiveness	0,00	0,33	0,00	27,71	0,17	4,62
17		Cost	23,13	0,33	7,71			
18		Asset Management	60,00	0,33	20,00			
TOTAL SCORE								54,29

The measurement of the XYZ SME supply chain performance based on the process approach is **54.29**. It includes the average category. Then, the value of supply chain performance measurement based on performance can be known by calculation in table 4. A low value from the measurement results can be defined as the following explanation. "Plan", SME doesn't have any forecasting method yet which can improve "plan supply chain", "plan source", "plan make", "plan deliver", and "plan return" activities. "Source", delayed material sourcing can effect the production process which has below 40 scores compared to performance indicator. "Make", managing the cost was the biggest impact for low-make value, knowing that XYZ SME still doesn't have any warranty cost and other problem from the production schedule and time. "Deliver", the biggest problem came from the managing labor to ship requirements, XYZ SME should manage it efficiently. "Return", a problem came from return authorized and deal with customer withdrawal for return activities. "Enable", XYZ SME still not have any specific and systematic supply chain performances, therefore the enable activities needs covered by the owner.

From the final score result, it can be seen that the highest score is asset performance. Higher score of asset performance indicates that the utilization of production facilities and the turnover of inventory is quite good. This is in line with research from [12] that asset management strengthens the performance especially for small-medium enterprises and inventory level of the company could increase asset performance.

Table 4. Final Score XYZ SME based on Performance

NO	ATTRIBUTES	SCORE	PROCESS	FINAL SCORE
1	Reliability	248,92	4	62,23
2	Responsiveness	299,90	6	49,98
3	Agility	9,09	1	9,09
4	Cost	184,44	4	46,11
5	Asset Management	240,00	3	80,00

On the other hand, the lowest attribute in performance score is agility. This indicates a lack of good ability to respond to market changes or changes in production capacity. The low level of agility can be seen from the lack of flexibility in the use of labor that cannot be transferred during overload work. Agility is one of key success to compete, especially companies that produce innovative product [13;14]. The fashion industry requires strategic flexibility and manufacturing flexibility [15] because it can affect agility supply chain performance. Flexible strategies affect directly and significantly while manufacturing flexibility does not. Examples of applying agile and responsive models are quick response, describe shorter, more flexible, demand driven, and based on process integration and networking [16]. In addition, supplier innovation is also needed through information sharing so that it can facilitate collaboration in the supply chain [17].

Reliability, responsiveness, and cost are a performance that have average scor. Responsiveness in manunfacturing companies can be upgraded by considering fundamental element that are spread into 4 paths, namely (i) Drivers, (ii) Enablers, (iii) Measures, and (iv) Impacts [18]. It is very important for manufacturing companies to have all elements of responsive in order to survive. Reliability is the ability to perform tasks as expected. Reliability focuses on the predictability of the outcome of a process [2]. Based on the study of 218 manufacturing SME in Texas [12], it can conclude that reliability strengthens the performance. Reliablity in this research is measured using metric forecast accuracy, on-time, the right quantity, and the right quality. The value of reliability is in the average category. This shows that the accuracy of manufacturing in planning, procurement, and delivery needs to be improved.

Costs in performance measurement are all costs incurred to operate the supply chain [2]. In this study, the low-cost performance shows that the cost of manufacturing is not yet efficient. The costs in the supply chain, especially those related to logistics for delivery and return, must consider the relationship of cost of transportation with the selection of shipping modes and the relationship between cost and distance and the time of delivery. Where all these indicators are linearly related as examined [19]. The company needs to consider this model to improve costs, especially the logistics costs of this company.

The crucial subjects in SCOR are process modeling and performance measurement [20] that different from many companies to gain competitive advantages. Therefore, this research was conducted by making a supply chain model and then measuring supply chain performance. So, XYZ SME can improve their performance based on their low performance and benchmark with other companies to enhance performance in accordance with the company that has the best performance.

5. Conclusions and Recommendations

Based on the measurement or supply chain assessment using the SCOR 12.0 method, some conclusions are obtained as follows. It define the value of each XYZ SME process in a based on plan; source; make;

deliver; return; and enable; are 50; 71,08; 57,58; 55.05; 64.44; 27,71. This value indicates that the best process that is owned is "source" and the lowest is "agility". Value for the overall supply chain performance and process of XYZ SME is 54.29. Based on all performance attributes possessed, the best value is asset management by 80, and the lowest value is agility with a value of 9.09. Based on the table of performance indicators, the performance value of the XYZ SME supply chain is included in the "Average" category. It still needs improvement in many sectors. Some of them are making forecasting on the production process, making collaborate with suppliers, managing by reducing production costs, logistic cost, clarity in product returns, and managing supply chain performance systematically covered by the owner. Moreover, XYZ SME Supply Chain Score can be made for another leather manufacture industry as a benchmark to improve supply chain performance.

REFERENCES

[1] Miradji MA 2014 Analisis Supply Chain Management pada PT. Monier di Sidoarjo *Balance Economics, Business, Mgt, and Acc. J.* **10** p 19.

[2] APICS, Quick Reference Guide SCOR (Supply Chain Operations Reference) 12.0, Chicago: APICS, 2017.

[3] Risqiyah IA, Santoso I 2015 Risiko Rantai Pasok Agroindustri Salak Menggunakan Fuzzy FMEA *Jurnal Manajemen & Agribisnis* **14**(1).

[4] Hendayani R, Pangestika S 2015 Analisis Kinerja Suplier pada Usaha Kecil Menengah (UKM) Jamur Tiram di Lampung Timur *Jurnal Studi Manajemen dan Bisnis* 2.

[5] Chin TA, Hamid ABA, Rasli A, Baharun R 2012 Adoption of Supply Chain Management in SMEs, *Social and Behavioral Sciences* **65** pp 614-619.

[6] Prakash S, Soni G, Sandeep, Rathore A 2013 Supply Chain Operations Reference (SCOR) Model : An Overview and a Structured Literature Review of Its Application *Int. Conf. on Smart Technologies for Mech. Eng.* (STME-2013) (p. 55), Delhi: Delhi Technological University.

[7] Vistasusiyanti, Kindangen P, Palandeng ID 2017 Analisis Manajemen Rantai Pasokan Spring Bed pada PT. Massindo Sinar Pratama Kota Manado *Jurnal EMBA* **5** 893-900.

[8] Zhou H, Benton WC, Jr., Schilling DA, Milligan GW Supply Chain Integration and the SCOR Model *J. of Business Logistics* **32** 332-344.

[9] Lima-Junior FR, Carpinetti LC 2019 Predicting Supply Chain Performance based on SCOR metrics and multilayer perceptron neural networks *Int. J. of Production Economics* **212** pp 19-38.

[10] Wahyuniardi R, Syarwani M, Anggani R 2017 Pengukuran Kinerja Supply Chain dengan pendekatan Supply Chain Operation References (SCOR) *Jurnal Ilmiah Teknik Industri* **16** pp 123-132.

[11] Hvolby HH, Trienekens J 2002 Supply chain planning opportunities for small and medium sized companies *Computers in Industry* **49**(1) pp 3-8.

[12] Dissanayake CK, Cross JA 2018 Systematic mechanism for identifying the relative impact of supply chain performance areas on the overall supply chain performance using SCOR model and SEM *Int. J. of Production Economics* **201** pp 102-115.

[13] Fisher ML 1997 *What Is the Right Supply Chain for Your Product?* Harvard Business Review pp. 105-116.

[14] Harris GA, Componation PJ, Farrington PA 2010 An exploration of Fisher's framework for the alignment of supply chain strategy with product characteristics *Eng. Mgt. J.* **22** (4).

[15] Chan A.T.L., Ngai EWT, Moon KKL 2017 The effects of strategic and manufacturing flexibilities and supply chain agility on firm performance in the fashion industry *European J. of Op. Research*, **259** (Issue 2), (2017) Pages 486-499.
[16] Čiarnienė R, Vienažindienė M 2014 Agility and responsiveness managing fashion supply chan, *Procedia - Social and Behavioral Sciences* **150** pp 1012 – 1019.
[17] Kim M, Chai S 2017 The impact of supplier innovativeness, information sharing and strategic sourcing on improving supply chain agility: global supply chain perspective, *Int. J. of Production Economics* **187** pp 42-52.
[18] Ebrahim Z, Ahmad NA, Muhamad MA 2014 Understanding responsiveness in manufacturing operations *Int. Symp. on Research in Innovation and Sustainability 2014 (ISoRIS '14)* 15-16 October 2014, Malacca, Malaysia.
[19] Fancelloa G, Schintua A, Serraa P 2018 An experimental analysis of Mediterranean supply chains through the use of cost KPIs *Transportation Research Procedia* **30** pp 137–146.
[20] Delipinar GE, Kocaoglu B 2016 Using SCOR model to gain competitive advantage: a literature review *Procedia - Social and Behavioral Sciences* **229** pp 398-406.

Acknowledgments

This research was supported by DPPM (Direktorat Penelitian dan Pengabdian Masyarakat) Universitas Islam Indonesia, Yogyakarta, Indonesia and Industrial Engineering Department, Faculty of Technology Industry, Universitas Islam Indonesia, Yogyakarta, Indonesia.

IOP Conf. Series: Materials Science and Engineering **697** (2019) 012025 doi:10.1088/1757-899X/697/1/012025

The Development of Sustainable Product Design Method for Sustainable and Successful New Products

H Alli, M S S Mohd Rashid, R Sulaiman, R Che Me, K M Kamarudin

Department of Industrial Design, Faculty of Design and Architecture, Universiti Putra Malaysia, 43400 Serdang, Selangor, Malaysia
*E-mail: halli@upm.edu.my

Abstract. Currently, most of the product designers have insufficient source in integrating the potential meanings when they are developing a new product. Also, it has become extremely challenging in verifying the variables of a product that could lead towards sustainable and successful products in the market. The development of a new product is not only to cater to the users' satisfactions, but additionally, it should attract more users. In this paper, the Sustainable Product Design Method (SPDM) is introduced as a process to identify and verify the variables of the users' emotional responses and perceived product quality, and element of sustainability in the early stage of the design process. The SPDM is a design guideline attempt to establish the sustainable product requirements (SPR) towards the sustainable and successful products. To prove the practicability of this method, one of the successful products was chosen as a case study. It has been found that the method built is valuable in facilitating the product designers in the early stage of product design and development process. The validation signified from this process is not only contributed towards product sustainability but also increase on its success in the market.

1. Introduction

The design of a product regulates from the user first insight and determines how it can communicate about its function. The users' perception of a product has been influenced by the product variables. Variables of a product have significant factors that influenced the users' perceptions particularly to evaluate on its physical and aesthetics elements [13, 15, 16]. The design of product will also generate the users' interpretations regarding the attribute of the product [7, 15] and would be directly applied to product choice [14]. Users' emotions have been an essential element of product design study since that late 1980s. An emotional response is also important in product interaction that provides a strong influence between users' behaviour and product attachment. Apart from emotions, perceived quality also contributed to evaluation regarding the product superiority to meet every individual's expectations. This refers to indirect intention among product features and users' orientation during decision-making process, helping the user to receive information about the product specifications. Researchers have recognised that perceived product quality as also a cognitive response towards a product which influences the product purchase intention [11, 5], referring to the indirect intention among product features and user orientation which affected by internal or external motivations of quality of a product through some perceived quality points during decision making process [12]. Many users including the product designers and manufacturers, unintentionally only concentrated of physical appearance (visceral level) and functionality (behavioural level) of a product that are easily replenished, then on an emotional part (reflective level) to produce and maintain continuing emotions [12, 1]. The users also do not get emotionally attached to products because they are excluded from the design and manufacturing processes. As a conclusion, the product designers are still facing the difficulty to integrate a meaning in the future products because they are not simply creating products, but also the are making users to want to be responsive towards their emotions a long with designing long-lasting products. They are also struggling in integrating a potential meaning to combine users' emotional attachment with good product characteristics [16, 4]. This has caused the users do not being able to assess and appreciate the products as the meaning that is being communicated is not clear. Product designers need a certain guideline that may be used at the early stage of the design process or in product assessment studies. But there is no specific methodology that could be referred to that incorporate both of the users' emotional responses and perceived quality characteristics for sustainable product as well as increase the probability its success in the market.

The purpose of this study is to develop a new sustainable product design method. The method is a design guideline for product designer and product developers, which the process of this method is incorporated the variables of emotional responses, perceived product quality and elements of sustainability in the early stage of product development. The significance of this method will be likelihood in contributing on product sustainability and successful in the market.

1.1 Sustainable and Successful New Products

The sustainable product design is not essentially about new technologies, but about reconsidering how to meet the need for growth and decreasing negative impacts at the same time. The idea of sustainable is becoming progressively important all over the world with the involvements of the incorporated approach indicators that link a community's society, economy and environment. There are three dimensions of sustainable product development, which have connection with each other that are economic, environment and social [9]. The combination of three elements in product design could reduce the negative impact on society, economic and environment aspects. Table 1 shows the important of sustainability elements in product design.

Table 1. The important of sustainability elements in product design

Element	Description
Social Sustainability	It performs specific function that includes perceptions of safety, equity, eco-prosumption and sustainable urban forms.
Economical Sustainability	It emphasises on the development of a higher quality product based on sustainable innovation, product durability, better price and cost, and effects on resources and environment.
Environmental Sustainability	It pushes manufacturers to be more responsible for the disposal and recycle of products.

Earlier studies have found that the best factor to distinguish the success of the new product from failure is a superior product in the eyes of the user. New product development could not be managed successfully without a clear understanding of users' needs and wants [2, 12]. A successful product is identified as having characteristics, meeting user needs, highly visible, and provides useful benefit or attributed easily by the user [8], and required to integrates other requirements such as market needs and environmental consideration [16, 3, 2]. The collaboration of the user and product designer in this process is required to produce a good and marketable product [15, 16], and very effective source in generating a new innovative product. In the process, the characteristics of a new product could be established in the initial stage of the design process, where the new product ideas are created based on the user wants and needs, and market demand.

1.2 Product Development, Process and Method

The introduction of a new product offers the opportunity for product developers to increase its sales and enhance both competitive position and potential of surviving. In today's competitive environment, there are many product developers who want to develop shorter product time, with lower cost, high quality of product, and consequently, the signification of its [2, 4]. User satisfaction has become the source of innovation success [15, 16, 3]. In addition, users cannot only distinguish which products that works and which do not, they can often provide the answer on the ways the products do not meet the expectations. User can contribute in designing a product that satisfies the expectation, which could increase the product quality. It is important to have an understanding on the relationship between user and product that could be improved by creating something sustainable [6]. Integrating user emotional responses and perceived quality is an important aspect of product interaction, helping users to choose and get attached with the product. Product designers should be able to understand the importance of detaining users' emotional responses and perceived quality characteristic in the product development. Nowadays, there are no specific methodology could be referred that would incorporated both users' emotional responses and perceived product quality characteristics for sustainable products as well as increasing the probability of their success in the market. Hence, the frameworks to produce a sustainable product design should be given as a main priority for improving and have better opportunities to gain product success in the market.

The literature and empirical study implied the outcome of the users' contribution in product development and process, which is crucial to determine the sustainable product design specification. There are 28 sustainable product design specifications (SPDS) have been identified to contribute in sustainability product on 20 specifications contributed from user emotional response, where it was divided into two group, namely Emotional Aesthetics (EAs) and Emotional Function (EFe) [5]. In addition, eight perceived quality characteristics were also identified. Both aesthetic and function elements are the primary elements in the development of physical products. These elements contribute in fulfilling or becoming the sustainable product design specification, which is stimulated from the users' emotional responses and perceived quality characteristics. By integrating all these elements, the specifications of sustainable product could be fulfilled, and the probability of its success is heightened. Figure 1 shows the Sustainable Product Design Method. The SPDM is a process to identify the sustainable product design specifications. It also acts as a decision-making for verifying the priority elements of product design that would fulfill the user requirements, and which can be integrated with sustainable elements to produce a sustainable product.

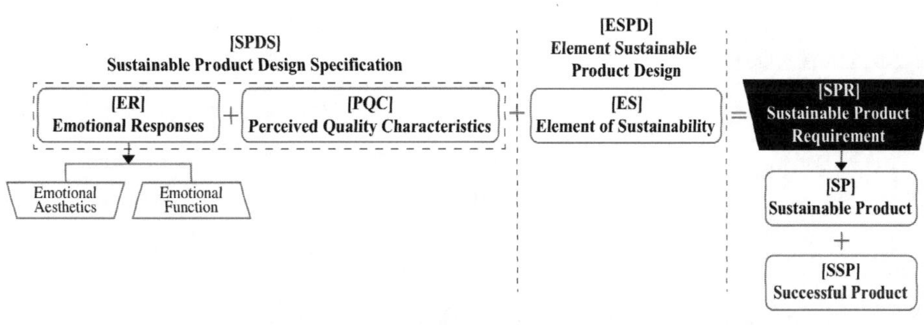

Figure 1. Sustainable product design method

Formula 1 also shown an assessment process in identifying and verifying the sustainable design specification (SPDS) and ultimately, establishing the sustainable product requirements (SPR). The process is divided into two steps. The first step is set in identifying the sustainable product design specification, where both elements of emotional aesthetics (EAe) and emotional function (EFe) need to be incorporated with the product quality characteristics (PQC) to determine a sustainable product design specification (SPDS). Meanwhile, the second step is integrating the sustainable product design specification (SPDS) with the elements of sustainable product design (ESPD), and thus, it will be determined a sustainable product requirements (SPR). Towards the end, selected sustainable product requirements (SPR) should contribute to the sustainability and successful of a product

$$\text{Formula:}([EAe] + [EFe] + [PQC]) = SPDS^{①} \tag{1}$$
$$([EAe] + [EFe] + [PQC]) = (SPDS + ESPD) = SPR^{②}$$

2. Method

An investigation conducted involving of a smartphone as a case study to validate a sustainability factors that could increase the success of a product. It supposed to meet the users' recognition by using the design guideline that was introduced. The product was selected based on the product reputation and sustainable factors in the market since the product were introduced until the present day. A survey was conducted using iPhone 6 model as a focus and involving the existing users and former users for product validation. The smartphone was selected, as the investigation from statistics that shows iPhone 6 model is the most iPhone adoption until September 2017. The snowball sampling method is used to identify the population, which the exact number is unknown or limited knowledge from some iPhone 6 model users. Figure 2 show the Sustainable Product Design Method process. The sustainable product requirements are identified from a mapping of 20 product specifications and nine elements of sustainable product design, which has been recognised by the user. The requirements are sets from the high priority and low priority value.

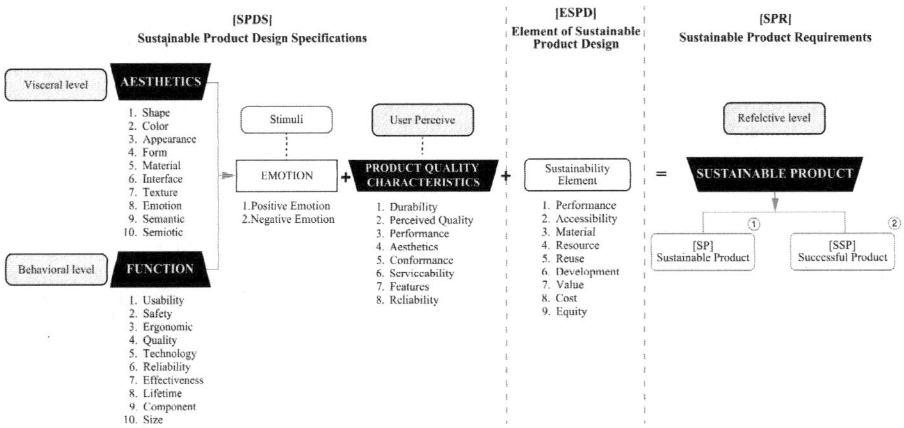

Figure 2. Sustainable product design method process

The validation process was divided into three phases; 1) Determine the users' emotional responses regarding the aesthetics elements and function elements of the product. 2) Attempt to find the product quality characteristics based on the user requirements to the product success. 3) Find the possibility factor of product sustainability. The quantitative data was used to gain an in-depth understanding of the propose investigation. The data from empirical is analysed using SPSS software.

3. Results and discussions

Sixty (60) product users were involved in this survey. Table 2 shows users' emotional responses of iPhone 6 model. It shows that most users show fascination (mean=3.78 point) emotion regarding the iPhone 6 model, while some of users' experience feel anger (mean=2.25 point).

Table 2. Emotional responses of iPhone 6 model

iPhone 6	N	Min	Max	Mode	Median	Mean	Std. Deviation	Sum
Fascination	60	1.00	5.00	4.00	4.00	3.7833	1.13633	227.00
Joy	60	1.00	5.00	4.00	4.00	3.6000	1.18178	216.00
Happiness	60	1.00	5.00	4.00	4.00	3.7167	1.18023	223.00
Admiration	60	1.00	5.00	4.00	4.00	3.5167	1.15702	211.00
Pride	60	1.00	5.00	4.00	4.00	3.4500	1.12634	207.00
Satisfaction	60	1.00	5.00	4.00	4.00	3.5333	.98233	212.00
Desire	60	1.00	5.00	3.00	3.00	3.3000	1.19745	198.00
Disgust	60	1.00	5.00	2.00	2.00	2.1000	1.10008	126.00
Anger	60	1.00	5.00	1.00	2.00	2.2500	1.31000	135.00
Disappointment	60	1.00	5.00	1.00	2.00	2.0000	1.07357	120.00
Shame	60	1.00	5.00	2.00	2.00	2.1333	1.17122	128.00
Fear	60	1.00	5.00	1.00	2.00	1.9833	1.04948	119.00
Sadness	60	1.00	4.00	1.00	2.00	1.9833	.96536	119.00
Boredom	60	1.00	4.00	1.00	2.00	1.8500	.89868	111.00

Figure 3 presents the important of aesthetics elements of iPhone 6 model based on the user requirements. The elements selected with the highest percentage are appearance (84.2%), color (83.3%), shape (82.9%), form (80.8%), material (72.1%), texture (66.7%), interface (64.2%), emotion and semantic (57.5%), and semiotic (52.1%).

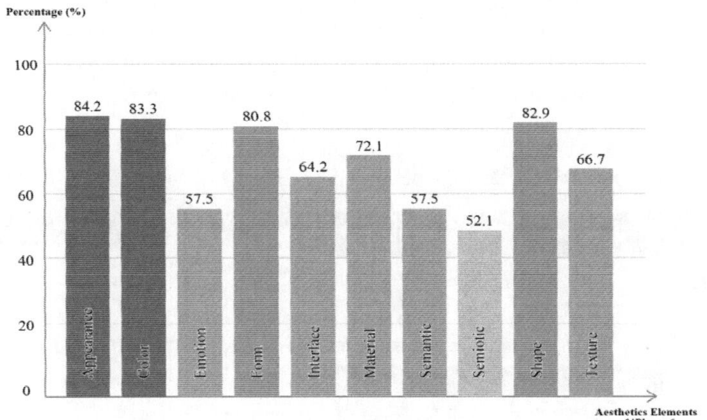

Figure 3. Aesthetics elements on iPhone 6 model

Figure 4 displays the important of function elements of iPhone 6 model based on the user preferences. The main elements selected with the highest measurement are usability (85.4%) followed by safety (83.3%) quality (82.5%), ergonomic and technology (81.7%), reliability (80.8%), effectiveness (77.1%), lifetime (68.8%), components (66.3%) and lastly size (64.6%).

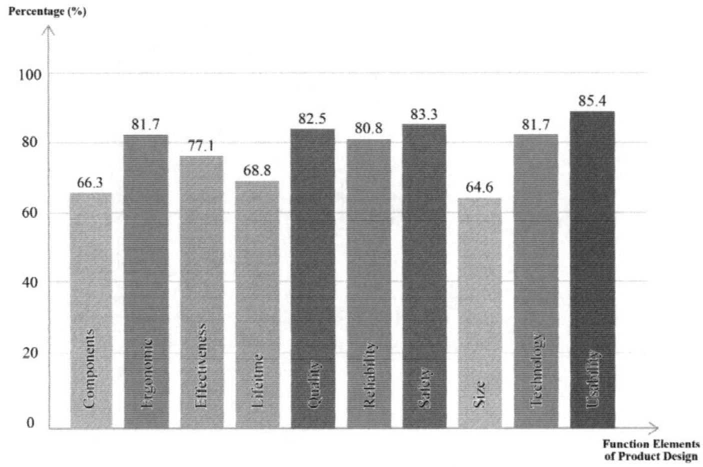

Figure 4. Function elements of iPhone 6 model

Product design elements are also considered as the specification of product. Table 3 shows the sustainable product design specification (SPDS) of successful iPhone 6 model. The very important aesthetic elements for iPhone 6 as identified are appearance, color, shape, form, meanwhile the function elements are stated as usability, safety, quality, ergonomic, technology and reliability. Meanwhile, it also shows perceived quality characteristics of the successful iPhone 6 model. The results indicated that most users preferred durability as their main priority followed by performance, features, aesthetics, perceived quality, reliability, conformance and serviceability.

Table 3. Sustainable product design specification (SPDS) of successful iPhone 6 model

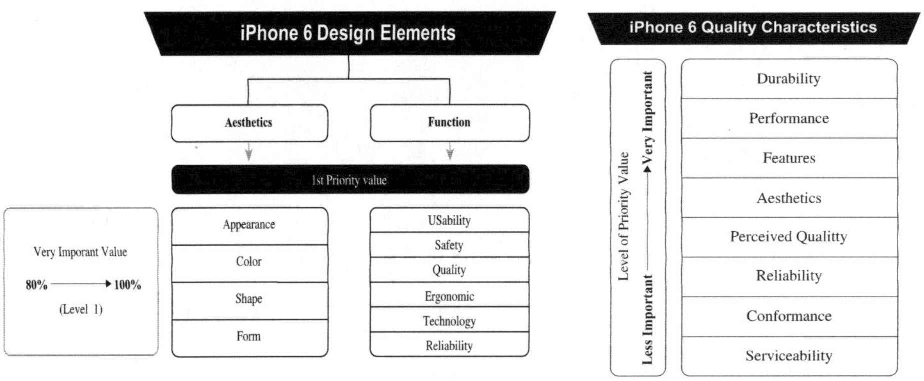

Emotional Responses (ER) Perceived Quality Characteristics (PQC)

Table 4 displays the elements of sustainable product design (ESPD) of iPhone 6 model. Users decide that accessibility as their top priority followed by performance, material, resource, reuse, value, cost, and equity.

Table 4. The elements of sustainable product design (ESPD) of the successful iPhone 6 model

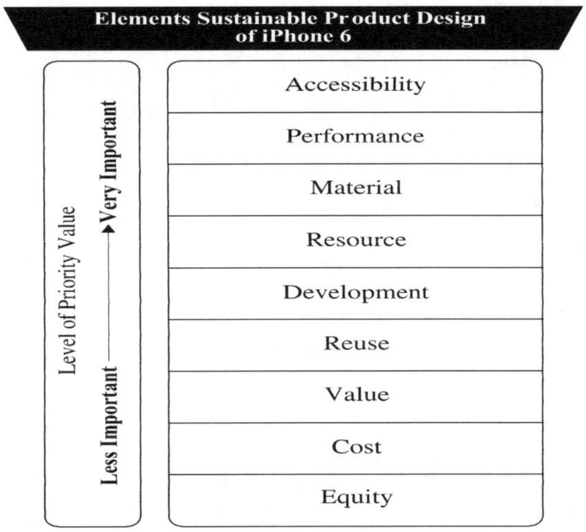

Element of Sustainability (ES)

The introduction of new design guideline attempts to assist the product designer to produce a good product. The development of design guideline or known as Sustainable Product Design Method (SPDM) is essential in the early concept development stage, which will enable the product designer to determine and establish the sustainable product requirements for a new product. This design guideline attempts to provide a framework to assist the product designer during the development of a new product. A new design guideline could help in verifying and establishing the sustainable product requirements (SPR) in the market. It also can help to improve the effectiveness in creating product ideas and quality in design process. Therefore, the systematic process is designated as a new decision-making tool for product designer during the development of a new product in the early stage of product development process. Meanwhile, the factors influencing the production for great products are not only contributed from the elements of aesthetics and function, but other aspects also needed to be recognised such as emotions, perceived quality and elements of sustainability. All the elements need to incorporate and integrate with a systematic process.

The validation results of the product sustainability and success of iPhone 6 model have found that there are four (4) elements of product design from aesthetic (EAe) and six (6) function (EFe), followed by eight (8) perceived quality characteristics (POC) and nine (9) elements of sustainability (ESPD) as indicated as the Sustainable Product Specification (SPR). Figure 5 shows sustainable and successful product elements of iPhone 6 model. Three important elements from aesthetic (appearance, shape, and, form) and function elements, (usability, safety, quality, ergonomics, technology, and reliability) are identified as product design specification assessed from iPhone 6 model. Most of the users show fascination level regarding the iPhone 6 model appraisals. Regarding the product quality characteristics, most users prefer iPhone 6 model choose durability as the main priority followed by performance, features, aesthetics, perceived quality, reliability, conformance, and serviceability. The highest priority of sustainability of iPhone 6 products is accessibility followed by performance, material, resource, development, reuse, value, cost, and equity.

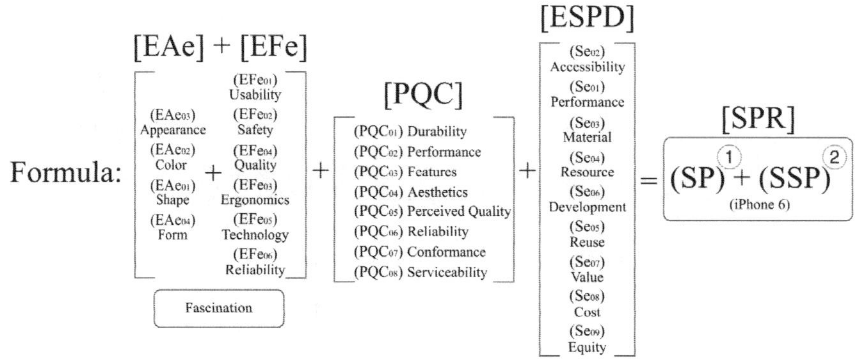

Figure 5. Sustainable and successful product elements of iPhone 6 model

In addition, the formula below also indicated that the specification for iPhone 6 model for its sustainability and its successful product. The equations of the product sustainable and successful of iPhone 6 model are stated as below:

[EAe03 + EAe02 + EAe01 + EAe04] +
[EFe01 + EFe02 + EFe04 + EFe03 + EFe05 + EFe06] +
[PQC01 + PQC02 + PQC03 + PQC04 + PQC05 + PQC06 + PQC07 + PQC08] +
[Se02 + Se01 + Se03 + Se04 + Se06 + Se05 + Se08 +Se09] = **SPR**

4. Conclusion

The development of a new design guideline acknowledged as Sustainable Product Design Method (SPDM) is successful implemented. The method is a process that incorporated the variables of emotional responses, perceived product quality, and the elements of sustainability, which is an attempt to increase the product sustainability and success in the market. Through this method, the product designer and product developer can easily verify and establish the sustainable product requirements in the early stage of the product development process. The findings for sustainable and successful products in the market was valid through the results from the data gathering using the iPhone 6 model as subject, using the SPDM. All the elements found have contributed to the product sustainability and as well as was increased its success in the market. As a conclusion, users' emotional responses and perceived product quality has significant of product design sustainability.

Acknowledgements

The authors gratefully acknowledge the help from the Research Management Centre (RMC), Universiti Putra Malaysia for their helpful support and providing the grant. The authors also wish to express appreciation and gratitude to everyone who has been involved and contributed knowledge, idea and experience as respondents in this study.

References

[1] Aftab M, Agustin H 2017 *Chinese Journal of Mechanical Engineering*. **30(5)** 1058-1068.
[2] Alli H 2018 *International Journal of Sustainable Tropical Design Research and Practice*. **11(1)** 23-28.
[3] Alli H 2014 *Sustainable Tropical Environmental Design Exhibition (STEdex)*. Serdang: UPM Press, 248-253.
[4] Alli H 2018 *International Journal of Sustainable Tropical Design Research and Practice*. **11(2)** 52-56.

[5] Alli H, Mohd Rashid M S S 2018 *World of Multidisciplinary Research & Application Conference*, Kota Kinabalu, April 16-17.
[6] Baetens J 2015 *London: Bloomsbury.*
[7] Bloch PH 2013 *Journal of Marketing*. **59(3)** 16-29.
[8] Cooper R G, Kleinschmidt E J 2011 *New York: Marketing Classics Press.*
[9] Flint R W 2013 *New York: Springer Science Business Media.*
[10] Gogoi B 2013 *International Journal of Sales & Marketing*. **3(2)** 73-86.
[11] Jaafar S N, Latip P E 2012 Asian *Journal of Business and Management Science*. **2(8)** 73-90.
[12] Norman D A 2007 *London: Hachette*.
[13] Rompay T J L, De Wries P W, Bontekoe F, Tanja-Dijkstra K 2012 *Psychology and Marketing*. **29** 919–928.
[14] Sundar A, Arroniz I, Machleit K 2014 *AMA Winter Educators' Conference Proceedings*. **2582** C-83.
[15] Taha Z, Alli H, Abdul-Rashid S H 2012 *Applied Mechanic and Materials*. **215-216** 484-488
[16] Taha Z, Alli H, Abdul-Rashid S H 2013 *Advanced Materials Research.* **739** 678-683.

IOP Conf. Series: Materials Science and Engineering **697** (2019) 012026 doi:10.1088/1757-899X/697/1/012026

Improvement of storage system upright piano cabinet using class based storage

S.L. Alfarokhi[1], Qurtubi[1*] and S. Miranda[1]
[1]Department of Industrial Engineering, Universitas Islam Indonesia
55584 Yogyakarta, Indonesia
[*]E-mail: qurtubi@uii.ac.id

Abstract. The objective of this research is to reduce the process distance on storage and retrieval (S/R) by suggesting storage system of piano's cabinets (piano's parts) on the certain storage shelf. The object of the research is department of setting cabinet that obligated to bridge and organize cabinet's supply from department of painting to department of assembly. Based on the planning and actual delivery of piano cabinets to department of assembly, it is found that only 33,3% deliveries were accomplished as recorded by data on April 2018. One of the causes is internal handling problem at the department of cabinet setting that takes longer time due to unorganized cabinet storage. After all data of "in-out stock" are resumed, later it will be processed by employing class-based storage method for material's grouping. The data indicated that cabinet with material class A is dominated by top 4 requested models of piano and the comparison between initial storage system layout and the proposed one shows the difference on total distance of retrieval activities for 210 kilometers during 5 months of working. It could be concluded that the purpose of storage system could reduce the retrieval distance by 68%.

1. Introduction

PT. XYZ is one of the big companies in Indonesia that has established almost 40 years. PT. XYZ produces 2 classic pianos, which are upright piano (UP) and grand piano (GP) that marketed internationally. The international market company should maintain its high standard in production process, started from standard of quality, cost and the speed of production flow as well as its deliverance.

The problem that emerged in this research is the product's flow of a department that considered as warehouse in PT. XYZ and functioned as temporary place (terminal) for cabinets (often called as piano's parts) that have produced by other departments to be later distributed to department of assembly for assembling process. Warehouse of PT. XYZ is named as department of cabinet setting since it is designated to set the incoming piano's cabinets.

Based on Table 1, it can be notified that the everyday data of delivery on April 2018 was not accomplished as daily plan. It shows that the target only reached 10 out of 40 or only fulfil 33,3% from total delivery planning.

According to field experts, in this case vice chairman of the group and foreman, there are two problems that interrupted the deliverance, which are: supply from previous department and handling of piano's cabinet that emerged as internal problem. The experts assumed that supply problem is considered as complicated and complex, hence, it requires access to all departments for managing that. While for internal problem, which is handling, is recommended by the experts for further research. This issue is emerged by the unorganized cabinets of UP in department of setting. So, it takes extra time for operators to put the cabinets on the storage shelves as well as retrieve the cabinets to be filled in special pushing shelf to department of assembly. Furthermore, longer search for cabinet location was generally occurred due to uncertain location. These issued should be minimized. measurements.

Table 1. Data plan and actual on case and side orders

Date	Case		Side		Target Unit Set	
	Plan	Actual	Plan	Actual	Case	Side
02-Apr-18	83	70	44	44	Not Achieved	Achieved
03-Apr-18	77	78	84	84	Achieved	Achieved
04-Apr-18	86	81	81	81	Not Achieved	Achieved
05-Apr-18	92	87	94	91	Not Achieved	Not Achieved
06-Apr-18	82	88	93	97	Achieved	Achieved
09-Apr-18	93	89	85	75	Not Achieved	Not Achieved
10-Apr-18	82	85	113	110	Achieved	Not Achieved
11-Apr-18	84	73	107	103	Not Achieved	Not Achieved
12-Apr-18	83	76	96	83	Not Achieved	Not Achieved
13-Apr-18	83	65	93	67	Not Achieved	Not Achieved
16-Apr-18	82	71	97	95	Not Achieved	Not Achieved
17-Apr-18	86	61	98	92	Not Achieved	Not Achieved
18-Apr-18	96	85	117	92	Not Achieved	Not Achieved
19-Apr-18	96	66	101	88	Not Achieved	Not Achieved
20-Apr-18	85	57	93	90	Not Achieved	Not Achieved
24-Apr-18	110	67	89	89	Not Achieved	Achieved
25-Apr-18	94	54	87	87	Not Achieved	Achieved
26-Apr-18	92	58	93	83	Not Achieved	Not Achieved
27-Apr-18	94	71	85	74	Not Achieved	Not Achieved
30-Apr-18	47	19	71	71	Not Achieved	Achieved

There are many references and previous researches regarding to class-based storage. A research by reference [1] defined the function of class-based storage to divide products to several classes, in which each class is situated at the certain area in the warehouse. This method is suitable for warehouse with flexibility and high storage activities. Other researcher, reference [2] argued that class-based storage could improve the accuracy in retrieval process and reduce the time wasted during that process. Besides, reference [3] have proven that class-based storage could enhance the speed accessibility on product's movement conducted by operators.

Previous research as studied by reference [4] performed the observation on the best shelf for storing the shuttle and retrieval system by using class-based storage method. Shuttle Based Storage and Retrieval System was the latest technology on automatic system in terms of storing and retrieval, which was employed in the warehouse with high transaction, yet the crane could not possible to achieve the speed standards in reaching designated shelf. Warehouse utilisation assessment is also applied in this research as well as modelled it to identify initial assessment. After the implementation of class-based storage method, it showed that the warehouse with high transaction was the suitable one for the method. Therefore, the storage warehouse required lesser lifting activities (lifts) and less investment.

Reference [5] discussed about the productivity improvement on products' order with manual retrieval and multi-level shelves. Storage warehouse with thousands of items will surely face difficult problems especially with the addition of new warehouse facilities, designated to stock numbers of items and materials. It leads to factors that should be considered such as the retrieval process, warehouse's dimension, material deliverance system, demands' trend, product characteristics and others. It was also discussed that the utilization of class-based storage method is a possible solution for existing warehouse's issues with distance and time as assessment's measurement. The result of research was effective implementation on storage system as stated above by adjusting certain storage system type to order variation from consumers. In this case, it could be stated as the implementation of class-based storage method.

Other things that related to warehouse problems were studied by reference [6] by conducting layout optimisation of three-dimensional order picking warehouse. In this research, it was explained that the layout in warehouse is highly related to important aspects as material handling, cost for warehouse and storage capacity. Therefore, algorithm was established to determine lane depth, numbers of storage level, lateral depth and dimension of warehouse. It was designated to minimize the material handling cost and minimize the space. In establishing the algorithm, several parameters are applied to help designers in designing the warehouse, so it will not be fixed to certain values as well as assist the manager in decision making.

2.Research methodology

This research was conducted from March 2018 in PT. XYZ by performing observation at department of setting cabinet for existing activities. Beside observation, interview was also carried out to related parties, in this case vice chairman of field group and foreman. Primary and secondary data are also required to support the research. Primary data cover the layout condition at department of setting cabinet, which are overall dimension, space area of each part and the order of cabinet storage. While, secondary data include company's data, which is In Out Stock activity data at this department. Those data are required to propose better cabinet storage system.

This research is initiated with problem identification at the department of setting cabinet that supported by literature studies. Hence, the problems could be well-maintained. The identification of problems that arised could be overcome by performing observation on existing system in this department. So, the requirement of system could be identified. Later, better cabinet storage system could be proposed, crucial actions could be retained to improve system, etc.

Next process is data collecting, for both primary and secondary data. Direct field observation was taken as primary data, such as interview with experts to reinforce data obtained, as well as field data collection about initial layout at department of setting cabinet, such as area coverage, storage area, distance among existing parts and the order of cabinet storage on the cabinet's storage. After all information are obtained, data of operator's step direction in retrieval process for the cabinet of certain piano's model are also be gathered to be later converted in distance unit (meter). While, the secondary data will be In Out Stock activity data of the existing cabinets to identify the movement of the cabinets.

From all obtained data, data processing with class-based storage is conducted, by clustering the materials in warehouse by considering their certain similar characteristics, such as similarity in types, size or preferences of consumers. Later, data derived from In Out Stock activity will be organized from the cabinets that have the highest activity to the lowest, respectively. Next, the cabinets are classified into three groups, which are: class A, designated for materials/piano's cabinets which has the highest activity, followed by class B and C.

The grouping of class materials/UP Piano's cabinets at the department of setting cabinet that already gathered is proposed for the new layout system of cabinet storage in order to optimize handling activities. System of cabinet placement on the shelf is based on the earliest formed groups, class A is in line with point I/O or the exit point of piano's cabinets, followed by cabinet on class B and class C, respectively. Two layouts of cabinet storage system before and after definitely have differences and comparable.

3. Results and discussion
After the process of observation and data collecting, several information is obtained to propose the storage system on piano's cabinet, hence the handling activities at the department of setting cabinet could run well. Following are the results and discussions.

3.1. Department of setting cabinet
Department of setting cabinet is the key department in regulating previous department, which is department of buffing. Later on, existing cabinets will be organized to intended models to be assembled in department of assembly. The system is started from department of supply, department of buffing that including small buffing (for small size cabinets) and panel buffing (for larger cabinets). Those cabinets will previously be processed by quality control (QC). Later, the cabinets need painting process, which includes unstain painting, emulsion for PE (Black) cabinet, PM (Red) and PW (Wood-textured), as well as matte and glossy white for PWH (white). Besides, before conducting setting process and cabinet organizing to certain model, drilling process is inserted that designated especially for fall back cabinet, while screen logo process is carried out for fall center cabinet. After all cabinet being painted, the finished cabinets are directly stored as one in the shelf-set as as proposed by the model (Fig. 1). While the other cabinets that unfitted to shelf-set due to excessive number will be stored in the storage shelf (Fig. 2).

From the intended system, apparently the cabinets that entering department of setting cabinet will skip the process storage process on the storage shelf for excessive cabinets since they are considered as waste that should be reworking, the shelf-set is sent to Assembly division. The intended system suggests the cabinets to be stuffed in the shelf-set and directly sent to Assembly division. Yet, previous production lines, such as buffing, sanding after spraying, basic sanding and wood works experiencing several issues that will cause uncertainty and unsynchronized number of cabinets inputted to department of setting cabinet with the request from Assembly division. It causes pile of unready cabinets that being stored in the storage shelf.

Figure 1. Modeled shelf-set of piano

Figure 2. Storage shelf for excessive cabinet for panel and small cabinet

Figure 3. Storage shelf for excessive small cabinets

3.2. Cabinet storage layout (Initial)

From the layout of initial storage (Fig. 4), it could be seen that the cabinet placement is still based on the type of cabinets instead of piano's model. Hence, to reach cabinets designated for piano model B, it needs the retrieval process to several storage shelves. Then, to identify the initial distance condition of retrieval for piano's cabinets, 4 types of pianos are taken, in which B1 PE, B2 PE, B3 PE and U1J PE. Those types are mostly produced. Following is the distance calculation based on steps of operator and later will be converted to meter unit.

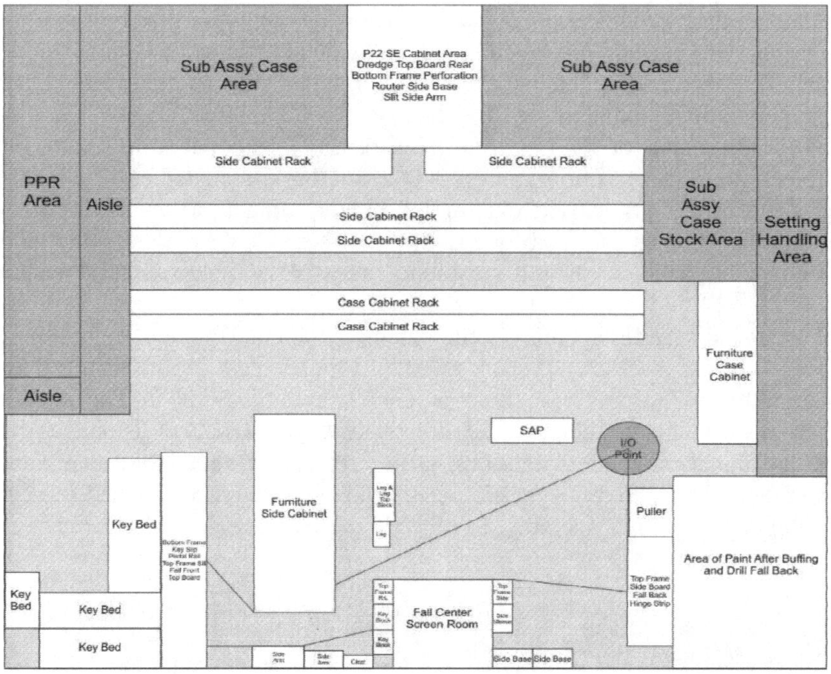

Figure 4. Layout setting cabinet for retrieval distance piano model B1 PE

Description:
= Area Setting Cabinet
= Not Area Setting Cabinet

Table 2. Retrieval distance of piano cabinets B1 PE

Model	Cabinet	Distance (cm)
B1 PE	Fall Back, Hinge Strip	484,4
B1 PE	Fall Center	557,5
B1 PE	Key Block	170
B1 PE	Side Arm R/L	225,1
B1 PE	Bottom Frame, Rail Pedal, Fall Front	590,7
	Return to I/O Point	1568,8
	Total	3596,5

Later, above data could be used to calculate 3 other models, which are model B2, B3, dan U1J. After retrieval distance for 4 models of piano is obtained, the total distance will be multiplied with production demand of designated piano model, or in PT. XYZ it is called as Plan Scheduling Index (PSI), for 5 months demand from January 2018 to May 2018. It is calculated to recognize the total distance conducted by operators to prepare cabinets of piano to shelf-set. It is necessary to figure out the total retrieval distance, in this case by excluding numbers of In Out activities frequency, but instead by employing PSI or piano's demands for 5 recorded months. It is determined by considering the difference of In Out activities frequency for each model, due to uncertain inputted cabinets to department of setting cabinet. Hence, PSI is functioned better as multiplier of retrieval distance to compare initial storage system layout with the proposed one. Following is the total retrieval distance for 4 models of piano.

Table 3. Retrievel distance of piano's upright cabinet (initial)

Model	Distance (cm)	PSI Piano 5 months (January - May 2018)	Total Distance (cm)	Total Distance (m)
B1 PE	3596,5	3696	13292664	132927
B2 PE	4113,8	1906	7840903	78409
B3 PE	4056,3	1323	5366485	53665
U1J PE	4056,3	1209	4904067	49041
Total				314041

From above 4 models, it could be notified that total distance of cabinet is 314041 meters or 314 km for 5 months. It considered as too much for retrieval activity.

3.3. In out stock data at the department of setting cabinet.

In determining classes for existing cabinets at this department as well as to organize the cabinets on the storage shelves, it takes In Out data for cabinets or Storage and Retrieval (S/R) at the existing stock condition. The data were taken from January 2018 to May 2018. Following are the data of In, Out and Stock at the department of setting cabinet.

Table 4. Data of in out frequency on piano's upright cabinet

Month		Total and Frequency		
Model	Cabinet	Total Frequency	Average of Frequency	Average of Stock
B1 PE	KEY BLOCK	334	67	56
B1 PE	FALL BACK	314	63	58
B1 PE	S.ARM R	314	63	13
B1 PE	HINGE STRIP	314	63	45
B1 PE	S.ARM L	314	63	13
B1 PE	BOTOM	313	63	7
B1 PE	P.RAIL	313	63	7
B1 PE	FALL FRONT	310	62	6
B1 PE	F.CENTER	302	60	1

Above data is presumed as entire data that show incoming and outgoing piano's cabinet at the Department Setting Cabinet and will be added on the column of 5 months total frequency. Data on the Table 4 has already organized started from the cabinet that has the highest In and Out activity to the lowest.

3.4. Class grouping.
Piano's cabinet levelling on Table 4 could be classified into 3 classes, which are Class A, B and C. In which Class A is dominated by 84% of all in out frequency, while Class B takes 11% of In Out frequency and Class C is left with 5% from all In Out activities. Below is the classification of cabinet's in-out activities.

Table 5. In-out activities class grouping for piano upright cabinet

Months		Total Frequency	Procentage of Utilization (%)	Total Procentage of Utilization (%)	Total Item (%)	Class
Model	Cabinet					
B1 PE	KEY BLOCK	334	2,92%	84%	42%	A
B1 PE	FALL BACK	314	2,75%			A
B1 PE	S.ARM R	314	2,75%			A
B1 PE	HINGE STRIP	314	2,75%			A
B1 PE	S.ARM L	314	2,75%			A
B1 PE	BOTOM	313	2,74%			A
B1 PE	P.RAIL	313	2,74%			A
B1 PE	FALL FRONT	310	2,71%			A
B1 PE	F.CENTER	302	2,64%			A

Above data is presumed as entire data. It indicates that Class A is filled with 4 models of piano, which are B1, B2, B3, and U1J, in which the production and the demands for those models are quite a lot. Hence, to identify frequency percentage of In Out activity and numbers of material/cabinet of each class, it could be seen from Table 6 below:

Table 6. Class percentage of piano's upright cabinet at department setting cabinet

Class	Numbers of Cabinets' Types	Total Percentage of Utilization (%)	Percentage of Numbers of Items (%)
Class A	59	84%	42%
Class B	33	11%	24%
Class C	47	5%	34%
Total	139	100%	100%

3.5. Cabinet storage layout (Proposed).
After grouping the piano's cabinets into three classes, later it could be proposed the layout system of piano's cabinet storage that supported by the proposed design of cabinets' storage shelf. Later, it could be employed based on classes' classification. Below is the proposed design of cabinets' storage shelf, as shown by Fig. 5.

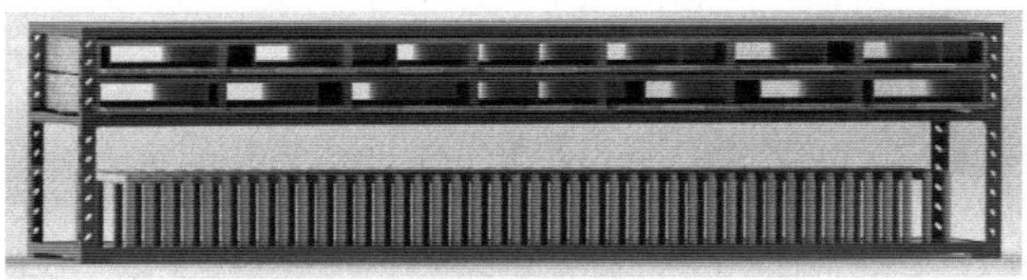

Figure 5. Proposed cabinets' shelf

The shelf is the combination of storage shelf designated for small cabinets (made from woods and located at the upper position) with the storage shelf by bigger size as generally called as panel cabinets (made from iron and located at the lower position). This shelf could help in placing the cabinets in accordance with the class grouping.

The proposed layout system of cabinet storage could identify the steps of operator that later will be converted in meters. Next it will be calculated with production demand (PSI) for 5 months (January to May 2018). The calculation is made to identify the distance of operators for the recorded 5 months. Below is the proposed layout for storage system of cabinets.

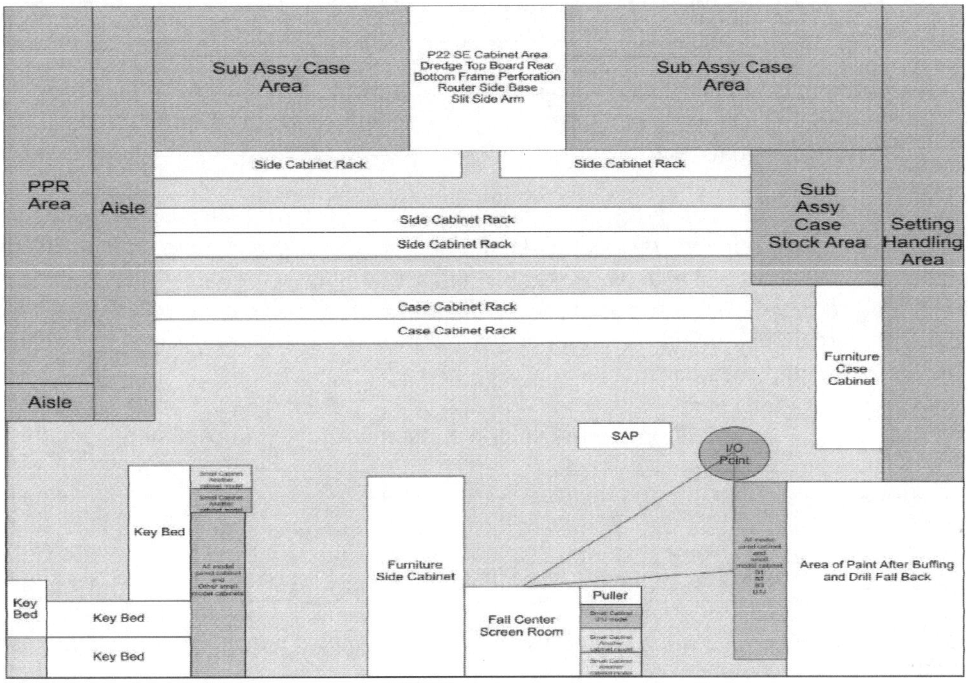

Figure 6. Proposed layout of cabinet setting retrieval distance for piano model B1 PE

= Class A
= Class B
= Class C

Table 7. Proposed retrieval distance for piano's cabinets B1 PE

Model	Cabinet	Distance (cm)
B1 PE	Fall Back, Hinge Strip, Key Block, Side Arm R/L, Bottom Frame, Pedal rail, Fall Front	399,2
B1 PE	Fall Center	557,6
	Return to I/O Point	716,4
	Total	1673,2

Table 8. Retrieval distance on piano's cabinets upright (proposed)

Model	Distance (cm)	PSI Piano 5 Months (January - May 2018)	Total Distance (cm)	Total Distance (m)
B1 PE	1673,2	3696	6184147	61841
B2 PE	798,4	1906	1521750	15218
B3 PE	798,4	1323	1056283	10563
U1J PE	1290,9	1209	1560698	15607
		Total		103229

From 4 mentioned models, it can be concluded that the total distance of cabinets' retrieval is 103229 meters or equal to 103 km for 5 months activities. The difference of this condition, if compared with initial layout system is quite high.

3.6. The comparison of layout system of initial cabinets' storage with the proposed one.
After conducting data collection on the initial condition of layout system, it can be notified the placement of cabinets on the storage shelf. Later the distance that should be performed by the operators in conducting the retrieval can be identified, so as the proposed layout that has established based on In Out Stock activity at the department of setting cabinet. Following is the comparison between initial layout condition with the proposed one.

Table 9. Initial layout comparison with the proposed one

	Initial Layout	Proposed Layout
Numbers of Retrieval on Piano's Cabinets (m)	314041	103229

It can be seen clearly that the retrieval distance on the storage shelf for initial layout condition shows overall distance of 314.041 meters or almost equal to 314 km, while on the proposed layout, the distance covers 103.229 meters or almost equal to 103 km. Hence, the difference of two layouts is 210.812 or almost equal to 210 km. In percentage, it can be said that the reduce of retrieval distance reaches 68%. The difference is quite big considering the calculation that designated for 5 months. The results may higher if it applies for one-year period (12 months). It can be resumed that the proposed layout is able to reduce retrieval distance, so it is expected to be able to improve and optimize the warehouse activities at the department of setting cabinet.

4. Conclusion and suggestion
Based on the result of the research, it could be concluded that the storage system of cabinets for piano UP (Upright) at the department of setting cabinet could be performed by classifying the cabinets that previously grouped based on types of cabinets' similarity to be grouped based on piano's model. The existing grouping will be divided in three classes, which are class A, B and C. It needs to be concerned that Class A is filled with cabinets of piano model B1, B2, B3, and U1J. While, Class B and C are filled with cabinets from other models. Besides, those cabinets are stored on the shelves with proposed design to maintain good storing. The shelf consists of small shelf for small cabinets and main shelf for storing panel's cabinets, which are placed vertically or horizontally.

Suggestions are advised for further research to perform wider research that could reach other aspects that possible for a research, such as studying the problems on uncertain cabinets supplies, complete layout problem at department of setting cabinet, considering other factors that influence handling duration in the process of storage and retrieval (S/R), and others.

IOP Conf. Series: Materials Science and Engineering **697** (2019) 012026 doi:10.1088/1757-899X/697/1/012026

References

[1] Hausman WH, Schwarz LB and Graves SC 1976 *Management Science.* **22** 629–638
[2] Brynzér H and Johansson MI 1996 *Int. J. Prod. Economics.* **46** 595–60
[3] Vickson RG and Lu X 1998 *European J. Op. Res.* **105** 18–28
[4] Ekren BY, Sari Z and Lerher T 2015 *IFAC-Papers OnLine.* **48** 1152–1154
[5] Chan FT and Chan H 2011 *Expert Systems with Applications.* **38** 2686–2700
[6] Rakesh V and Adil GK 2015 *IFAC-Papers OnLine.* **48** 1155–1160

IOP Conf. Series: Materials Science and Engineering **697** (2019) 012027 doi:10.1088/1757-899X/697/1/012027

Review on bio-inspired algorithms approach to solve assembly line balancing problem

N Sulaiman[1*], J Mohamad-Saleh[2], N R H Md-Haron[3], Z A Kamaruzzaman[3]

[1] Faculty of Electrical & Electronic Engineering Technology, Universiti Malaysia Pahang, 26600 Pekan, Pahang, Malaysia.
[2] School of Electrical & Electronic Engineering, Universiti Sains Malaysia, 14300 Nibong Tebal, Penang, Malaysia.
[3] Faculty of Industrial Management, Universiti Malaysia Pahang, 26300 Gambang, Pahang, Malaysia.
[*]E-mail: azliza@ump.edu.my

Abstract. Bio-inspired algorithms that have been introduced by mimicking the biological phenomenon of nature have widely implemented to cater various real-world problems. As example, memetic algorithm, EGSJAABC3 is applied for economic environmental dispatch (EED) optimization, Hybrid Pareto Grey Wolf Optimization to minimize emission of noise and carbon in U-shaped robotic assembly line and Polar Bear Optimization to optimize heat production. The results obtained from their research have clearly portrayed the robustness of bio-inspired algorithms to cater complex problems. This paper highlights the efficiencies of bio-inspired algorithms implemented to cater problem relate to assembly line balancing. This kind of problem is very crucial to counter since it involves minimizing the time of the machines and operators or cost that required optimal task distribution. The outcome of this paper shows the effectiveness of bio-inspired algorithms in solving assembly line balancing problem compared to traditional method.

1. Introduction

Bio-inspired algorithms (BIAs) are approaches that have been developed based on the biological phenomenon of nature [1]. BIAs can be categorized as metaheuristic method and swarm-intelligence-based (SI) algorithms are among the most prominent classes in BIAs classes. This is because they have been discovered the exhibit the capability to solve various optimization problems including optimizing economic environmental dispatch [2], optimal reactive power dispatch [3] and cutting parameters optimization [4]. With the increasing of the complexity of problems, BIAs seems to be the most effective methods to solve those problems [5]. One of the problem that been meticulously studied by researchers in literature is assembly line problem.

Assembly line, which is normally the last step of production that involves final assembly of the products. An assembly line generally consists of several workstations placed in sequential order. Each of the workstation is in charge to complete certain specific jobs [6]. Hence, it is a must to make the best use of the efficiency of the assembly line by minimizing the time of the operators and machines for each workstation. This is the challenge faced by the researchers in catering the problem as it is a combinatorial problem due to the correlation between each job in each workstation. In other words, the idle time of the operators caused by the different loads in each workstation are to be minimized as well as required optimal task distribution [7].

This paper highlights the implementation of BIAs to cater assembly line problem. The main objective is to analyse the efficiency of BIAs as a problem solver to solve assembly line problem in which later can be used as a guidance for manufacturing companies specifically. The paper is organized into 4 section. Then, few prominent BIAs are highlighted to provide ideas on what is BIAs itself. Section 3 then provide the review on the application of BIAs in solving assembly line problem before concluding remarks are presented in section 4.

2. Bio-Inspired Optimization Algorithms (BIAs)

There are thousands of BIAs that have been established by optimization researchers to solve various optimization problems. The most important element in developing BIAs is to have balanced exploitation and exploration process of the algorithm. Having balanced exploration and exploitation process, it means that the algorithm has the capabilities as a robust algorithm. Among the most prominent BIAs are:

2.1. Ant colony optimization (ACO) algorithm

ACO algorithm was suggested by Dorigo et. al in 1996. ACO is adopted the cooperative communications among ants facilitated by pheromones. Pheromones is an artificial trail that works as a guide for the ants to their destinations. It works by using collective behaviour which means that as the trail is followed by more ants, the trail become attractive to be followed. Besides that, ACO uses constructive greedy heuristic besides having positive feedback and distributed computation [8].

2.2. Particle swarm optimization (PSO) algorithm

The other BIAs is PSO algorithm. PSO has been created by Shi and Eberhart in 1998 is based on the collective behaviour of creatures such as fish schooling or birds flocking. The creatures or call as particles fly or swarm in the population with the aim to reach the optimum position and velocity. The algorithm works in a sequence. First, within the search space, the fitness of each particle is being evaluated. Then, the algorithm updates the individual and global-best particle in the population. Lastly, the velocity and positions each particle is being updated [9].

2.3. Artificial bee colony (ABC) optimization algorithm

ABC was being introduced by Karaboga in 2005. This algorithm adopted the foraging behaviour of honey bees. There are three types of bees in the population which are employed-bees, onlooker-bees and scout-bee. Employed-bees and scout-bee are in charge with the exploration process while onlooker-bees are in charge with the exploitation process. In other words, ABC is possessing both exploration and exploitation capabilities. First, after initialization, each employed-bees have been assigned with food sources that represent the potential solutions. Then, the employed-bees shared the fitness values information of each of the food source with the onlooker-bees in the hive through waggle dance as a medium of communication. Onlooker-bees select only the best-so-far food source for further exploring to find the best food source. Then, the food source that cannot be updated through predetermined time, its respective employer-bee will be abandoned. New bee will be recruited to replace the abandoned employed-bees prior to stabilize the number of the population. The new bee is referred as scout-bee [10].

2.4. Raven roosting optimization (RRO) algorithm

RRO algorithm has been proposed by Brabazon et al. in 2016. This global search algorithm was stimulated from the foraging and social roosting behaviour of raven. Generally, ravens live and socialize in groups of 200 to ten thousand during the non-breeding seasons called roost. Hence, they interacted among each other to perform task like foraging for food sources [11]. The algorithm starts with random selection of a roost in the search space. Then, the fitness value of the position of each raven is calculated after each population of the members is being placed randomly in the search space. Next, the raven with the finest solution is selected as leader. After the selection, a portion of the population follows the leader to gain the food source while the others take a flight to their best-so-far personal position to forage. In each cycle, raven will stop its flight at the location that is better location than the current best position's bird, and forage. The processes of evaluating and updating the location are continued until the stopping criterion is met [12].

2.5. Modified ABC variant (JA-ABC5) algorithm
Another example of optimization algorithm is modified artificial bee colony (ABC) named as JA-ABC5. JA-ABC5 is introduced by Sulaiman *et. al* in 2015. Basically, JA-ABC5 is the extended and modified version of two other ABC; JA-ABC3 [13] and JA-ABC4b [14]. There are four modifications that have been implemented into the standard ABC algorithm. The first modification happens after initialization phase which is the insertion of new stages, aimed to improve the exploitation process of the algorithm by improving the average fitness of the population. The first stage identifies few poor food sources and update the poor food sources around global-best food source. The process has improved the exploitation capability of the algorithm. This is because the recent population now consist of fitter food sources. Also, the random selection of food sources has created diverse population as well [15].

3. BIAs Approach in Assembly Line Balancing Problem
There are not many BIAs found in the literature that have been applied to particularly solve assembly line problem. Most of researchers are fond to use other methods to solve the problem such as exact methods [16][17], constraint programming [18], goal programming and many others. One of the example is the application of bees algorithm (BA) and ABC algorithm to solve one type of assembly line balancing problem by Tapkan et. al in 2016. In their research, they have introduced assembly line balancing problem with parallel two-sided that associates the strength of two-sided and parallel and lines. Besides that, they have also included walking times that is necessary to be considered when involving large assembly line system. This is because the larger the assembly system, the longer the walking distances between each production line. In the end, they have concluded that both BA and ABC algorithm have shown better performance than other existing optimization algorithms which is Tabu search algorithm [19].

Another example is the implementation of PSO algorithm in robotic assembly line balancing problems by Janardhanan *et. al* in 2017. In their research, their objective is to minimize the cycle time. PSO has been applied to solve two configurations of robotic assembly line which are straight and U-shaped configurations. PSO has shown excellent performance in computational time in comparison to other algorithms [20].

The other example is the implementation of hybrid algorithm based on genetic algorithm to solve cost-oriented robotic assembly line problem (cRALBP) by Pereira *et. al* in 2018. In their research, they are more focusing on cost-oriented assembly line problem. They have proposed a hybrid algorithm to cater two case studies. For the first case study, the memetic algorithm is able to create feasible solution to the problem by encoding individuals as ordered lists of tasks. For the other case study, the method gives suggestion for an effective local search technique and give guidance on the strong NP-hardness of a number of assembly line balancing problems [21].

Besides that, in 2018, Babazadeh *et. al* have introduced an algorithm called as an enhanced non-dominated sorting genetic algorithm II (NSGA II) to cater straight and U-shaped assembly line balancing problems. Before applying the algorithm to solve the problem, they have tested the algorithm on several benchmark functions. The obtained results have clearly portrayed the excellent performance of the algorithm in comparison to others. Later, after implementing the algorithm on the straight and U-shaped assembly line balancing problems, they have concluded that their proposed algorithm has shown the capability in dealing with the problem that consists of several conflicting objective functions. Lastly, they have mentioned the gap that can be fulfilled by other researchers which is to apply the algorithm to solve problem with high complexity such as mixed-model assembly line balancing problem since they are more focusing on single-model [22].

Later, in 2019, Şahin and Kellegöz have developed a new hybrid method for multi-manned workstations that involved solving a resource constrained assembly line problem. The new hybrid method, which is the hybridization of PSO with a special constructive heuristic has been used for that purpose and it is being compared with tabu search and cuckoo search algorithm. Lastly, they have concluded that their proposed hybrid BIAs are able to produce results that have acceptable deviations from the lower bounds [23].

Then, in 2019 as well, Zhang *et. al* have designed hybrid pareto grey wolf optimization (HPGWO) algorithm to solve assembly line problem that having multi-objective because it involves with minimizing cycle time, noise and carbon emission. They have also compared the performance of their designed BIAs with other five algorithms. The results showed their proposed algorithm is able to achieve promising results in minimizing cycle time and noise and carbon emission [24].

4. Conclusion

Based on the above review, it can be concluded that, although there are not many literatures found regarding the implementation of BIAs in solving assembly line problem, the efficiency of the BIAs in solving the problems tremendously show a good sign that BIAs is perfect choice as the problem solver. Hence, more study on the applications of BIAs to optimize assembly line balancing problem should be conducted to further analyze and prove the effectiveness of BIAs as the best method at solving assembly line balancing problem.

Acknowledgements
The authors acknowledge UMP Research Grant project No. RDU180370 and Universiti Malaysia Pahang and Universiti Sains Malaysia for their financial and facilities support.

References

[1] Ahmad M, Ali M G, Alireza F and Pendar S 2013 *International Journal of Bio-Inspired Computation* **5** 19-34.
[2] Sulaiman N, Mohamad-Saleh J and Abro A G 2016 *ARPN Journal of Engineering and Applied Sciences* **11(18)** 10814-10819.
[3] Subbaraj P and Rajnarayanan P N (2009) *Electric Power Systems Research* **79(2)** 374-381.
[4] Yildiz A R 2013 *Information Sciences* **220** 399-407.
[5] Sulaiman N, Mohamad-Saleh J and Abro A G (2018) *Engineering Applications of Artificial Intelligence* **74** 10-22.
[6] Álvarez-Miranda E and Pereira J (2019) *Comput. Oper. Res.*
[7] Vishnu Raj A S, Mathew J, Jose P and Sivan G (2016) *Procedia Technology* **25** 1146-1153.
[8] Dorigo M, Maniezzo V and Colorni A (1996) *IEEE Transactions on Systems, Man, and Cybernetics, Part B: Cybernetics* **26(1)** 29-41.
[9] Kennedy J and Eberhart R (1995) *Proceedings IEEE International Conference on Neural Networks* Perth, WA.
[10] Karaboga D (2005) *Technical Report* TR06.
[11] Torabi S and Safi-Esfahani F (2018) *Swarm and Evolutionary Computation* **40** 144-154.
[12] Brabazon A, Cui W and O'Neill M, *Soft Comput.* **20** 525.
[13] Sulaiman N, Mohamad-Saleh J and Abro A G (2015) *AIP Conference Proceedings* **1660(1)** 050037.
[14] Sulaiman N, Mohamad-Saleh J and Abro A G (2017) *International Journal of Bio-Inspired Computation* **10(2)** 99-108.
[15] Sulaiman N, Mohamad-Saleh J and Abro A G (2015) *The Scientific World Journal* **396189**.

IOP Conf. Series: Materials Science and Engineering **697** (2019) 012027 doi:10.1088/1757-899X/697/1/012027

[16] Pereira J and Álvarez-Miranda E (2018) *Omega* **78** 85-98.
[17] Borba L, Ritt M and Miralles C (2018) *European Journal of Operational Research* **270(1)** 146-156.
[18] Bukchin Y and Raviv T (2018) *Omega* **78** 57-68.
[19] Tapkan P, Özbakır L, and Baykasoğlu A (2016) *Applied Soft Computing* **39** 275-291.
[20] Janardhanan M N, Nielsen P and Ponnambalam S G (2016) *Distributed Computing and Artificial Intelligence* **474**.
[21] Pereira J, Ritt M and Vásquez Ó C (2018) *Computers & Operations Research* **99** 249-261.
[22] Babazadeh H, Alavidoost M H, Fazel Zarandi M H and Sayyari S T (2018) *Computers & Industrial Engineering* **123** 189-208.
[23] Şahin M and Kellegöz T (2019) *Computers & Industrial Engineering* **133** 107-120.
[24] Zhang Z, Tang Q and Zhang L (2019) *Journal of Cleaner Production* **215** 744-756.

IOP Conf. Series: Materials Science and Engineering **697** (2019) 012028 doi:10.1088/1757-899X/697/1/012028

Designing Disaster Recovery Plan of Data System for University

Difana Meilani[1], Ikhwan Arief[1], Mulya Habibitullah[1]

[1]Industrial Engineering Department, Faculty of Engineering, Universitas Andalas, Padang, Indonesia
*E-mail: difana@eng.unand.ac.id

Abstract. According to the QS World University Rankings, Andalas University is a 3-star university and is in ranked 12th of the best universities in Indonesia based on the Indonesian Ministry of Research and Higher Education criteria. The university is located in three different areas which are Padang, Payakumbuh, and Dhamasraya. Currently, the university's Information and Communication Technology Development Unit which is responsible in managing the information and communication technology has no formal documentation of response procedures in an event of an emergency. Being used by some 26,702 students and staffs, the system is potentially corrupted by a few possible disasters like viruses, hackers, and server downs. Therefore it is crucial for the organization to plan a documented security action to protect essential archives and application software in anticipating the catastrophes. This study developed a Disaster Recovery Plan (DRP), as a document to act on accidents which may occur in occur in information system failures. It is a plan for rapid recovery from the disaster or the emergency response so that the organization does not perceive the impact given by the information system failure.

1. Introduction

Recently, organizations are often faced various types of disruptions that could affect to the organizations. The effects of each disruption might be different. Traditionally, Business Continuity Planning (BCP) and Disaster Recovery Planning (DRP) as the main contingency plans are carried out separately in different time horizons within organizations. To face disruptive events that may happen, proactive planning of internal and external resources of the organization should be prepared so that the disasters will coped with effectively and efficiently [1],[2]. A disaster is defined as the occurrence of any event that causes a significant disruption in IT capabilities [3].

The disaster recovery plan will provide some activities to do in response to a disaster. This plan will provide effective and efficient ways to recover from any disaster that affect IT system in organizations. Disaster loses can be minimized by having a good disaster recovery plan for every business subsystem and operation within an enterprise. Every business disaster has one or more causes and effects. The causes can be natural, human or mechanical in origin, ranging from simple case such as a tiny hardware or software component's malfunctioning to universally recognized events such as earthquakes, fire, and flood. Effects of disasters range from small interruptions to total business shutdown for days or months, even fatal damage to the business [4].

Information Technology (IT) system has become vital to every aspect of business [5],[6]. IT Disaster Recovery Plan must contain requirements and steps that need to be taken in response to disaster that affect IT System in University (in this case is Andalas University). The fundamental goal of disaster

recovery plan is allowing basic business functions to resume and continue until system can be restored into its previous disaster functions [7]. This Disaster Recovery Plan is designed to ensure the continuation of vital business processes in the case that a disaster occurs [8]. An effective solution that can be used to solve the problems in IT system. After disaster, the organizations should provide disaster recovery plan to restore all required frame vital business processes [9].
IT disaster recovery plans provides step-by-step procedures for recovering disrupted systems and networks, and help them resume normal operations. The goal of these processes is to minimize any negative impacts to company operations. The IT disaster recovery process identifies critical IT systems and networks; prioritizes their recovery time objective; and delineates the steps needed to restart, reconfigure, and recover them. A comprehensive IT DR plan also includes all the relevant supplier contacts, sources of expertise for recovering disrupted systems and a logical sequence of action steps to take for a smooth recovery [9].

Disaster recovery plan will have an outcome as a disaster recovery plan document. This document will be the primary source in term of emergency when the disaster happened and attact the IT system. The disaster recovery plan documents will provide procedures to restore all the lost in IT system. Therefore, the document should be readable, simple and the instructions are detailed. The purpose of the document must be clearly stated in the introduction, defining the objectives the plan intends to achieve.

Specifically for information technology operations and management, it is namely ISO / IEC 27031. It provides guidance on the concepts and principles behind the role of information and communications technology (ICT) in ensuring business continuity. The International BCM (Business Continuity Management) Standard developed by ISO / TC 223 summarizes the BCM approach to prevent, react and recover from incidents. Activities involved in BCM are incident preparedness, operational continuity management, Disaster Recovery Planning (DRP) and risk mitigation that focus on improving the resilience of the organization and preparing it to react effectively to incidents and recover pre-determined timeframes [10], [11].

Andalas University is a 3-star university and it is in ranked 12th of the best universities in Indonesia based on the Indonesian Ministry of Research and Higher Education criteria. The university is located in three different areas which are Padang, Payakumbuh, and Dhamasraya. The university's Information and Communication Technology Development Unit which is responsible in managing the information and communication technology and in providing applications that are used for the academic community. Based on its responsibility, the unit is the center of information and storage of all important files or data that must be maintain so the system of Andalas University runs continuously. Currently, it has no formal documentation of response procedures in an event of an emergency. If a disaster happens, it can cause lost and damage of all information. Furthermore the organization and lecturer system can be paralysis. The purpose of this research is to design a Disaster Recovery Plan for Information and Communication Technology Development Unit of Andalas University.

2. Method

The data and information being collected were related to the problems faced in information systems, especially regarding handling in disasters. In addition, data deals with designing a Disaster Recovery Plan were also being collected. The Data and information regarding the Information and Communication Technology Development Unit were obtained from direct observation, interviews, and discussions with related experts. The stages of the survey were carried out with literature related to this study. Literature studies were carried out by reading theories on books and journals which concern in Disaster Recovery Plan and also collecting information from the online sources through internet.

IOP Conf. Series: Materials Science and Engineering **697** (2019) 012028 doi:10.1088/1757-899X/697/1/012028

The method used in this study was ISO 27031. This was chosen since it offers several advantages such as:

1. ISO is a standard with a middle to upper area coverage
2. ISO is suitable for use in IT management
3. The framework can be used to manage technology and information infrastructure in organizations, and ways to provide the best service for users of information technology.
4. The framework is used since it can help them at the highest level of the organization to achieve and fulfill their legal, regulatory and ethical requirements in terms of the use of IT organizations.

The design of DRP began by analyzing the running system and identifying threats that occurred, both from the outside and originating from within. After obtaining a risk assessment one of steps in risk management process then risk analysis is carried out for determining the level of each risk [12]. The risk assessment has been validated and verified by the expert. Determination of risk level is done by using the Failure Mode and Effect Analysis (FMEA) method [13], [14]. Then the Business Impact Analysis (BIA) is carried out. It is a systematic process to determine and evaluate the potential effects of an interruption to critical business operations as a result of a disaster, accident or emergency [15].

BIA is analyzed using qualitative and quantitative approach. The qualitative analysis is for obtaining an assessment from the point of view of the system manager and user regarding the critical assets at the Information and Communication Technology Development Unit of Andalas University. While, the quantitative analysis is for getting assessment from the point of view of the productivity of the existing system and, it is also to sharpen the result of qualitative method. In the quantitative approach, Recovery Point Objective (RPO) and Recovery Time Objective (RTO) are two important parameters of a disaster recovery or data protection plan which are being calculated [16], [17], [18]. These are objectives which can guide enterprises to choose an optimal data backup plan.

Finally, the design of the content of the Disaster Recovery Plan is based on the standard of ISO/IEC 27031. At the end, the validation process is done by the stakeholders at the Information and Communication Technology Development Unit of Andalas University to prove that the DRP document has already fulfilled the needs of the unit.

3. Finding and Result

The Risk Assessment stages based on ISO 27031 are as follows:

1. Identify the potential failure mode.
2. Determine the potential impact on ICT services; the severity of each failure mode.
3. Identify the frequency of failure modes that the organization has prior experience, as well as ease of monitoring and detection of failure modes.
4. Identify indicators that will provide signals or information of failing components. .
5. Identify the direct and indirect events that are related, and will change the state of each indicator.
6. Identify existing controls that prevent important components of failure, or can detect those failures.
7. Identify related data sources, and possible methods for detecting changes in indicator values.
8. Identify whether suitable risk reduction or elimination of controls can be applied to prevent such ooccurrences.

At this stage, risk identification was carried out from the literature which specifically threatens the information systems. The potential risks to the information system from previous research are shown in Table 1.

Table 1. Risk based on Literature and Previous Research

	No	Item /Attribute Risk Agent

IOP Conf. Series: Materials Science and Engineering **697** (2019) 012028 doi:10.1088/1757-899X/697/1/012028

Indicators / Dimensions		
Nature / Environment	1	Fire
	2	Flood
	3	Earthquake
	4	Lightning
	5	Storm
	6	Dew
Nature / Environment	7	Heat radiation
	8	Temperature varies
	9	Dust / dirt
	10	Humidity
Human	11	Device theft
	12	Information is accessed by unauthorized parties
	13	Leaks of data / internal information of the company / institution
	14	Data and information do not match the facts
	15	Misuse of access rights / User ID
	16	Former users / employees still have access to information
	17	Unauthorized physical access
	18	Data loss
	19	Human error
	20	Risk of human-caused damage such as cybercrime, terrorism, piracy and vandalism
System and Infrastructure	21	Hardware failure / damage
	22	Server down
	23	Overheat
	24	The network connection is lost
	25	System crash
	26	Overcapacity
	27	Overload
	28	Data corrupt
	29	Backup failure
	30	Failed to update
	31	Poor network quality
	32	Obsolete technology
	33	Risk of damage due to problems
	34	power / voltage

The risks obtained from the literature were validated and verified by the expert. In this validation and verification, the expert can add, reduce and improve the risk data points to be relevant to the Information and Communication Technology Development Unit of Andalas University. Furthermore, the expert gave the points of the risks relevant to the Information and Communication Technology Development Unit of Andalas University and the causes. Table 2 shows the risks after the validation from the expert.

Table. 2 Risks after Validation

IOP Conf. Series: Materials Science and Engineering **697** (2019) 012028 doi:10.1088/1757-899X/697/1/012028

Indicators / Dimensions	No	Item / Attribute Risk Agent
Nature / Environment	1	Fire
	2	Earthquake
Human	3	Temperature varies
	4	Humidity
	5	Device theft
	6	Former users / employees still have access to information
	7	Data loss
	8	Human error
	9	Risk of human-caused damage such as cybercrime, terrorism, piracy and vandalism
	10	Late Data Backup
System and Infrastructure	11	Hardware failure / damage
	12	Server down
	13	Network connection disconnected ISP
	14	Data is corrupt
	15	Backup failure
	16	Failed to update
	17	Power / voltage
	18	Blacklist from Certain Sites
	19	The internal network is disconnected

Description of Color:
(white) : The attribute chosen by Expert
(grey) : Additional attributes from Expert

Table. 3 Table of Levels for Severity

Effect	Explanation	Non Productive Time	Level
Harmful without changes	Can harm the server without warning	> 6 x 24 hours	10
Dangerous but there is a warning	Can endanger the server with a warning first	>5 x 24 jam - >6 x 24 hours	9
Very high	Failure to completely interfere with the server	>4 x 24hours - >5 x 24 hours	8
High	Failure interferes with 50% of server work	>3 x 24hours - >4 x 24 hours	7
Medium	Failure interferes with 25% of server work	>2 x 24 hours - >3 x 24 hours	6
Low	Failure interferes with 10% of server work	> 24 hours - >2 x 24 hours	5
Very low	Failure affects server work	> 12 hours - >24 hours	4
Small	Failure has a minor effect on the server	> 6 hours - >12 hours	3
Very small	Failure has a negligible effect	> 3 hours - >6 hours	2
There is no	Failure has no effect	> 0 hours - >3 hours	1

The level for Occurrence is shown in Table 4

Table 4. Table of Level for Occurrence

IOP Conf. Series: Materials Science and Engineering **697** (2019) 012028 doi:10.1088/1757-899X/697/1/012028

Probability of failure	Number of Failures Per Year	Possible Failure Rat	Level
Very high: inevitable	>500	> 1 in 2	10
	366 – 500	1 in 3	9
Height: Repeated failure	300 – 3365	1 in 8	8
	250 – 300	1 in 20	7
Moderate: failure seasonal	150 – 249	1 in 80	6
	50 – 149	1 in 400	5
	10 - 49	1 in 2000	4
Low	5 - 9	1 in 15000	3
	1 - 4	1 in 150000	2
Remote: rarely happening	<1	1 in 1500000	1

The level for Detectability is shown in Table 5

Table 5. Table of Level for Detectability

Detection	Explanation	Level
Almost impossible	It is almost impossible to check to detect failure	10
Very few possibilities	Very little possibility for checking large failures	9
Little possibility	It's less likely to check to detect a failure	8
Very low	Check has a low chance of detecting failure	7
Low	Check for possible low detection	6
Enough	Check will likely detect failure	5
High enough	Check is likely to detect failure	4
High	Checking has a great chance of detecting failure	3
Very tall	Check almost certainly can detect failure	2
Almost certain	Checking can definitely detect failure	1

After obtaining the level of each FMEA criterion, the Risk Priority Number (RPN) was then being calculated.

Example :

Risk Priority Number = Severity x Occurrence x Detectability (1)

RPN = 7 x 5 x 6

= 210

From the value of the RPN, it could be seen that the risk rating has the most potential for failure of the Information and Communication Technology Development Unit of Andalas University. The highest RPN value was the most potential risk. The RPN calculation result is shown in Table 6.

Table 6. RPN Calculation Result

No	Item /Attribute Risk Agent	RPN	Rank
1	Fire	180	2
2	Earthquake	210	1
3	Temperature varies	125	5
4	Humidity	100	7
5	Device theft	175	3
6	Former users / employees still have access to information	175	3
7	Data loss	96	8
8	Human error	80	9
9	Risk of human-caused damage such as cybercrime, terrorism, piracy	150	4
10	Late Data Backup	100	7
11	Hardware failure / damage	150	4
12	Server down	120	6
13	Network connection disconnected ISP	96	8
14	Data is corrupt	80	9
15	Backup failure	96	8
16	Failed to update	80	9
17	Power / voltage	96	8
18	Blacklist from Certain Sites	125	5
19	The internal network is disconnected	100	7

The next step is the Business Impact Analysis, which was a step to determine RTO and RPO values for each critical server in the Information and Communication Technology Development Unit of Andalas University. RTO value was needed to see how long the server are cut off during repairing if a disaster occurs that causes the server to be cut off. The RTO assessment is shown in Table 7.

Table 7. RTO Assessment

RTO (Recovery Time Objective)	
RTO is defined as how long a user tolerates a loss before regaining use of the application. Simply put: "Recovery that is 0 minutes away - immediate recovery / no interruption in service"	
0	Recovery within 0 hours - 0 days
1	Recovery within 24 hours - 1 day
2	Recovery within 48 hours - 2 days
3	Recovery within 72 hours - 3 days
4	Recovery is greater than 72 hours -> 3 days

While the RPO assessment is to see how much data is allowed to disappear if a disaster occurs. This RPO value is very influential on the backup method of the server that is in the Information and Communication Technology Development Unit of Andalas University. The RPO Assessment is shown in Table 8.

Table 8. RPO Assessment

RPO (Recovery Point Objective)	
"RTO is defined as how much user data will be lost. If the system that supports your business process fails, the data can be recovered and how many losses can be received?	
0	No data is allowed to be lost
1	In the last 8 hours regardless of when the disturbance occurred
2	All data entered since the last backup must be re-inputted
3	It can take up to one week of lost data to be reconstructed

The value of RTO and RPO were determined by distributing questionnaires. The distribution of the questionnaire was filled out by one person from each division, namely the network division, the data collection division, and the multimedia division. The questionnaires were distributed to three people with the aimed to obtain accurate data on the state of the Information and Communication Technology Development Unit of Andalas University. To see the criticality of the server, it can be seen from how long the server is cut off. The RTO Value on each the Information and Communication Technology Development Unit of Andalas University Server is shown in Table 9.

Table 9. RTO value on each Information and Communication Technology Development Unit of Andalas University Server

RTO 0	**RTO 1**		**RTO 2**	
Data Base SIA	SIA	E Office	Administrasi	LP3M
Jurnal 1	FAPERTA	Evaluasi Pendidikan	Akademik	Pasca
Jurnal 2	Buku	FATETA	Alat Lab	RSP Unand
UNAND.AC	Development	Feder PDPT	Berkas	Seminar
	DNS Server	Fekon	Dosen.Unand	SIA Dikti
	E Juornal	FHUK	E Books	SIKS
	I-Learn 1	FIB	Educonf	Sister
	I-Learn 3	FKM	Eorfernece	UKT
	Katalog Pustaka	FMIPA	Farmasi	BAPSI
	Kerja Sama	FTUA	Faterna	Blog Mahasiswa
	LPPM	IKA	FISIB	I-learn Depan
	LPTIK	JNTE	FJI	Karir
	Server Pustaka	Litbang Aset	Fkep	PPID
	Sim Keuangan	LPSE Unand	FKG	Share Point
	Simpeg	Repo Sitory	FTI	Staf Unand
	Analytic	Scolar	Galery	Teknik Elektro
	BIOS	Sim Kehadiran	I-Learn 2	Teknik Industri
	Digilib	Simalyar	Intake D3	Teknik Lingkungan
		SNMPTN	Journal Mahasiswa	Teknik Mesin
			Kemahasiswaan	Teknik Sipil

The RPO value on each the Information and Communication Technology Development Unit of Andalas University server is shown in Table 10.

Table 10. RPO Value on each Information and Communication Technology Development Unit of Andalas University Server

RPO 0	RPO 1			RPO 2
UNAND.AC	I-Learn 1	FJI	Eorfernece	Kerja Sama
Jurnal 1	Feder PDPT	SNMPTN	Faterna	Development
Simpeg	Sister	LPSE Unand	RSP Unand	LPPM
Jurnal 2	I-Learn 3	Alat Lab	Administrasi	IKA
SIA	Katalog Pustaka	Digilib	Karir	Analytic
	Sim Keuangan	Journal Mahasiswa	Fekon	Kemahasiswaan
	Sim Kehadiran	SIA Dikti	E Books	Galery
	Litbang Aset	Simalyar	SIKS	PPID
	E Juornal	Scolar	Pasca	Staf Unand
	JNTE	LPTIK	FISIB	Seminar
	Server Pustaka	FIB	Buku	Educonf
	Data Base SIA	FMIPA	FKG	Teknik Sipil
	FAPERTA	FATETA	LP3M	Teknik Industri
	Evaluasi Pendidikan	FKM	Berkas	Teknik Elektro
	BIOS	Intake D3	Akademik	Teknik Mesin
	DNS Server	FHUK	BAPSI	Teknik Lingkungan
	I-Learn 2	FTUA		Share Point
	Repo Sitory	Fkep		Dosen.Unand
	E Office	FTI		I-learn Depan
	UKT	Farmasi		Blog Mahasiswa

In planning DRP, a backup server location was needed to maintain data security. The location's indicators is referred to ISO 27031, which is compared to the circumstances of each location. The alternative locations and considerations is shown in Table 11.

Table 11. Alternative Locations and Considerations

No	Location Considerations	Alternative location		
		Padang	Payakumbuh	Dharmasraya
1	Being in a safe area but within an affordable distance from the location it will serve	√	√	√
2	Minimum> 40 km from the data center		√	√
3	Be outside the mitigation radius		√	√
4	Not in the geological fault line			
5	Small potential risks from disasters		√	√
6	Availability of local staff and vendors	√	√	
7	Availability and quality of electricity / battery	√	√	√
8	Nearby Fiber Routes	√	√	√
9	Information technology can function at that location	√	√	√
10	Not prone to conflict and terrorism	√	√	√
11	Tax Incentive	√	√	√
	Total	7	10	9

The design of the DRP document for the Information and Communication Technology Development Unit of Andalas University are based on the ISO/IEC 27031. The standard content that must be in the DRP document has been set at ISO / IEC 27031 as follows; Purpose and scope, Roles and responsibility, Calling Plans, Response and Recovery Plan, Application Recovery Team. The plan must be documented so that competent personnel can use it in the event of an incident. Documentation must define the overall framework in which the recovery plan is prepared, which includes; Overall strategy, Critical services based from RTO and RPO values, Timeline for recovery, Recovery team and their responsibilities before and during the disaster. The plan must be documented so that competent personnel team can use it in the event of an incident. The documentation consist of Purpose, Coverage based on the BIA, Availability requirements determined by the business for the availability of services and related technologies, Information security requirements, the appendix. The Appendix consist of Inventory of information systems, applications, and databases; Review of network infrastructure and server names; Inventory of hardware and software system; Service level contracts and agreements. Last the documentation the main ICT supplier. After designing the DRP document, the document validation is carried out. The document was validated by Mr. Hendra Gunawan as chairman of the Information and Communication Technology Development Unit of Andalas University, Mr. Prama Wahyudi as the multimedia division, Mr. Rudi as the network division. The results of the validation by the validator show that the DRP document designed is in accordance with the needs of the unit.

4. Conclusion

This research has developed Disaster Recovery Plan (DRP) documents for data systems in the Information and Communication Technology Development Unit of Andalas University. In the DRP documents consist of quick response procedures and activities in the event of a disaster. This DRP documents will be used in the event of a disaster not for daily operational activities. In the DRP document there is a risk analysis. It analysis what potential destroy the data system by using FMEA method. This DRP documents also have RTO and RPO values that will be used to save the system and the location of backup service which later can be used as a backup server during the repairing of the main server. Suggestions for further research on DRP covers documents and also requiring costs.

5. Acknowledgment

Authors would like to thank the Engineering Faculty of Universitas Andalas for providing publication grant 2019 with contract number 017/UN.16.09.D/PL/2019.

6. References

[1] N. Sahebjamnia, S. A. Torabi, and S. A. Mansouri, "Integrated Business Continuity and Disaster Recovery Planning : Towards Organizational Resilience," *Eur. J. Oper. Res.*, vol. 242, no. 1, pp. 261–273, 2015.

[2] V. Lozupone, "Disaster Recovery Plan for Medical Records Company," *Int. J. Inf. Manage.*, vol. 37, no. 6, pp. 622–626, 2017.

[3] E. Andrade, P. T. Endo, R. Lima, E. Andrade, P. T. Endo, and R. Lima, "Disaster Recovery Solutions for IT Systems: A Systematic Mapping Study," *J. Syst. Softw.*, 2018.

[4] R. B. Duffey, "Power Restoration Prediction Following Extreme Events and Disasters,"

Int. J. Disaster Risk Sci., vol. 10, no. 1, pp. 134–148, 2019.

[5] H. Chae, C. E. Koh, and K. O. Park, "Information Technology Capability and Firm Performance : Role of industry," *Inf. Manag.*, vol. 55, no. October 2017, pp. 525–546, 2018.

[6] D. Meilani, A. Andiningtias, and D. Fatrias, "Decision Support System For Inventory Control of Raw Material," in *5th International Conference on Industrial Engineering and Applications (ICIEA)*, 2018, pp. 2–6.

[7] H. R. Soufi, S. A. Torabi, and N. Sahebjamnia, "Developing a Novel Quantitative Framework for Business Continuity Planning," *Int. J. Prod. Res.*, vol. 0, no. 0, pp. 1–22, 2018.

[8] Faisal, "The Backup Recovery Strategy Selection to Maintain the Business Continuity Plan," in *International Conference on Disaster Management (ICDM)*, 2018, vol. 229.

[9] H. Abdel and R. Mohamed, "A Proposed Model for IT Disaster Recovery Plan," *Int. J. Mod. Educ. Comput. Sci.*, vol. 4, no. April, pp. 57–67, 2014.

[10] M. Niemimaa, J. Järveläinen, M. Heikkilä, and J. Heikkilä, "Business Continuity of Business Models : Evaluating the Resilience of Business Models for Contingencies," *Int. J. Inf. Manage.*, vol. 49, no. April 2018, pp. 208–216, 2019.

[11] S. A. Torabi, R. Giahi, and N. Sahebjamnia, "An Enhanced Risk Assessment Framework for Business Continuity Management Systems," *Saf. Sci.*, vol. 89, pp. 201–218, 2016.

[12] R. A. Hadiguna, "Decision Support Framework for Risk Assessment of Sustainable Supply Chain Decision Support Framework for Risk Assessment of Sustainable Supply Chain," *Int. J. Logist. Econ. Glob.*, no. January 2012, 2012.

[13] S. Vazdani, Gh. Sabzghabaei, S. Dashti, M. Cheraghi, R. Alizadeh, and A. Hemmati, "FMEA Techniques Used in Environmental Risk Assessment," *Environ. Ecosyst. Sci.*, vol. 1, no. 2, pp. 16–18, 2017.

[14] M. Yazdi, "Improving failure mode and effect analysis (FMEA) with consideration of uncertainty handling as an interactive approach," *Int. J. Interact. Des. Manuf.*, vol. 13, no. 2, pp. 441–458, 2019.

[15] P. Radanliev *et al.*, "Future Developments in Cyber Risk Assessment for the Internet of Things," *Comput. Ind.*, vol. 102, pp. 14–22, 2018.

[16] H. E. Miller and K. J. Engemann, "Business continuity management in data center environments," *Int. J. Inf. Technol. Syst. Approach*, vol. 12, no. 1, pp. 52–72, 2019.

[17] V. Gupta, P. K. Kapur, and D. Kumar, "Prioritizing and Optimizing Disaster Recovery Solution using Analytic Network Process and Multi Attribute Utility Theory," *Int. J. Inf. Technol. Decis. Mak.*, vol. 18, no. 1, pp. 65–86, 2019.

[18] E. Andrade and B. Nogueira, "Performability Evaluation of a Cloud-Based Disaster," *J. Grid Comput.*, vol. 17, no. 3, pp. 603–621, 2018.

IOP Conf. Series: Materials Science and Engineering **697** (2019) 012029 doi:10.1088/1757-899X/697/1/012029

The Development of SORREL Systematic Roselle Harvesting System for Efficient Transfer Process

E A Ali[1], K M Kamarudin[1*], R Che Me[1], R Sulaiman[1] and H Alli[1]

[1]Department of Industrial Design, Faculty of Design & Architecture, Universiti Putra Malaysia, 43400 UPM Serdang, Selangor, Malaysia

*Email: manami@upm.edu.my

Abstract. Roselle (Hibiscus sabdariffa L.) belongs to the family of Malvaceae or Hibiscus the Malaysian national flower. In Malaysia, this plant is first introduced in the state of Terengganu as one of the alternatives to replace tobacco plants in sandy soils. Studies conducted on Roselle plantation in Malaysia identified too many repetitive steps in harvesting and deseeding process, as well as separating different grades of Roselle flowers and its logistics. The paper focuses on the improvement of harvesting and deseeding Roselle flowers, aiming to hybridize both processes into one effective process. The reason of developing such combination is to prevent frequent transferring of Roselle flowers to avoid damages, at the same time increasing the quality of harvest. This study also considers embedding few appropriate but simple technologies for systematic harvesting, making sure the Roselle flowers collected are in top quality and in best hygiene by user-friendly, ergonomic and safe to use processes.

1. Introduction

In Malaysia, Roselle is locally called “Asam Paya”, literally means sour for “Asam” and swamp for “Paya”. Other names are “Asam Susur” (sour-rail) or “Asam Kumbang” (sour-beetle), or famously called “Ribena Malaysia” [1] used mainly for cordial drinks, juices, jam, sour pickled, piccalilli and halwa. The important component in commercialization of Roselle is on it’s calyx, a layer surrounding the fruit’s seed (capsules). The flower calyces can be used as beverages, dried to be soaked in water for cold drinks, or boiled in water and taken as tea. It also has some medicinal properties [2, 3], especially its Malic acid which known to increase body’s energy, promoting younger skin and other benefits [4]. In Malaysia, the plant was introduced by farmers in Terengganu as a new agriculture industry in 1993 by the Agriculture Department of Terengganu. The Roselle plant is one of the alternatives to replace the tobacco plants in the area of BRIS (Beach Ridges Interspersed with Swales) soil and it also become a new commercial plantation experiment in Malaysia [5]. Currently, Roselle type UMKL-1(Terengganu) and UMKL-2(Arab) [6] are planted actively in the state of Kedah, Pahang, Perak and Johor. The objectives of carrying the research on Roselle harvesting and deseeding are to identify potential improvement of both processes. Second objective is to develop a new product that can efficiently help in both processes by applying available simple technology as well as embedding ergonomic elements to make the product easy and safe to use.

IOP Conf. Series: Materials Science and Engineering **697** (2019) 012029 doi:10.1088/1757-899X/697/1/012029

2. Problems

To ensure the quality of Roselle at its best, the Roselle harvester and deseeder have to handle Roselle calyces carefully, avoiding damages and maintains its freshness and cleanliness. This research focuses on Roselle plantation in the state of Johor, Malaysia. Although specific plantation is chosen, but in general, any Roselle plantation that practices the same harvesting and deseeding method may experience similar problems. There are two main problems associated with harvesting Roselle at the Johor Plantation: 1) frequent transferring of Roselle flowers, and 2) inefficiency of Roselle deseeding process. Current practices of harvesting Roselle at the Johor plantation is that the flower collectors uses reusable items such as used paint bucket, used gunny sacks and used large baskets to collect and transfer Roselle flowers frequently from place to place before delivering them to the deseeding centre (Figure 1). At the deseeding centre (Figure 2), workers again transfer the Roselle flowers on the table, then deseeding the flowers with a deseeding tool, which again damaging the flowers as well as creating hygienity concerns.

In order to maximize the quantity of good and high quality Roselle flowers, minimizing the transferring process of Roselle flowers starting from picking Roselle from its plant up to the Federal Agricultural Marketing Authority (FAMA) of Malaysia collection center should be established. Minimising transfer will avoid Roselle from getting damaged and will affect its quality. An effective harvesting method should be introduced to ease the process of travelling the harvest and managing workers.

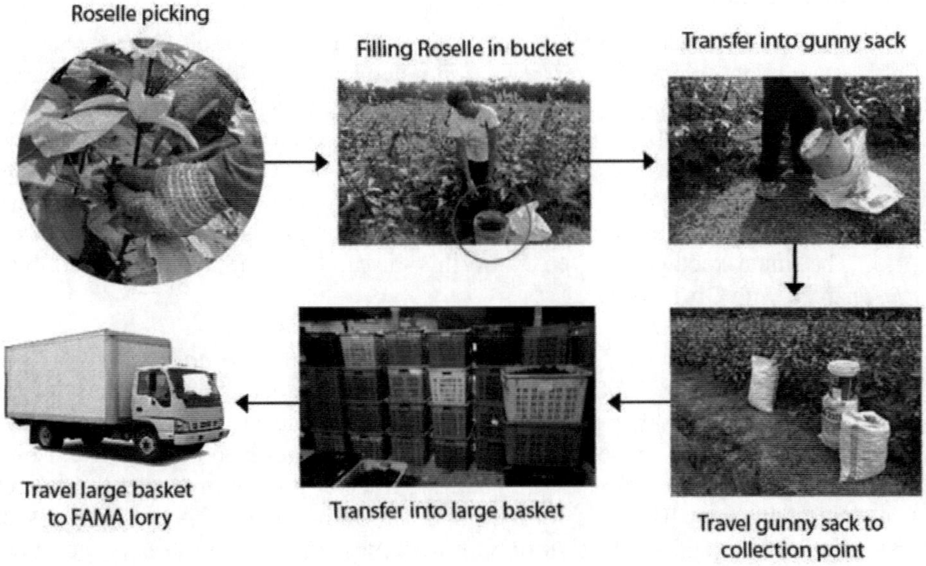

Figure 1. Harvesting process at Roselle plantation at Kg. Teluk Rimba, Sg. Mati, Muar, Johor, Malaysia.

Figure 2. Roselle deseeding process at Kg. Baharu Senggarang, Batu Pahat, Johor, Malaysia.

3. Materials and Method

There are primary and secondary groups of data collected for this research. Primary data are collected through field research, observations, interviews, and video analysis. Meanwhile, secondary data comes from literature reviews, internet articles and online sharing video reviews. The investigation was carried out between 11th Sept 2017 to 30th June 2018, at two venues both at Johor state, Malaysia: the Roselle plantation at Kg Teluk Rimba, Sungai Mati at Muar district and the collecting centre, situated in Kg. Baharu, Senggarang at Batu Pahat district. At the Roselle plantation, the areas involved in planting up to 2000 plants of Roselle is around 3 hectares. The plantation was originally a paddy field (Figure 3:a). The main method delivered for this research is the field research by participant observations, named with Ob.1 and Ob.2.

3.1 Ob.1: Roselle Plantation at Kg Teluk Rimba, Sungai Mati, Muar, Johor

Based on the observations, all workers are using reusable products as main tools for collection of harvest (Figure 3:b). The reusable products are used paint buckets, used gunny sacks, and large baskets. To start the harvesting process, workers use the paint bucket to collect the Roselle flowers at the Roselle plants. After the bucket is full, they need to transfer it to gunny sack at certain checkpoints. They will repeat the process back and forth from 8 am until 5 pm, with lunch breaks daily during harvesting season, a time and energy consuming activities. Filled gunny sacks then brought to the plantation center for another transferring process, into large baskets. Supervisors of the plantation calculate the weight of the large baskets for worker's wage payments purposes. All workers are paid based on the weight of the Roselle flowers collected per day. The flowers are transported using FAMA's refrigerated lorry, where a suitable temperature of 10°c maintained to keep Roselle flowers fresh. Time taken to travel from the Roselle plantation to the deseeding centre located in Kg. Baharu Senggarang takes up to two and a half hours.

Figure 3. a) The Roselle plantation besides paddy field. b) Harvesting Roselle flowers activity.

3.2 Ob.2: Deseeding center at Kg. Baharu Senggarang, Batu Pahat, Johor
At the deseeding center (Figure 4: a), all collected Roselle flowers are poured onto table for deseeding workers to starts separating Roselle's capsules from Roselle calyces. The existing deseeding tool (Figure 4: b) is made out of stainless steel rod with pointy sharp end to punch out the capsule from the flower. The workers will deseed Roselle until the rod is full, about 10 flowers before transferring them into a bucket. During deseeding, workers often found damaged flowers, capsule stuck at the tip of the tool and Roselle juices all over gloves occurred frequently (Figure 5). The stained gloves raises issues on hygienity and again, transferring deseeded Roselle into large basket repeating.

Based on observations done at both venues, there have been numerous transferring activities. Too frequent transfer can affect the quality of Roselle flower and also its hygienity. When the quality already affected, it can lower the grade of the Roselle harvest simultaneously lowers the grade and its pricing per kilogram.

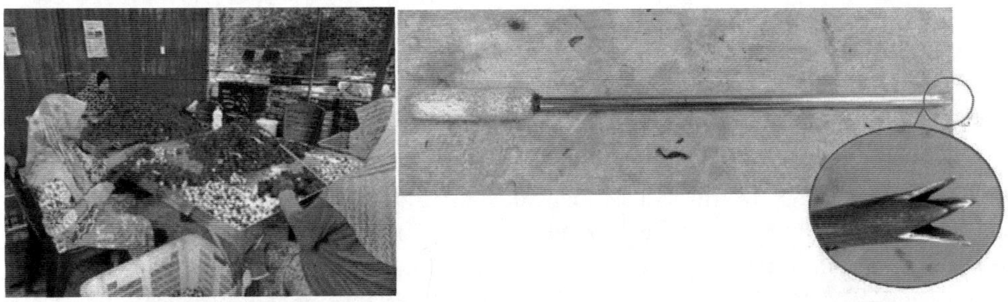

Figure 4. a) Workers deseeding Roselle flowers. b) Deseeding Tool.

Figure 5. Roselle damages due to frequent transfers and ineffective deseeding tool.

4. Solution Development

The development of this project involved several stages using ADDIE method (Figure 6). ADDIE is an adapted model based on Instructional Systems Design consisting of Analysis, Design, Development, Implementation and Evaluation [7]. ADDIE act as the main structure of this research starting from idea to the validation of the proposed solution. The "Analyze" part of ADDIE is done by participant observations and interviews at both venues, meanwhile the "Design" part of ADDIE is done at the Department of Industrial Design, Faculty of Design & Architecture, Universiti Putra Malaysia.

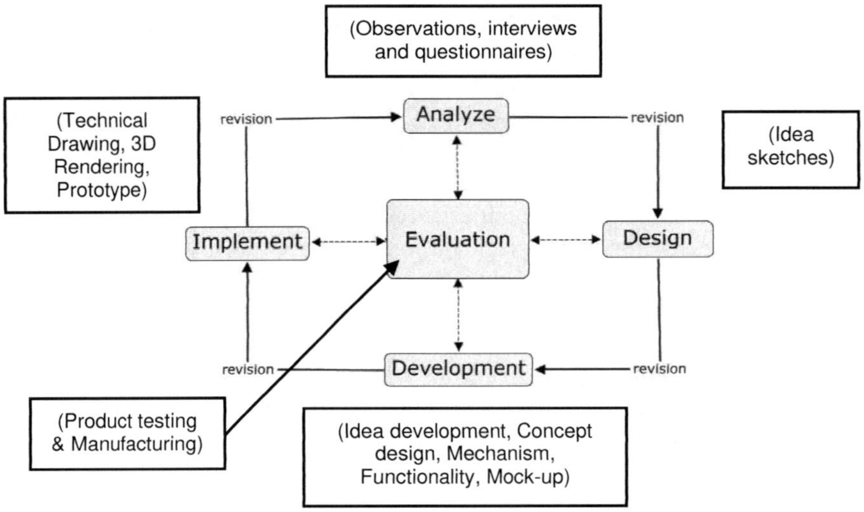

Figure 6. The ADDIE Model [8], with additions of research activities conducted on each process.

4.1 Product Design Specifications (PDS

At the "Development" stage of ADDIE, apart from abstraction and generalization of the problem process that considers safety in product concept design [9], the Product Design Specification (PDS) [10] model is used to capture several elements for design consideration that suits for harvesting and deseeding Roselle, particularly for Johor plantation. Elements shown in Figure 7.

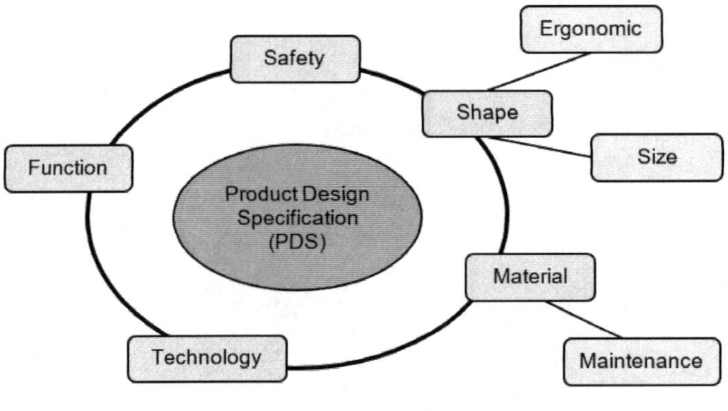

Figure 7. Product Design Specification (PDS)

The elements from the PDS model are elaborated as:

i. Safety: deseeding Roselle flowers must be safe, avoiding any sharp shapes in contact with hands.
ii. Shape: the shape of harvesting and deseeding tools should be simple and easy to operate.
   a. Ergonomic: the shape of both tools should be comfortable and promotes hygienity.
   b. Size: sizing of harvesting and deseeding product should fit comfortably with Malaysian anthropometric.
iii. Material: Should be durable for long term usage but low in cost, using waterproof material so that it is easy to wash.
   a. Maintenance: referring to material types.
iv. Technology: Efficient and systematic harvesting and deseeding even logistics requires suitable IoT technologies.
v. Function: The functions of the new product should be simple and easy to use.

In the development of the harvesting product design, few steps including ideation sketch, several 3-dimensional (3D) mock-ups and a product prototype are involved. Apart from the process of design development, establishments of IoT (Internet of Things) technology is also included in the prototype for the purpose of systematic management of harvesting and logistics of Roselle flowers between Roselle plantation and FAMA. After "Implementing" step, all the necessary information from "Analyze" part done at both venues, solution to systematic harvesting is developed, named "SORREL".

5. Development of SORREL

SORREL is a harvesting system aimed to ease the process of harvesting Roselle and reducing the transferring process drastically. In previous practice, transferring happened at least 6 times between buckets, gunny sacks and large baskets, not including travelling the harvest from place to place. SORREL suggests transferring happened only once and transferring happened again, at minimum recurrences at FAMA for grade sorting. The solution is equipped with QR code system, linking the code scanning with weighing machine and wage payments apps. FAMA will collect all filled bags and return empty bags for workers to continue harvesting until Roselle harvesting season ends.

5.1 SORREL Harvesting Bag

The harvesting bag (Figure 8), similarly look like a laundry bag, consists of handles both on the left and right side. The handles makes it easier to carry when its full, and the workers can easily change to another bag to continue harvesting the flowers. At first, workers will scan the QR code of empty bag, to secure the assignment of the bag to the worker. The workers will collect and deseed Roselle flowers at the same time, and put the deseeded Roselle into the bag until the bag is full. The bag can accommodate up to 5 kilograms and workers just have to bring the filled bag onto weight scan machine provided at the plantation's weighing centre. The bag has both upper and lower opening, only to be operated at FAMA centre, where transferring deseeded Roselle flowers are much easier and uses less work. To secure the bag from spilling, the upper part of the bag uses zip while the bottom part uses buckle system. At the centre, scanning of the bag's accurate weight linked with workers ID eases the process of facilitating wages to the workers. All Roselle filled bags will be collected by FAMA workers straight to FAMA sorting centre. The material used for the harvesting bag is from tarpaulin material, waterproof, easy to clean and high durability. The material also does not allow hot environment as to sustain the quality of Roselle flowers. The bag equipped with padding support on the shoulder, to reduce discomfort on the body when load is applied in the bag.

Figure 8. The SORREL harvesting bag, equipped with QR Code stickers.

5.2 SORREL Deseeding Tool

The deseeding tool (Figure 9) consists of three parts: the hollow part for the knife, the part for the thumb and the handle. The knife is a detachable components, makes it easier to replace, clean and store. The second part is the thumb section, which is important for the purpose of removing the capsule of Roselle flowers. The thumb section have a push point where it only requires little force when deseeding. The handle focuses on the ergonomic shape for user's comfortability when using it in long hours. It has a gear system inside which makes the product easier to handle by the worker. The design of this product is based on the validation responses at the Roselle plantation.

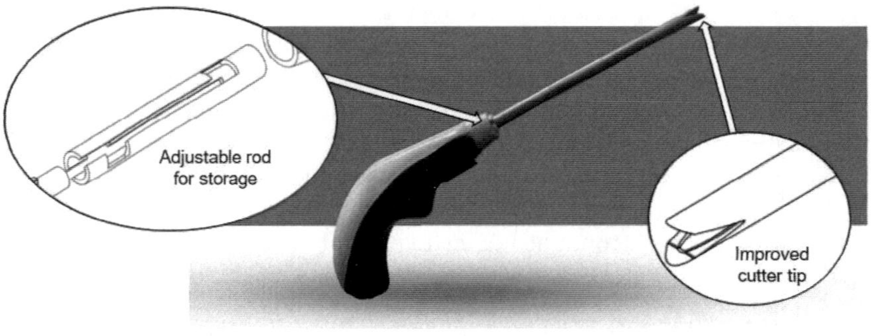

Figure 9. The SORREL harvesting+deseeding tool.

Although the tool looks simple, the design is more focused on the ergonomic feature; easier to hold and to clean.

6. Efficient Transfer

Systematic transfer process is important to ensure effortless process and maintains the freshness and hygienity of Roselle flowers. The tools used must be suitable for Roselle picking and deseeding activity, avoiding damage or defect on the calyces that will decrease its quality and lowers its grade. As shown in Figure 10, the workers just need to transfer Roselle only once: from harvesting the flowers into the SORREL harvesting bag. There are no more Roselle transferring after that from the workers. Deseeded Roselles are kept inside the harvesting bag and travels to weighing machine, into pallets and into FAMA lorry straight to FAMA centre. Before, FAMA lorry have to travel to Roselle plantation and deseeding center and return back to FAMA. Now with SORREL system, FAMA lorries only have to travel between its center and Roselle plantation. This resulted to time and petrol savings.

IOP Conf. Series: Materials Science and Engineering **697** (2019) 012029 doi:10.1088/1757-899X/697/1/012029

The travel flow of the harvesting bag of Roselle are shown in Figure 11. At FAMA center, transferring will happened at minimum frequency, involving sorting Roselle's grade, cleaning and logistics to Roselle food factories. By this improves system, the quality and freshness of Roselle are kept in maximum.

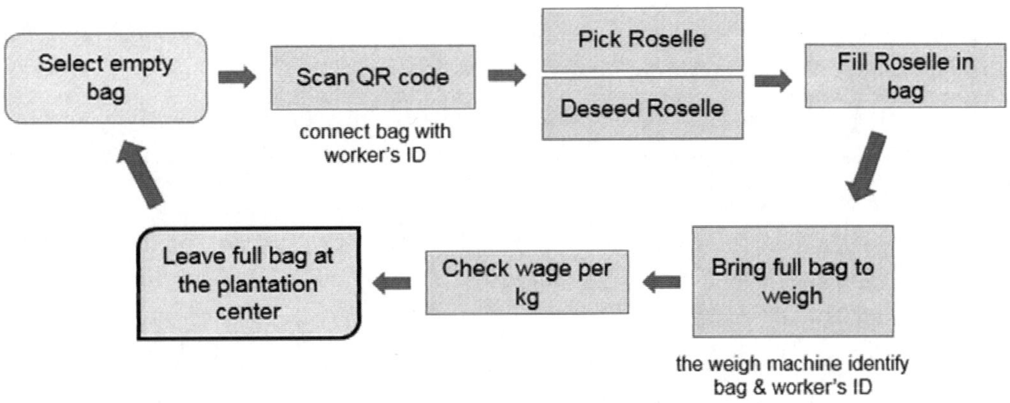

Figure 10. Workers workflow when using SORREL harvesting process.

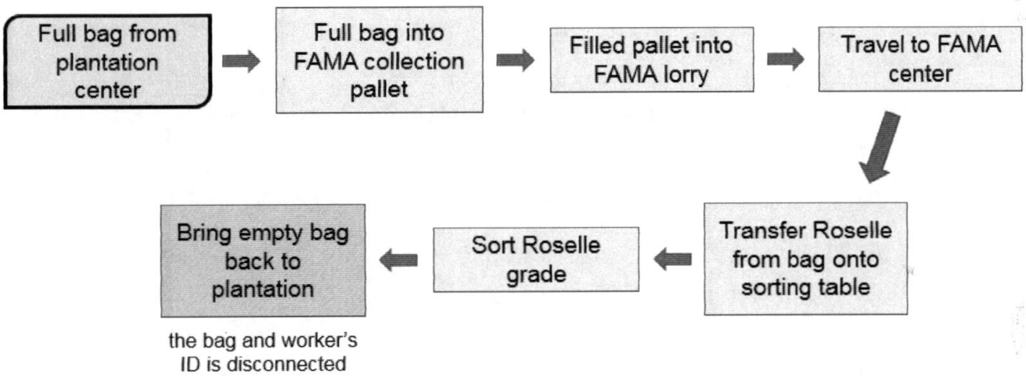

Figure 11. The SORREL harvesting bag movement flow.

7. Conclusion

Any plant harvesting should have a systematic process flow so that the collection of the harvest is a smooth process, furthermore maintains the harvest freshness and quality. After research done on both venues in Johor, an effective harvesting, deseeding, wage payments and logistics of Roselle flowers are developed with an improved simple system, fulfilling the needs of the workers to complete the work at the minimum time and place. Therefore, it can reduce the cost for travelling and still maintaining the same amount of workers, provided they all can work at the plantation. The researchers are currently working on the improvement of the conceptual design, as the study presented in the paper is at preliminary stage at the time. In the future, validation of SORREL and proof of concept will be carried out to strengthen the proposed system. SORREL system consisting of harvesting bag, deseeding tool and SORREL apps is copyrighted under UPM.

References

[1] Mohd-Esa N, Hern F S, Ismail A 2010 Yee C L. Antioxidant activity in different parts of roselle (Hibiscus sabdariffa L.) extracts and potential exploitation of the seeds *Food Chemistry,* **122(4)**, 1055-1060.

[2] Nkumah O C 2015 Phytochemical analysis and medicinal uses of Hibiscus sabdariffa *International Journal of Herbal Medicine*, **2(6)** 16-19.

[3] Wahyuningsih S, Sukandar E Y, Sukrasno 2016 Antihyperuricemia activity of the ethanol extract of Roselle calyx and its fraction (Hibiscus sabdariffa Linn) on male wistar rats *Int. J. Pharm. Pharm, Sci.* **8** 278–280.

[4] Al-Wandawi H 2015 Organic acids composition of different parts of the medicinal plant–roselle (Hibiscus sabdariffa), *International Journal of Biological and Pharmaceutical Research* **6(10)** 808-813.

[5] Naimah R, Nur Amirah Y, Adzemi M A, Wan Zaliha W S 2014 Quality and growth development of Roselle grown on BRIS soil in relation to regulated deficit irrigation, *Journal of Tropical Plant Physiology* **6** 23-34.

[6] Susanti D, Osman M, Draman N N 2012 Natural antioxidant contents in the seeds of four Roselle Malaysian varieties and a Kenaf: a comparison *Journal of Herbs, Spices & Medicinal Plants* **18(3)** 240-245.

[7] Branson R K 1978 The Interservice Procedures for Instructional Systems Development, *Educational Technology* **18(3)** 11-14.

[8] Information on https://commons.wikimedia.org/wiki/File:ADDIE_Model_of_Design.jpg

[9] Kamarudin K M, Ridgway K, Ismail N 2016 Abstraction and generalization in conceptual design process: Involving safety principles in TRIZ-SDA environment, *Procedia CIRP* **39** 16-21.

[10] Pugh S 1991 *Total Design: Integrated Methods for Successful Product Engineering*, (Wokingham: Addison-Wesley)

Acknowledgement

Special thanks goes to Dr. Latinah Mohamad from Pusat Kesihatan Universiti Putra Malaysia, Mr. Ridzuan Nizar of Aslah Supply & Services company, Mr. Zainudin of the Roselle plantation, Mr. Luqman of the Roselle deseeding center, and all workers at both Roselle plantation and deseeding center involved in giving valuable information and experiences for this research.

UniSZA as the National Centre of Design for Disability in Malaysia

Marzuki Ibrahim[1,a], Muhammad Sophist Ahmad[2,b], Imran Abdullah[3,c], Mohd Hisham Omar[4,d], Ahmad Fairuz Ariff[5,e], Sheikh Ahmad Tajuddin Sh. Yusoff[6,f],

[1, 2, 3, 4, 5, 6]School of Industrial Design, Faculty of Innovative Design and Technology, Universiti Sultan Zainal Abidin (UniSZA), Gong Badak Campus, Terengganu, Malaysia

Abstract. This paper explains the initial steps of establishing the national centre of design for disabled people in Malaysia. As we all know, there is still no evidence of the existence of a creative centre for industrial product design in the country, especially for this special group. Through the initiative from the School of Industrial Design located in Universiti Sultan Zainal Abidin (UniSZA), we take a commitment to channel societal significances through designing, developing and producing innovative product design for various categories of disabilities in this country, such as physically disability, mentally disability, vision impair, learning disability, speech disability and others. It will be done entirely in a centre known as National Centre of Design for Disability (CDD), which highlights the standards and quality of design, where the outcomes meet the satisfaction of the end users through its function, ergonomics, safety, comfort and aesthetics for them to not ashamed to join their routine activities in the open public environment. The CDD improves to enhance the existing equipment alongside providing advice and assistance for the state-of-the-art research outcomes.

Keywords: National Centre of Design for Disability; *Tetraphochomelia* Child; Special Products and Equipment; School of Industrial Design

1. Introduction

The Activities involving social services to disabled people in our country have been around since the 20th century. The establishment of Social Welfare Department in 1946 was an initial step towards helping disabled people for the business continued until 1960s. Since 1950s, several NGOs have been established like Malaysia Association of Blind in 1951, Spastic Centre in 1960, National Society for the Deaf in 1961, Cheshire Home in 1964, and Bethany Home for Epileptic Children in 1966 [1].

The addressing of the Malaysian government regarding well-being for disabled people was reviewed in some strategies and programs to ensure that Malaysian disabled people enjoy their lives more comfortably. For instance, the raising of caring Malaysian societies has outlined in Malaysia's 2020 vision [5] as a stepping stone for disabled people to get along with the local community without shame or embarrassment [6]. In addition, the Malaysian government has set up a panel of advisors for this special group on their needs, reviewing equipment in their programs through joint venture between the government and private sectors, identifying any specific related issues, and suggesting long last plans for activities in their daily lives [7].

2. The Existing Scenery

A random survey conducted in 1958 has shown that disabled people in Malaysia represent 1% of the total population. Also, the Infosoc Malaysia 2002 e-forum reports stated that disabled people in Malaysia represent 0.42% of the total population [2]. However, the real numbers of people with disabilities in Malaysia is quite difficult to obtain due to statistical constraints [8]. This is due to the customary problems in the country, where parents are embarrassed to register their disabled children [9].

IOP Conf. Series: Materials Science and Engineering **697** (2019) 012030 doi:10.1088/1757-899X/697/1/012030

In 1995, the Social Welfare Department specified that there were 56,000 of disabled people in Malaysia have successfully registered thus after a year, the department took the initiative to collaborate with the local telecommunications companies to encourage more disabled people in their registrations [2]. As a result, there were 445,006 of disabled people in Malaysia have had registered in 2012 [3].

The Malaysian government has taken initiatives to protect the rights of people with disabilities through ratifying the Convention on the Rights of Persons with Disabilities (CRPD) in 2010. The Government has also adopted the Persons with Disabilities Act 2008 as well as has provided social protection services in the fields of health, rehabilitation, and education for children with disabilities guided by National Policy for Persons with Disabilities and also National Plan of Action for Persons with Disabilities [3].

Based on Table 1 below, since 2014 the Social Welfare Department stated that there were 531,962 of registered disabled people in Malaysia have been divided into Malay (324,668), Chinese (106,523), Indian (55,453), Peninsular Natives (2,289), Sabah Natives (25,872), Sarawak Natives (12,153), and others (5,004). On the other hand, Table 2 showed a breakdown of statistics for learning disabilities (188,911), physically handicapped (174,795), hearing disabilities (62,153), vision impair (50,827), mental (24,263), speech (3,988), and others (27,025). By breaking down the percentage of statistics, the data can be referred to in Figure 1. Also, the statistical breakdown of people with disabilities in Malaysia through the breakdown of each state can be found in Table 3.

Table 1 : Disabled People Registration by Ethnic Groups in 2014 (Source: [4])

Kumpulan Etnik / *Ethnic Group*	TAHUN 2014
Melayu / *Malay*	324,668
Cina / *Chinese*	106,523
India / *Indian*	55,453
PSm / *PSm (Peninsular Natives)*	2,289
PSb / *PSb (Sabah Natives)*	25,872
PSw / *PSw (Sarawak Natives)*	12,153
Lain-lain / *Others*	5,004
Jumlah/ *Total*	**531,962**

Table 2 : Disabled People Registration by Disability Categories in 2014 (Source: [4])

Kategori Ketidakupayaan / *Diasbilities Category*	TAHUN 2014
Penglihatan / *Vision Impair*	50,827
Pendengaran / *Hearing*	62,153
Fizikal / *Physical*	174,795
Masalah Pembelajaran / *Learning Disabilities*	188,911
Pertuturan / *Speech*	3,988
Mental / *Mental*	24,263
Pelbagai / *Others*	27,025
Jumlah/ *Total*	**531,962**

Table 3 : Disabled People Registration by Malaysians' States in 2014 (Source: [4])

Negeri / State	Tahun / Year 2014
Johor	69,824
Kedah	41,262
Kelantan	44,169
Melaka	24,547
Negeri Sembilan	28,036
Pahang	33,317
Perak	49,537
Perlis	7,719
Pulau Pinang	33,618
Sabah	33,494
Sarawak	25,202
Selangor	77,051
Terengganu	30,411
W.P. Kuala Lumpur	32,654
	1,121
Jumlah/ *Total*	**531,962**

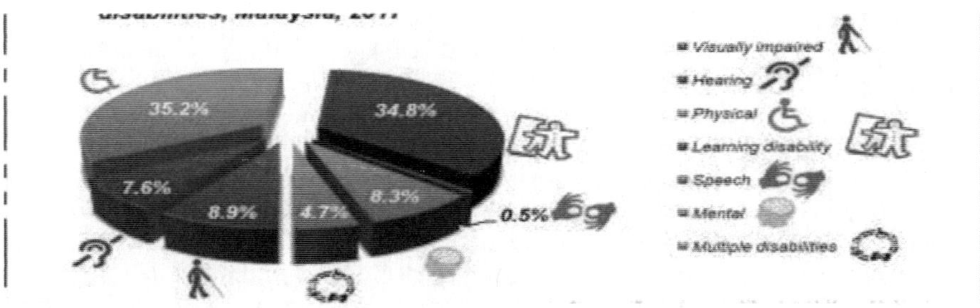

Figure 1 : Disabled People Percentage by Disability Categories (Source: [2])

3. Understanding Disability

In 2016, UNICEF Malaysia has conducted a study on disabilities that engaged 756 participants from Selangor, Kelantan, Sabah, and Sarawak covered a few approaches like surveys, discussions, workshops, and interviews. The findings of the study were used to advise the stakeholders to develop programs and systems that were more inclusive to this special group. Six out of 10 participants (58.4%) were under-informed or ill-informed about disability; 87% said they wanted to know more. Only 20% of the participants saw behavioural and mental conditions as disabilities. Most regarded children with learning disabilities, hyperactivity, or aggression as being badly behaved. Those with mental disabilities were labelled as crazy or stupid and were often made fun of. The finding was also saying that to light the lack of knowledge and misconceptions about disability and has resulted in stigma against people with disabilities and limited their access to their rights to be a part of society. Therefore, the stakeholders must not just taking a note of the data, but to create an environment that allows and encourages children with disabilities to be part of active society. Based on Figure 2 show that people were more accepting and understanding of people with physical disabilities, the conditions which were tangible to them such as loss of vision, inability to walk, total loss of hearing, or loss of limbs. About 80% of the participants recognised these physical limitations as disabilities [3].

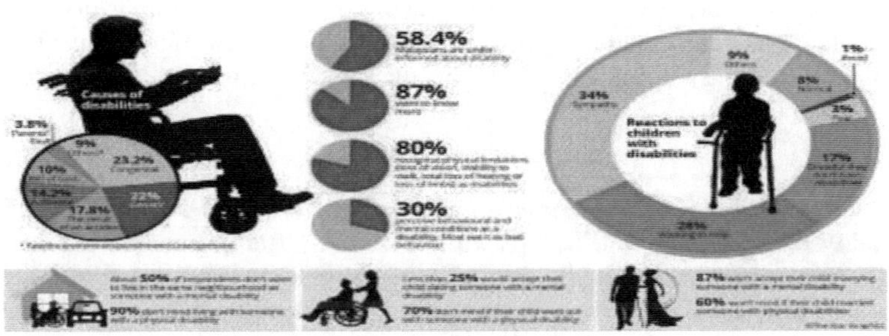

Figure 2 : How Malaysians Understand Disabilities (Source: [3])

4. The Strategies to Establish and Implement the National Centre of Design for Disability (CDD)

There are ten plans to establish and implement the national centre of design for disability under the management and administration of School of Industrial Design, UniSZA – (i) philosophy behind the establishment, (ii) design for disability, (iii) a pilot of special project, (iv) special interest group, (v) design vs. non-design experts; (vi) the rationales of formation; (vii) the values of universal design; (viii) the objectives of establishment; (ix) the facilities for implementation; and (x) the funding.

IOP Conf. Series: Materials Science and Engineering **697** (2019) 012030 doi:10.1088/1757-899X/697/1/012030

4.1 Philosophy behind the Establishment

The national centre of design for disability (CDD) @ UniSZA takes an initiative to contribute its expert for disabled people in Malaysia comprises various categories of disabilities, such as physically disabilities, mentally disabilities, vision impair, learning disabilities, speech disabilities, and multiple disabilities. It highlights the standards and quality of design, where the outcomes meet the satisfaction of the end users through its function, ergonomics, safety, comfort and aesthetics. So that end users are not ashamed to join their routine activities in the open public environment. As well, it also improves to enhance the existing equipment alongside providing advice and assistance for the state-of-the-art research outcomes. The CDD @ UniSZA strongly believes that disabled people are a part of the community and they should not be isolated or alienated. It is a tough duty to ensure a better life for them in the future. The CDD @ UniSZA realizes to design, develop, produce, and commercialise special products and equipment for disabled people that should be managed in systematic way.

4.2 Design for Disability: Reality or Illusion?

The CDD @ UniSZA has conducted several research projects on disabilities, particularly from category of physically disabilities. There are also scholars and students completed their projects and theses on these particular subjects. However currently, there is nothing to be commercialised due to the absence of a platform like a centre of excellence. The CDD @ UniSZA also thinks about the survival of research findings contained in theses that are quite disappointing due to the many valuable contributions that are only stored in the bookshelves.

4.3 A Pilot of Special Project with Tetraphochomelia Child

In realizing the goal, the CDD @ UniSZA come up with a strategic plan by having a special design project with a *Tetraphochomelia* child named Nordamia binti Daeng, a 12 years old girl. The *Tetraphochomelia* is a very rare case found in the world, which are only 20 cases registered patients globally. As the CDD @ UniSZA was directly involved in this new product development, the focus was in designing and manufactures a wheelchair that was specially designed for her, including producing a practical design that would cater to the needs of the *Tetraphochomelia* user, this project was carried out for nearly two years, beginning in 2016 and ending in 2017.

4.4 Special Interest Group (SIG)

In ensuring the smooth formation of a platform as a centre of excellence, the setting up a special interest group, or known as SIG is considered necessary. The SIG will gather as many experts from various related fields (outside the university) to work together professionally covering aspects of design, development, research, academic, intellectual property, licensing and transfer of knowledge, manufacturing, commercialization, marketing, logistics, and management.

4.5 Design vs. Non-Design Experts

Other than that, the CDD @ UniSZA will strengthen all teaching force from three sub-fields: Product Design, Transport Design and Furniture Design. The experts with PhD and Master degrees (local and foreign universities) with sufficient professional experience in academic and industry sectors – locally, nationally and internationally. The experts are also categorised as practicing industrial designer with actively engaged in research, consultancy, publications, and development of many industrial products and equipment for local and international markets, including producing new design management strategies to involve in the other related works of public interest. On the other hand, the CDD @ UniSZA is also welcoming collaboration of experts from non-design fields: Engineering, Social Sciences, Business and Management, Health Sciences, and Medicine.

4.6 The Rationales of Formation

There are five rationales of formation for CDD @ UniSZA, namely (i) to assess and focus on the needs of disabled people from design research process to commercialisation stages, (ii) to meet market demand for new or enhanced of special products and equipment, (iii) to become a hub for disabilities data and reference in Malaysia and Asia Pacific region; (iv) to act as an intermediate to play the important role between the governmental agencies, NGOs, industries, charitable organizations, institutions of higher learning and other bodies as well as the public who are willing to work together in helping the disabled, and (v) to provide advice and consultation in locating or enhancing any special products and equipment in Malaysia and ASEAN region.

4.7 The Values of Universal Design
The CDD @ UniSZA will complete all seven values of universal design as follows: (i) to equitable the use of outcomes that must be useful and marketable to any group of disabled people to avoid segregation or stigmatisation condition, (ii) to flexible the use of outcomes that must accommodate a wide range of individual preferences and abilities, (iii) to simple and intuitive the use of outcomes that must be easy and understandable to use regardless of the user's experience, knowledge, skills, or concentration levels, (iv) to perceivable the information of outcomes that must communicate necessary information effectively to the user regardless of ambient conditions or user's sensory abilities; (v) to tolerance the error of outcomes that must minimise the hazards and adverse consequences of accidental or unintended actions, (vi) to low physical the effort of outcomes that must be usable efficiently and comfortably with minimum fatigue or vibration, and
(vii) to size and space the approach and use of outcomes that must appropriate in size and space for user's approach, reach, manipulation, and use that regardless of the user's body size, posture, or mobility.

4.8 The Objectives of Establishment
In order to establish and implement CDD @ UniSZA, there are four objectives that need to be taken into account, which are (i) to design and manufacture special products and equipment for people with disabilities based on the values of universal design; (ii) to conduct and encourage the enhancement of design and manufacture special products and equipment for the current and the future generation of disabled people; (iii) to improve the standards and quality of special products and equipment by providing satisfaction to all targeted disabled users; and (iv) to increase public awareness of the role of design to meet the needs of people with disabilities.

4.9 The Facilities for Implementation
The facilities available in the CDD @ UniSZA are ranked amongst the best in Malaysia. Those include industrial design studios, model making workshops, and computer labs with 3D printing machines. The accessible software of design is such as CAID, AutoCAD, Rhinoceros, and Solid Works that used for design conceptual and surfacing. Meanwhile, a detailing designing is up to product analysis software such as NX and Master Cam which enable the lab to be a one stop design and development centre. Furthermore, the other extensively software from Desktop Publishing are such as Adobe Illustrator, Adobe Photoshop, Adobe Dreamweaver, and Macromedia Director would be very helpful in creating the front end product design presentation, either in output of hardcopy presentation panel or for the animation needs. The industrial design workshop is completed with all wood-working tools, whereas for the metal and plastic sections; the workspace is fully-equipped with a range of hand tools and machineries for forming, fabricating and assembling, from lightly to heavily jobs.

4.10 The Funding
One of the most important things in the establishment and implementation of CDD @ UniSZA is the need for funds. The contributions are indeed through the governmental channels, private sectors, NGOs, and also from public funds. The funds will be fully managed by UniSZA Malaysia.

5. Conclusion
It is the hope of everyone in School of Industrial Design, Universiti Sultan Zainal Abidin Malaysia that to enable this national centre of design for disability to be established in the near future. With the implementation of the CDD @ UniSZA will progress a step further in being a leader in a field that has not been tapped by anyone locally or globally. The setting up of CDD @ UniSZA if surveyed in terms of marketing has a wide and unique potential. The special products and equipment that are designed and developed will not only be marketed in the country, but will also encourage demand from the overseas market, which has to be catered for. Therefore, the establishment and implementation of CDD @ UniSZA will fulfil the needs of 159,547 of physically disabled people identified in 2017 in Malaysia and will also be in demand globally – through a wide range of the outcomes that includes (i) daily and routine needs like beds, chairs, bath, and toilet equipment; (ii) mobile equipment like wheelchairs and scooters; (iii) posture and carriage items like trolleys and stroke support systems; (iv) paediatric equipment like postural aids; and (v) small items like specially-designed toothbrushes, paintbrushes, stationaries, and cutleries.

IOP Conf. Series: Materials Science and Engineering **697** (2019) 012030 doi:10.1088/1757-899X/697/1/012030

References

[1] D. Jayasooria, and G. Ooi, Disabled peoples movement in Malaysia, Disability & Society Journal. 9 (1994) 17
[2] Information on http://www.jkm.gov.my/jkm/uploads/files/pdf/Buku%20JKM%202014.pdf
[3] Information on https://www.unicef.org/malaysia/sowc_sowc_13-situation_in_malaysia.html
[4] Information on https://www.dosm.gov.my/v1/index.php?r=column/cthemeByCat&cat=152&bul_id=NU5hZTRkOSs0RVZwRytTRE5zSitLUT09&menu_id=U3VPMldoYUxzVzFaYm N kWXZteGduZz09
[5] O.A. Rahman, Industrial Targets of Vision 2020: The Science and Technology Perspective, Pelanduk Publications, Kuala Lumpur, 1993
[6] A. Sarji, Malaysia's Vision 2020: Understanding the Concept, Implications and Challenges. Pelanduk Publications, Kuala Lumpur, 1993
[7] M. Yeoh, Management Strategies for Vision 2020: Developing Competitive Strategies for the Winning Edge, Pelanduk Publications, Kuala Lumpur, 1993
[8] M. Katie, and B. Juliet, Childhood Disability in Malaysia: A Study of Knowledge, Attitudes and Practices, UNICEF Malaysia, Kuala Lumpur, 2017
[9] Information on https://www.nst.com.my/news/2015/09/disabled-friendly-malaysia-more-needs-be-done-address-pwd%E2%80%99s-welfare

IOP Conf. Series: Materials Science and Engineering **697** (2019) 012031 doi:10.1088/1757-899X/697/1/012031

A review of accidents & incidents on Boeing and Airbus commercial aircraft's avionics-related system in two decades (1996-2015)

I J Baidzawi[1*], N Mohd Nur[1], S Ahmad Shukri[1], K D Mohd Aris[1], N L Asmadi[1] and M Mohamed[1]

[1]Universiti Kuala Lumpur, Malaysian Institute of Aviation Technology, Lot 2891, Jalan Jenderam Hulu, 43800 Dengkil, Selangor, Malaysia

*Email:inarni@unikl.edu.my

Abstract. Aviation accidents still hit the news even though the growth of technological advancement on commercial aircraft avionic systems has been impressive. Hence, one of the objectives of this study is to plot the time-based graph of commercial aviation accidents, with direct consequence from avionics instrumentations in the period of two decades, from 1996 to 2015. The second objective is to analyse two main aircraft manufacturers, Boeing and Airbus, in determining specific model of its kind that significantly involved in avionics-related instrumentation as one of the contributing factors that leads to the incidents or accidents. The third objective is to identify which avionics system that most frequently involved in aviation incidents, for both manufacturers. The final objective is to examine the main probable cause that has the highest percentage in those accidents within the said time frame. The method of collecting data is by doing comparative analysis from reliable official websites of four well-known bodies such as National Transportation Safety Board (NTSB), Federal Aviation Administration (FAA), Aviation Safety Network and Flight Safety Foundation. Results show that misfortune occurrences are directly associated with avionics within the said two decades, which mainly involved Autopilot and Flight Management System (FMS) (14% each), meanwhile the aircraft model Boeing 737 carries the highest percentage of avionics-related incidents or accidents. Nonetheless, 67% of the misfortune occurrences within the scope of study are mainly due to human error instead of technology.

Acronyms

ADIRU	Air Data Inertia Reference Unit
AOA	Angle of Attack
ECAM	Electronic Centralised Aircraft Monitor
FCS	Flight Control System
FMS	Flight Management System
FQIS	Fuel Quantity Indicating System
GPWS	Ground Positioning Warning System
INS	Inertial Navigation System
MSAW	Minimum Safe Altitude Warning (MSAW)
NDB	Non Directional Beacon
TOWS	Take Off Warning System

1. Introduction

It is known that air transportation is the safest mode compared to other types of transportation as data have shown that only one passenger fatality per 7.1 million air travellers for scheduled passenger airline service [1]. In fact, the International Air Transport Association (IATA) stated in 2010 that in the global scenario, airline accident rate was one accident for every 1.6 million flights [2]. Meanwhile for year 2018, the rate is slightly higher where the commercial airline industry was reported as having 1.35 accidents per 1 million flight or equivalent to one accident for every 740,000 flights [3]. The fact that air transportation is considered safe, does not change the reality that aircraft accidents and incidents still occur annually even though robust technological advancement is already implemented in the aircraft system. Hence, this study was intended to plot a time-based graph of commercial aviation accidents, which had avionics-related cases within the two decades from 1996 to 2015, for Boeing and Airbus airplanes. The second objective of this study is to determine the significant aircraft model from Boeing and Airbus, which frequently involved in these avionics-related accidents. In addition, it was also an interest to determine the specific avionics system mostly involved in those accidents. Lastly, is to identify the main causes of those avionics-related aviation accidents, within the mentioned time span.

2. Literature Review

The safeness of the air transportation comes from many sources, for example technological improvements where better and reliable aircraft systems are developed with advance avionics and steadfast engines. In addition, the enhanced pilot training, improved weather forecasting, enrichment of air traffic management as well as judicious air accidents investigation have also contributed to the establishment of stringent aviation regulations, giving big impact to this prudent environment [1].

Avionics is one of the key elements that significantly has impacted the overall aircraft systems, which apparently plays critical function in making sure smooth, safe and reliable operation of an aircraft. For large commercial aircraft, avionics involves communication, navigation, cockpit display as well as management of the overall aircraft systems. Manufacturers have designed the avionics system to be reliable enough so that it can detect any inconsistencies of input and output data from the original data setting and information, in making sure a practical aircraft operation as well as part of back up system in case of emergency. However, there are also other issues involved technically or human related with this intricate system.

As an example, Gawron [4] had produced a report listing down all the related accidents involving flight deck automation issues with regards to flight crew, from year 1972 to 2013. He found that the recurring issues correlated with automation were human poor attentiveness, skill deficiency, trust in automation and self-complacent. The findings suggested that in any new automation design, the cultural differences between nations also need to be taken into account when it involves training, certification and aircraft operations as these will directly affect the performance and how automation is used. In an investigation that was carried out after any aviation mishap, a report was released with possible causal factors together with several recommendations to prevent similar event from recurring. Hence, the national authorities and international aviation bodies have proposed as well as imposed stringent regulations when it comes to design issues to the aircraft manufacturers in order to prevent and minimize as well as learned some lessons from such occurrences.

In fact, a researcher had done a lot of study on the aviation accidents, came up with theories on what could have caused such events [5]. The well-known James Reason [5] stated that any high profile accidents in this world, not limited to aviation alone (eg. aerospace, nuclear, marine, rail) happened due to failed defences in every chain of interlink events that contributed to such disasters, rooted from the organisational environment where safety culture was not really inculcated in every level of company structure. Cooper [6] and Civil Air Navigation Services Organisation (CANSO) [7] described that safety culture involves three aspects: psychological, behavioural and situational. Nevertheless, in general view it was found that 70% to 80% of aviation accidents were attributed by human error [8].

IOP Conf. Series: Materials Science and Engineering **697** (2019) 012031 doi:10.1088/1757-899X/697/1/012031

3. Methodology

This study looked into air incidents and accidents of two aircraft manufacturers that have the highest market share for commercial aircraft, Airbus and Boeing [9] within two decades of time, from the year 1996 to 2015, specifically on their avionics systems. The method used in this study was by comparative study, through analysing the published secondary data from well-known and reliable official websites of aviation bodies such as National Transportation Safety Board (NTSB), Federal Aviation Administration (FAA), Aviation Safety Network and Flight Safety Foundation. Based on the published official investigation reports and data taken from these four websites, each case was scrutinized and compared for accuracy and consistency. Those cases were then compiled by arranging them according to date, year, location, aircraft operator, aircraft manufacturer & model (Airbus or Boeing), aircraft registration, its main cause of accident/ incident and in particular its specific avionics system involved. From this list of commercial aviation accident cases in twenty years' period, emphasis was given on the avionics-related incidents and accidents. This allows further examination on the overall picture of avionics system feasibility experienced in those two decades.

It is important to emphasize here that in any air disaster investigation one single cause does not lead towards the unfortunate incident. In reality, several interlink contributing factors do have discernible impact to a particular air disaster [8]. However, this study utilized the official reports in the mentioned official websites to identify the main probable cause. The accuracy of the data was mainly based on the information provided in those four websites. Therefore, any air disaster data which were not published in those websites, are not included in the discussion of this paper.

4. Results and Discussion

The data of aviation accidents from the official websites of NTSB, FAA, Aviation Safety Network and Flight Safety Foundation were compared to, cross referenced and compiled based on specific date, location, aircraft operator, aircraft manufacturer & model (Airbus or Boeing), aircraft registration, its main cause and whether it is avionics system-related. Based on the data obtained, Figure 1 presents the trend of the accidents/incidents of Airbus and Boeing aircraft in the period of 1996 to 2015.

Figure 1 shows the fluctuations on the rising of commercial aircraft accidents every year. Nevertheless, the cases that involved avionics instrument slightly decreased. The fact that fast development in aviation industry such as in technological innovation [10] has encouraged frequent use of aircraft as one of the main transportations.

Based on the data collected within the period of 1996 to 2015, there were 405 cases of commercial air accidents involving both aircraft manufacturers, based on those four official websites alone. 42 cases of them involving direct causal or related with avionics system.

Based on the same data, there were 314 total air accident cases for Boeing, which 25 (8%) out of them involved avionics-related cause. Meanwhile for Airbus models, a total of 91 cases were involved in general air disasters, and 17 (18.6%) of them involving avionics instruments as shown in figure 2.

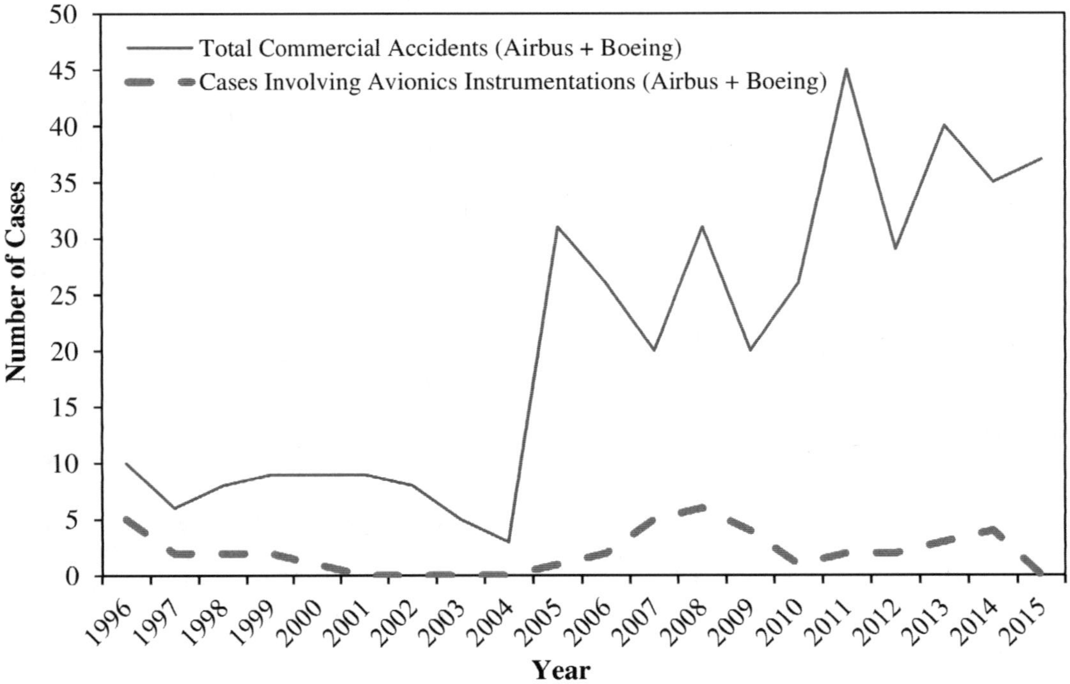

Figure 1. Trend of aircraft accidents & incidents (Airbus and Boeing) year 1996-2015

Figure 2 shows the details on accidents and incidents involving avionics instruments between these two major manufacturers for each year within the two decades, with 42 cases. Out of 42 cases related with avionics, 25 cases involved Boeing aircraft which reflects to 59.52% of the total cases. Airbus types with 17 cases (40.48%), which shows that avionics instrumentation in Boeing aircraft has significant contribution towards aircraft accidents compared to Airbus in those years. The highest cases of Boeing aircraft were in 1996 where those cases involved with autopilot, barometric altimeter, Flight Management System (FMS), Fuel Quantity Indicating System (FQIS) and static port. However, these cases do not represent the flaw of the avionics technology as an isolated system, but other contributory factors were tangled, as example the maintenance error, pilot error, design or environmental factors.

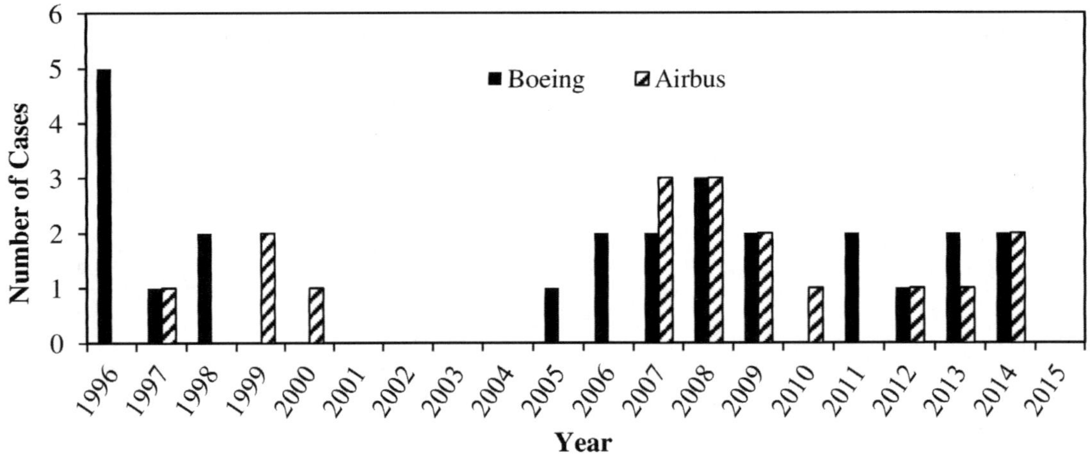

Figure 2. Comparison of Boeing & Airbus accidents/ incidents involving avionics instruments (1996 – 2015)

IOP Conf. Series: Materials Science and Engineering **697** (2019) 012031 doi:10.1088/1757-899X/697/1/012031

4.1 Specific Avionics Instruments

To look into details of the instruments involved in commercial aviation accidents and incidents in the year 1996 to 2015, the following figure 3 reflects the specific avionic systems involved.

Figure 3 shows that Flight Management System (FMS) and Autopilot are the most frequent instruments contributed to the air accident/ incident which involves both manufacturers, Airbus and Boeing. Each manufacturer had three cases for each type instrument system, covering 14% from the overall instruments as shown, within the said period of time under study. For the record, most of those cases happened during landing. The subsequent avionic systems which contributed the highest to the misfortune events were anti-skid (10%) and altimeter (7%) systems.

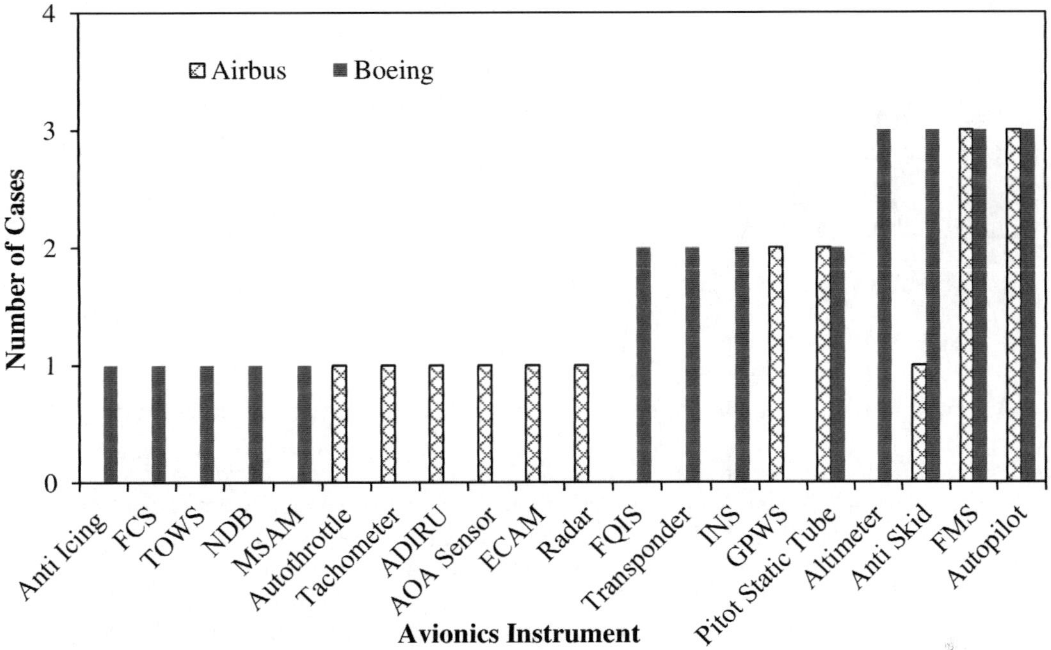

Figure 3. Avionics instrument related to aircraft accidents/incidents of Boeing & Airbus (1996 -2015).

4.2 Specific Aircraft Model

Figure 4 presents which aircraft model frequently involved in the avionics-related aviation accident. In general, it can be stated that within a total of 42 cases involving avionics-related incidents of Boeing and Airbus models in the said period (1996 - 2015), the prominent aircraft model was B737 which had 15 cases (35.7%) followed by A320 model which involved in 9 cases (21.4%). Hence, the B737 model involved in avionics-related accident was higher by 14.3% compared to A320. In fact, these two models are popularly utilized for single aisle aircraft design throughout the world especially for low cost carriers.

Nevertheless, Business Insider stated that, "From 1984 to the present (January, 2019) the A320 has outsold the B737 by 439 planes" [11]. This means that A320 could have higher flight operations in the aviation world for single aisle design, yet this model has less impact towards avionics-related issues in aviation accidents compared to B737 model.

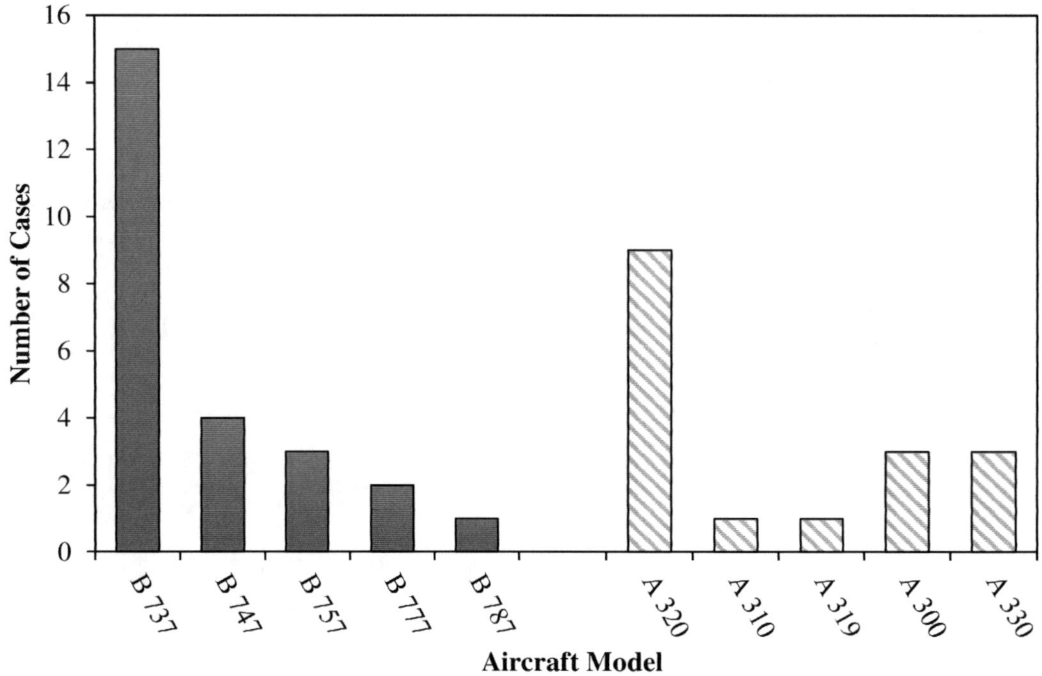

Figure 4. Aircraft models involved in accidents/incidents related to avionic instruments (1996-2015).

4.3 Main Contributing Factor towards Avionics-Related Aircraft Accidents/ Incidents

Many research on aviation accidents claimed that aviation accidents/incidents were caused by several contributing factors [1, 5, 8, 12, 13, 14, 15]. Technological advancement or avionics instrumentation could be one of the contributing factors to such misfortune cases [4, 8]. From the data that were collected and compiled, the following figure 5 shows the general summary of the main factors of the aviation accidents/incidents which related to avionics instrumentation, within the two decades of 1996 - 2015.

Figure 5 shows that there were three main categories that have caused the aircraft accident/incidents related to avionics system which mainly Flight Crew (Human Error) with 54%, Faulty Design / Technical Error contributed 33% and Ground Crew (Human Error) with 13%. If both groups of human error are combined, it accumulates to 67%, which is consistent with the study done by Shappell and his researchers [12]. However, as mentioned earlier it is not easy to simply identify the main cause of any accident as a single main factor since any mishap occurs due to interlink events, when all defences were failed to prevent the next occurrence as explained in Swiss Cheese model of James Reason [5]. Yet, this chart was concluded from the data collection based on published investigation reports of the recognized websites as stated earlier.

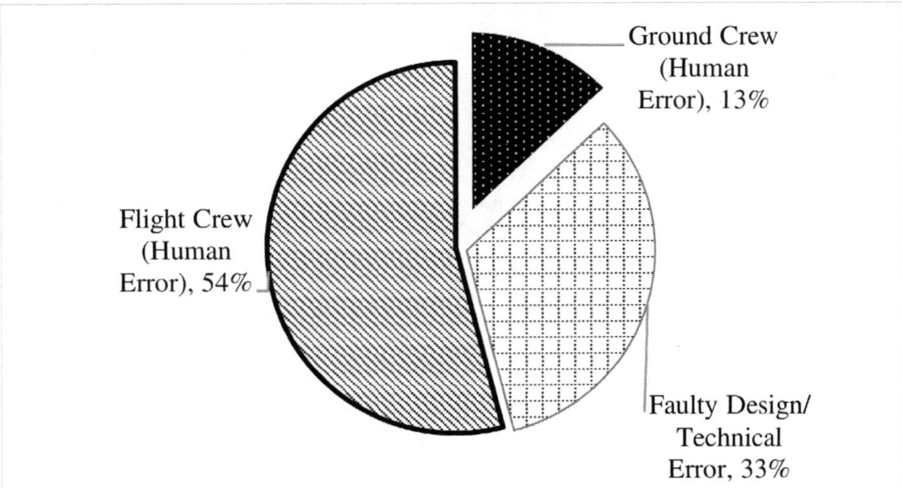

Figure 5. General summary of causes of avionics-related aircraft accidents & incidents (1996 -2015)

Thus, it can be pointed out that human error was still the prominent contributing factor towards aviation accidents, even though avionics instruments were auspiciously placed in the aircraft system. Wiegmann and Shappell [13] had specified five different perspectives of human error that led towards an aviation disaster. These were the limitations that we as human need to cope [12] where there are elements around us that cannot be controlled. In terms of automation system, we can only control the input while the outputs are processed by the system based on designed input data, which could have not responded efficiently as predicted in the preliminary stage that could create outcome disparity. The relationship between human expectations and systems efficiency could give unfavourable repercussion if the existing conditions were not sufficiently conducive, especially with external factors such as environmental stress, organisational hierarchy issues and many more [14].

Looking at the overall data presented in this study, most of the designed avionics systems were quite robust as can be seen from the reduction of air accidents (due to avionics-instrument), compared to overall commercial aviation accidents. There is always a grey line in identifying the actual main factor contributed to any accidents as mishap could happen due to multiple linked factors and not mainly as a result of one single cause. Johnson and Holloway [15] also emphasized where if human error was determined as the main cause of an accident, the organisational influence should also be looked into. In many cases, the organizational factors and accountability do give direct impact towards individual human error at work environment.

Another critical element is the process of how an accident investigation was carried out, which would reflect the findings of the probable versus contributory causes in an aviation accident investigation report. Shappell and his researchers [12] reported that human error components in aviation accident should be scrutinized further as any accident report could be inherently biased or incomplete.

In the meantime, Boeing and Airbus had forecasted that the global fleet is projected to essentially double in size in the near future to 42,730 [16] and 37, 400 [17] respectively by year 2037. Consequently, due to this projection, they need to do extensive research development to increase the safety of their aircraft systems. This includes inculcating the safety culture [6,7] within all related supply chain organisations associated with the manufacturers especially when the regulatory implementations are in place. Issues associated with pilots' additional training on any new upgraded avionics system, as an example, should not be compromised for the sake of cost reduction within all levels in the aviation industry hierarchy.

Moreover, market share competition between these two giant manufacturers should not jeopardize the passengers' safety and affect the airlines businesses with their rivalry on offering lower cost of technological improvements. This is crucial especially when newly implemented software requires additional obligatory safety features. Studies have shown that the cases of human errors in aviation do increase with the use of automation [14, 15], hence fail-safe measures should be taken into account when any new automation system is proposed by original equipment manufacturer (OEM).

5. Conclusion

In conclusion, as the fluctuation of the aircraft accident rises, avionics-related incidents seemed to decrease within the time frame under study. It was found that the avionics-related cases were mostly originated from Boeing B737 and Airbus A320 models from the whole avionics-related accidents. FMS and autopilot were the two main avionics systems that carry the highest contributory factors to incidents, followed by anti-skid and pitot-static system. Nevertheless, even with steadfast technological advancement, human error still carries the highest percentage of causal factor in aviation accidents.

References

[1] Oster C V, Strong J S, Zorn C K 2013 *Research in Transportation Economics* **43** 148-164.
[2] Hersman D A P Assuring safety in aviation's second century *Quebec 8th annual Assad Kotaite lecture at Montreal*, Retrieved from http://www.ntsb.gov/news/speeches/hersman/daph111208.html; 11 May, 2019
[3] International Air Transport Association (IATA) 2018 Safety performance data for commercial airlines, *International Airport Review*. Retrieved from https://www.internationalairportreview.com/news/81497/iata-2018-safety-performance-airlines/; 26 April 2019
[4] Gawron V 2019 Automation in aviation-accident analyses, *Mitre Technical Report*, January 2019.
[5] Reason J 1998 *Work & Stress: An International Journal of Work, Health & Organisations* **12:3** 293-306.
[6] Cooper M D 2000 *Safety Science* **36** 111-136.
[7] Civil Air Navigation Services (CANSO), Safety culture definition and enhancement process, 2008. Retrieved from https://www.canso.org/sites/default/files/Safety%20Culture%20Definition%20and%20Enhancement%20Process.pdf ;10 May 2019
[8] Wiegmann D A and Shappell S 2004 *A human error approach to aviation accident analysis*, (Great Britain: Ashgate Publishing Limited)
[9] Aviation Week Network, Industry data: 2016 vs. 2025 fleet market share: top 10 original equipment manufacturer, Aviation Daily, Retrieved from https://aviationweek.com/sitefiles/aviationweek.com/files/uploads/2016/06/17/avd_06_20_2016_cht1.pdf ; May 13, 2019
[10] Oliver Wyman, Global fleet & MRO market forecast commentary 2018–2028, Retrieved from https://www.oliverwyman.com/content/dam/oliverwyman/v2/publications/2018/January/20182028_Global_Fleet_MRO_Market_Forecast_Commentary_Public_Final_web.pdf; 12 May 2019
[11] Business Insider, The amazing story of how the Airbus A320 became the Boeing 737's greatest rival, Jan 2, 2019. Retrieved from https://www.businessinsider.com/airbus-a320-history-boeing-737-rival-2018-9 ; May 11, 2019.
[12] Shappell S, Detwiler C, Holcomb K, Hackworth C, Boquet A, and Wiegmann D A 2007 *Human Factors: The Journal of the Human Factors and Ergonomics Society* **49** 227
[13] Wiegmann D A and Shappell S A 2000 *The International Journal of Aviation Psychology* **11(4)** 341–357.
[14] Kahuho-Mwarari P 2014 Human Factors as an element in aviation accidents : lessons learnt from 'Human Failure', Retrieved from https://www.academia.edu/37594387/Human_Factors_as_an_element_in_aviation_accidents._Lessons_learnt_from_Human_Failure_2014; 11 May 2019
[15] Johnson C W and Holloway C M, On the over-emphasis of human 'error' as a cause of aviation accidents: 'systemic failures' and 'human error' in US NTSB and Canadian TSB Aviation Reports 1996-2003, Retrieved from

https://pdfs.semanticscholar.org/f128/85d1d7b479f2072d42fde11481ab737b4df7.pdf; May 11, 2019.

[16] Boeing, Boeing commercial market outlook 2018-2037, Retrieved from https://public.tableau.com/en-us/s/gallery/boeing-commercial-market-outlook-2018-2037 ; May 11, 2019.

[17] Airbus, Global market forecast, 2018. Retrieved from https://www.airbus.com › publications › media-day › GMF-2018-2037; July 30, 2019.

IOP Conf. Series: Materials Science and Engineering **697** (2019) 012032 doi:10.1088/1757-899X/697/1/012032

Preliminary analysis of hybrid laminate composite for aircraft radome application

M.Y. Haris[1], A. Fernández-López[2], K. D. Mohd Aris [1], Z. Muzafar[3], N.Z.M. Zuhudi[1], D.Laila[4] and K. Nurhafiza[5]

[1]Universiti Kuala Lumpur – Malaysian Institute of Aviation Technology, Lot 2891, Jalan Jenderam Hulu, 43800 Dengkil, Selangor, Malaysia

[2]Escuela Técnica Superior de Ingenieria Aeronáutica y del Espacio, Universidad Politécnica de Madrid, Plaza Cardenal Cisneros, 3, 28040 Madrid, Spain,

[3]Green Chemistry and Sustainable Research Cluster, Universiti Kuala Lumpur – Malaysian Institute of Chemical and Bio-Engineering Technology, 78000 Alor Gajah, Melaka, Malaysia, [4]Department of Aerospace Engineering, Universiti Putra Malaysia, 43400 Universiti Putra Malaysia Serdang, Selangor, Malaysia

[5]Research and Development Unit, Strategic Planning and Business Development, German-Malaysian Institute, 43000 Kajang, Selangor, Malaysia

Abstract. The preliminary analysis of laminate composite was conducted to analyse the optimal number of layers for aircraft radome application. In this work, the ESAComp software was used for preliminary analysis to predict the displacement during the flight operation. The ESAComp is a finite element software used for preliminary and conceptual design for composite layers. Flax fibre, glass fibre and epoxy resin were applied as the hybrid composite laminate design. The laminate construction consists of flax as the center laminate and sandwiched between glass fibre. Both fibres have the same weave pattern which is twill 2/2 and 200 gsm. The material properties for each materials were obtained from technical data sheet and used as input value for ESAComp. This software was used to analyse a single ply engineering constant by conducting fibre/matrix micromechanics analysis. The value of engineering constant of single ply for each material is then used for laminate analysis. It has range from 4,6 and 8 layers with 20%, 30% and 40% of fibre content on a flat panel with the size of 300 mm x 300 mm. The panel were fixed on each edges and under aerodynamic pressure load plus safety factor of two. Based on the result of analysis, minimum layer required is at least 6 layers if acceptance displacement is 0.1 mm. Therefore, experimental works on 6 layers hybrid laminate will be carried out to determine other requirement for aircraft radome application.

Keywords: Finite element analysis, ESAComp software, flax fibre, hybrid composite laminate, aerospace structure

1. Introduction

The Natural fibres have attracted the interest of material scientists, researchers and industries due to their specific advantages as compared to glass fibres [1]. The classification of natural fibres are broad as discussed by K. Mohanty [2] and intention of this paper is to focus on bast flax fibres. The flax fibres has an advantages such as low density, low cost, renewable and biodegradable [2]. Furthermore, most of natural fibre or specifically a flax fibre has a low dielectric property, which is suitable for aircraft radome application [3, 4, 5]. The purpose of aircraft radome is as a protection to weather radar and therefore it should has a good structural strength and radio wave can be transmitted effectively [6]. The

IOP Conf. Series: Materials Science and Engineering **697** (2019) 012032 doi:10.1088/1757-899X/697/1/012032

common material used for aircraft radome is a type E glass fibre [6]. Finite element analysis has been used for simulation in the aerospace and automotive industries. Both industries have recognised simulation as part of the design process to minimize the design costs and create more efficient structures. Prototyping and testing has to be performed to verify the design, nevertheless simulation is essential for standard practice throughout the design process [7]. Commonly, there are several finite elemant analysis software used for composite [8] and ESAComp is one of the softwares for analysis and design of composites at conceptual or preliminary stage [9, 10]. An aircraft radome wall from Beechcraft Duke 60 is used as a reference as shown in Figure 1 [11]. The Beechcraft Duke 60 aircraft radome wall construction is a thin walled radome or style c and it is fabricated from glass fibre and epoxy. The main purpose of laminate composite preliminary analysis of is to analyse the optimal number of layers for aircraft radome application using hybrid material which is flax fibre and glass fibre.

Figure 1. Beechcraft Duke 60

Table 1. Mechanical properties of flax fibre, glass fibre and epoxy resin [12,13]

Properties	Flax fibre	Glass fibre	Epoxy resin
Density (kg/m^3)	1500	2540	1100
Young modulus (GPa)	50	73	3.12
Shear modulus (GPa)	19.20	30	1.15
Poisson's ratio	0.40	0.18	0.35

Flax fibre (Figure 2a), glass fibre (Figure 2b) and epoxy resin as a matrix are used to produced hybrid laminate composite. Flax fibre is constructed at the center and glass fibre at the outer and inner layer. Both fibres have the same weaving pattern which is twill weave 2/2 and has weight per unit area of 200 gsm. Flax fibre and glass fibre were supplied from Easy Composite and Epox Amite epoxy resin from Smooth-on. The mechanical properties for each material are obtained from technical data sheet as showed in table 1.

(a)Flax fibre (b)Glass fibre

Figure 2. Weave pattern twill 2/2 and 200gsm

1.1 ESAcomp Analysis Setup

The ESAComp is a finite element software for composite structures with the focus on conceptual and preliminary design. Figure 3 shows an analysis flow for ESAComp.

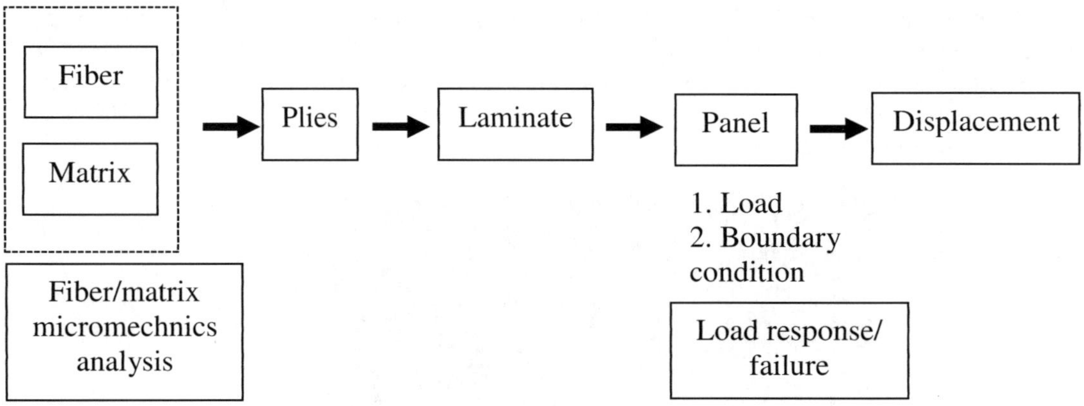

Figure 3. ESAComp analysis flow

The data in Table 1 is used in ESAComp for initial analysis to analyse the optimal number of layers for aircraft radome application. The analysis started with combined fibre and matrix to create single ply using fiber/matrix micromechanic analysis. Rules of mixture are used for ply analysis and fibre content ratio range from 20% to 40%. For laminate analysis, load response/failure option is used on flat panel with size 300 mm x 300 mm which are under pressure wind load and all edges are fixed. The assumptions of classical laminate theory are applied in the analysis to determine the displacement [14].

2. Results and discussion

The Fiber/Matrix micromechanics analysis was conducted to analyse a single ply engineering constants. The value of engineering constants were then used to determine properties of laminate. Table 2 below shows the result of value engineering constant for single ply based on rules of mixture calculation from ESAComp.

Table 2. Single ply engineering constant

Engineering constant	Glass/epoxy			Flax/epoxy		
	20%wt	30%wt	40%wt	20%wt	30%wt	40%wt
Density (kg/m^3)	1342	1426	1521	1250	1277	1304
E_1 (GPa)	10.50	14.89	19.86	10.93	15.09	19.43
E_2 (GPa)	4.10	4.54	5.17	4.50	5.29	6.42
E_3 (GPa)	4.10	4.54	5.17	4.50	5.29	6.42
G_{12} (GPa)	1.67	1.92	2.24	1.90	2.29	2.80
G_{31} (GPa)	1.67	1.92	2.24	1.90	2.29	2.80
G_{23} (GPa)	1.37	1.53	1.77	1.51	1.80	2.22
nu_{12}	0.332	0.321	0.309	0.342	0.337	0.333
nu_{13}	0.332	0.321	0.309	0.342	0.337	0.333
nu_{23}	0.5	0.48	0.458	0.49	0.47	0.449

The wind load value is calculated based on the situation of a radome during flight operation which is around 100 N. With additional a safety factor equal to two, the total wind load apply during flight operation is 200 N. Then, a composite laminate panel consist of 4, 6 and 8 layers at 20%, 30% and 40% fibre weight ratio is created. Figure 4 show an example result from ESAComp.

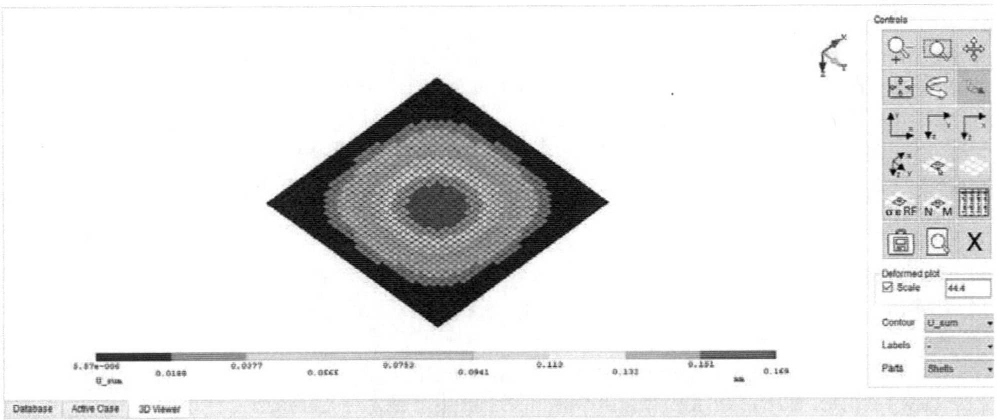

Figure 4. Result from ESAComp

The total displacement result for each configuration is tabulated in Table 3. If the displacement design value is around 0.1 mm, the minimum layer required is at least 6 layers. Furthermore, Figure 5 shows the number of layers increase, the displacement is decrease due to increase of thickness of composite panel. However, the experimental works still need to be done before to verified the result it can be used in aircraft radome construction.

Table 3. Total displacement result for laminate

Laminate	Displacement (mm)		
	20%wt	30%wt	40%wt
4 layers	0.307	0.552	0.767
6 layers	0.072	0.169	0.203
8 layers	0.027	0.072	0.081

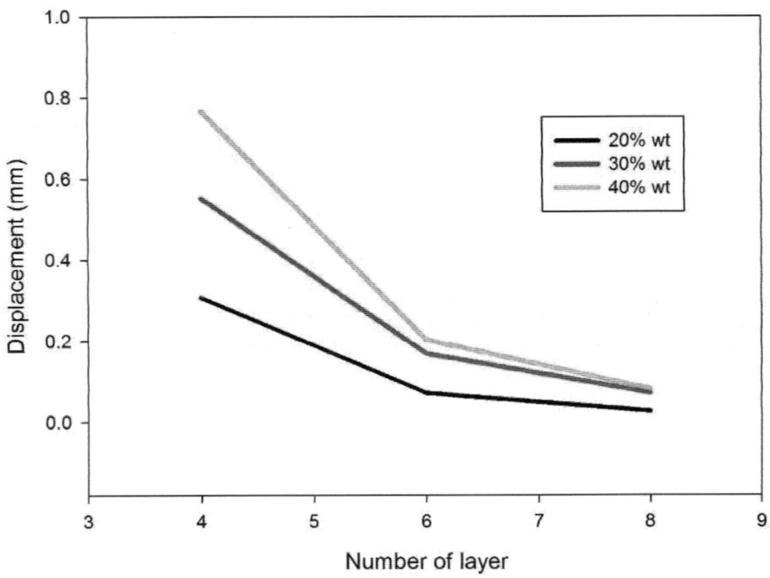

Figure 5. Displacement vs number of layer

3. Conclusions

The understanding of preliminary analysis on laminate composite has significant impact in decision making especially on laminate design. The ESAComp software can be used as a preliminary analysis to determine the suitability laminate design for further investigation. As the number of layers increase, the displacement is decrease. Meanwhile, the percentage displacement increase with increase the percentage of fibre weight ratio. This shows that the displacement decrease as the number of layer increase of the increase of stiffness.

References

[1] Varun M, Rahul S, Shishir S. Natural fiber-mediated epoxy composites - A review. *Composite Part B: Engineering 2016*; **99**: 425–435.
[2] Amar KM, Manjusri M, Lawrence TD, Susan ES, Bruce R, Georg H. Natural fibers, biopolymers, and biocomposites: An introduction. In: Amar KM, Manjusri M, Lawrence TD, editors. Natural fibers, biopolymers, and biocomposites: CRC Press Taylor & Francis Group, 2005, p 1-36.
[3] Maya JJ, Radhouan T, Rajesh DA, Abderrahim B, Laurent I. Effect of amphiphilic coupling agent on heat flow and dielectric properties of flax-polypropylene composites. *Composite Part B: Engineering 2010*; **43**: 526–532.
[4] Dragana DC, Jablan RD, Kovijka AA, Tatjana AM. Dielectric investigation of some woven fabrics. *Journal of Applied Physic 2009*; **106(8).**
[5] Haris MY, Halim Z, Zainudin ES, Mustapha F, Majid DL, Zahari R. Preliminary review of biocomposites materials for aircraft radome application, *Key Engineering Materials* 2011; 471–472: 563–567.
[6] Dennis JK. Analysis of radome-enclosed antennas, 2nd ed. Artech House Inc; 2010.
[7] Minh QN, David J, Javid B, Rodney ST, Murry LS. A review of explicit finite element software for composite impact analysis, *Journal of Composite Materials* 2006; **39(4):** 375–386.
[8] Peković O, Aleksandar S, Slobodan S, Srđan T, Zorana P. Contemporary software tools in the design process of composite structures. In: 5th International Scientific Conference on Defensive Technologies, Belgrade, Serbia, pp 117–122; 2012.
[9] Saarela O, Palanterä M, Häberle J, Klein M. ESAComp: A powerful tool for analysis and design of composite materials, In: International Symposium on Advanced Materials for Lightweight Structures, Noordwijk, The Netherlands, pp. 161–169; 1994
[10] Saarela O, Haberle J, Klein M. Composite analysis and design system ESAComp, Computational Mechanics Publication, USA, pp 23–30; 1994.
[11] BEECHCRAFT Duke 60 Series Maintenance Manual, Revision 1973.
[12] Composite Evolution, Biotex Flax 200g/m2 2x2 Twill, Data Sheet Revision 2018
[13] Smooth-on, EpoxAmite^{tm} 100 Epoxy Laminating System, 101718-JR Data Sheet, Revision 2019.
[14] ESAComp, ESAComp User's Guide, Release 4.6.

Acknowledgement

This research was supported by Short Term Research Grant Universiti Kuala Lumpur STR17051 and ESAComp software license under Universidad Politechnica de Madrid, Spain

IOP Conf. Series: Materials Science and Engineering **697** (2019) 012033 doi:10.1088/1757-899X/697/1/012033

Characterization of Banana peels waste adsorbent for preliminary study of methylene blue removal from aqueous solution

M Z A Zamri[1], N Y Yahya[1*], R S Ramli[1], N Ngadi[2] and M Widia[3]
[1]Faculty of Civil Engineering Technology, Universiti Malaysia Pahang, Lebuhraya Tun Razak, 26300 Gambang, Kuantan, Pahang Darul Makmur, Malaysia
[2]School of Chemical and Energy Engineering, Faculty of Engineering, Universiti Teknologi Malaysia, 81310 UTM Skudai, Johor Darul Takzim, Malaysia
[3]Faculty of Industrial Sciences and Technology, Universiti Malaysia Pahang, Lebuhraya Tun Razak, 26300 Gambang, Kuantan, Pahang Darul Makmur, Malaysia
*E-mail: yahida@ump.edu.my

Abstract. This paper presents the removal of methylene blue (MB) from aqueous solution using adsorbent produced from Banana peels waste (BPW). Banana peels waste adsorbent (BPWA) was prepared via carbonization in laboratory muffle furnace for 1 h. A preliminary study was performed on the adsorption reaction by varying the initial concentration of MB, adsorbent dosage, contact time, pH and temperature of the aqueous solution. Under the best conditions obtained, the maximum percentage removal of MB is approximately 99 %. The potential of BPWA as dyeing wastewater adsorbent was proven by Scanning Electron Microscopy (SEM) analysis, N_2 adsorption-desorption analysis and Fourier Transform Infrared Spectroscopy (FTIR). From the results, the synthesized adsorbent shown a good percentage removal of MB. In addition, the optimum conditions for the adsorption study are 1 g of adsorbent in the 1 g L^{-1} of MB initial concentration, for 1 h at pH 8 in 65 °C. This study reveals potential of BPW in removal of MB dyes from aqueous solution however, further studies are required to establish the applicability of the synthesized adsorbent for the treatment of wastewater from textile industry.

1. Introduction

Wastewater containing dyes is produced by various industries such as textile, food, paper, plastic and cosmetic [1,2]. Effluent discharge containing dyeing wastewater led to negative effects to the environment and human [1,3]. Thus, it is crucial to develop an eco-friendly and effective wastewater treatment and management for this type of water pollution [4]. Numerous approaches are currently being used for treatment of industrial and municipal wastewater, including adsorption, membrane filtration, electroflotation, flocculation and coagulation as well as biological treatment [5,6]. Nevertheless, as various contaminants commonly present in such wastewaters, combination of more than a few approaches in wastewater treatment process is necessary. In treatment of dyeing wastewater, adsorption is more practical than the other processes, as this method involves low cost material and facile operation [7]. Development of new adsorbents that are hydrophobic, porous, cost-effective and possess high adsorption capacity is necessary to ameliorate the negative impacts of dyeing wastewater on the environment and human [8,9]. Banana peels waste (BPW) is one of the potential materials for dyeing wastewater treatment. Banana or is scientific name Musa is one of local fruit in Malaysia. Banana has benefits such as provide nutrient, helps the digestive system and improve insulin sensitivity. It was used to produce a product such as jams, chips, and juice. Banana peels could be prepared as an adsorbent for dye removal due to its chemical composition which contained a high amount of hemicellulose, pectin, cellulose, lignin and contained various polar functional groups such as phenolic acid groups, carboxylic and hydroxyl [10,11]. 40% of total weight of banana comes from the peel. However, the peels are generated as a waste product and dumped as solid waste [12]. Production of adsorbent from BPW reduces the amount of agricultural waste that ends up in landfills as well as this value-added biomass is not harmful to the flora and fauna. In this study, BPW was synthesized through carbonization process. The adsorbent preparation method is cost effective which can replace the high cost of conventional adsorbent preparation. The problem faced by conventional adsorbent was, it required complicated pre-treatment that use high amount of chemical and required high usage of energy. By utilizing this resource and preparation method, it will be able to reduce the amount of chemical the usage of energy. The preliminary study of adsorption reaction was carried out to study several adsorption parameters such as initial concentration of MB, adsorbent dosage, contact time, pH and temperature of the aqueous solution toward the percentage of MB removal. The result of this preliminary study is important to evaluate the significant of the material used and the preparation method to be improved in the future.

2. Materials and method

2.1. Materials and chemicals

Banana peels waste (BPW) was collected from local area near campus. Methylene blue (MB) was obtained from Sigma-Aldrich.

2.2. Synthesis of Banana peels waste adsorbent (BPWA)

Banana peels waste (BPW) was collected and subsequently washed with distilled water until the dirty completely removed. Next, the BPW was cut into small pieces and dried in a hot air oven at 105 °C for 24 h. Then, the dried BPW was crushed into granules and sieved to 200 μm particle size. Finally, the powder was carbonized in muffle furnace at 550 °C for 2 h to developed porosity and high adsorption efficiency of the synthesized adsorbent.

2.3. Preparation of methylene blue (MB) dye solution

Aqueous methylene blue (MB) stock solution with concentrations of 100 mg·L^{-1} was prepared by mixing 100 mg of MB powder with 1 L of distilled water. To prevent decolourization by direct sunlight, the stock solution was stored in a dark place before being used.

2.4. Characterization of banana peels waste adsorbent (BPWA)

The functional groups of banana peels waste adsorbent (BPWA) was characterized using Fourier Transform Infrared (FTIR) Spectroscopy (IRTracer-100 Spectrophotometer, Shimadzu Cooperation, Japan). The spectra were recorded from 400 to 4000 cm^{-1}. The surface morphology of BPW adsorbent was observed using Scanning Electron Microscopy (SEM) (Hitachi SU8020). The surface area and pore size of were measured by a N_2 Adsorption-Desorption Isotherm Analyzer (NAD) model Micromeritics 3 Flex Surface Characterization Analyser. Sample was degassed at -196.15 °C under nitrogen gas flow for 4 h prior to analysis.

2.5. Batch mode adsorption study

The prepared banana peels waste (BPW) adsorbent was evaluated for removal of methylene (MB) from aqueous solution through batch equilibrium adsorption studies. The effects of various process parameters, such as initial dye concentration (0.1-1 g L^{-1}), adsorbent dosage (0.1-1 g), pH of the solution (3-10), contact time (10-60 min) and temperature (25-70 °C) on the adsorption capacity of MB were evaluated. Batch experiments were performed using 50 mL glass vials with 50 mL dye solution and 250 rpm stirring on a shaking incubator. At the end of each experiment, solid and liquid phases were separated from each other by filtration process. The remaining concentration of MB in solution after adsorption on BPW adsorbent was determined using UV-Vis spectroscopy (Shimadzu 1800) at 630 nm. The percentage removal of MB dye, %R was calculated by Eq. (1) as follow:

$$\text{Percentage removal of MB dye, } \%R = \frac{C_o - C_e}{C_o} \quad (1)$$

where C_o and C_e denote the initial and equilibrium concentration ($mg \cdot L^{-1}$) of MB dye.

3. Results and discussion

3.1. Adsorption study

Figure 1 shows the parametric effects on the percentage removal of MB after the adsorption study. For the effect of initial dye concentration on percentage removal of MB (Figure 1(a)), the graph shown that the capacity of the adsorption of the BPWA increase with the increase in the initial concentration of MB which can be explained by the increase of the adsorbate (MB ions) on the free active sites (pores in the adsorbent surface). This is because, at high initial concentration, the driving force to overcome all resistances between ions of aqueous and solid phases is also high, yet increasing the uptake. Also, increasing the initial ions concentration of MB increases the number of collisions between the dye ions and the adsorbent, hence enhancing the adsorption process [13]. On the other hands, Figure 1(b) demonstrates the effect of the adsorbent dosage on the percentage removal of MB. It is obvious that, the percentage removal of MB increases significantly with the increase in the amount of adsorbent from 0.1 g to 0.3 g. This can be due to greater availability of surface area for adsorption reaction to occur which leads to the introduction of more binding sites during the reaction [14]. An increase in the percentage removal was also observed when the dosage was further increased from 0.3 g to 0.5 g. Any further addition of adsorbent beyond this amount did not cause significant change in adsorption performance. This may be due to the overlapping of adsorption sites as a result of overcrowding of adsorbent particles [15]. For all adsorbents dosage ranging from 0.5 g to 1 g, maximum removal of all studied MB removals was obtained at the adsorbent dosage of 1 g with up to approximately 99 % of MB removal. Moving on to Figure 1(c), the effect of pH on the adsorption of methylene blue was not significant due to the low dosage of adsorbent which 0.1 g. This can be explained by the fact that at low values of the initial pH, the surface of the adsorbent would be surrounded by the ions H^+ inducing repulsive forces and hence decreasing the interaction of the ions of the MB with the sites of the adsorbent. Thus insignificant in the adsorption rate at low pH value can be due to the high concentration and the mobility of the ions H^+. Conversely, at high pH, the concentration of H^+ decreases and the total surface of the adsorbent become negatively charged, enhancing the sorption of the ions dye, by means of electrostatic interactions. As a result, the highest percentage removal of MB was at pH 8 with approximately 25 %. This findings is similar to that reported by [16]. It also can be seen from Figure 1(d) that by increasing the contact time from 10 min to 30 min the percentage removal increases. After that, the percentage removal was decrease from 30 min to 50 min. This can be attributed to the availability of sites for the sorbate. In addition, a very high adsorption driving force at the beginning resulted in a higher adsorption rate. After the initial period, slower adsorption may be attributed to the slower diffusion of molecules into the interior pores of the adsorbent, and the molecules subsequently occupy the positions within the adsorbent framework. For effect of temperature on percentage removal of MB, BPW adsorbent shows higher percentage removal of MB as the temperature increase. When the temperature increase, it will decrease the solubility of MB and increase the adsorption of MB.

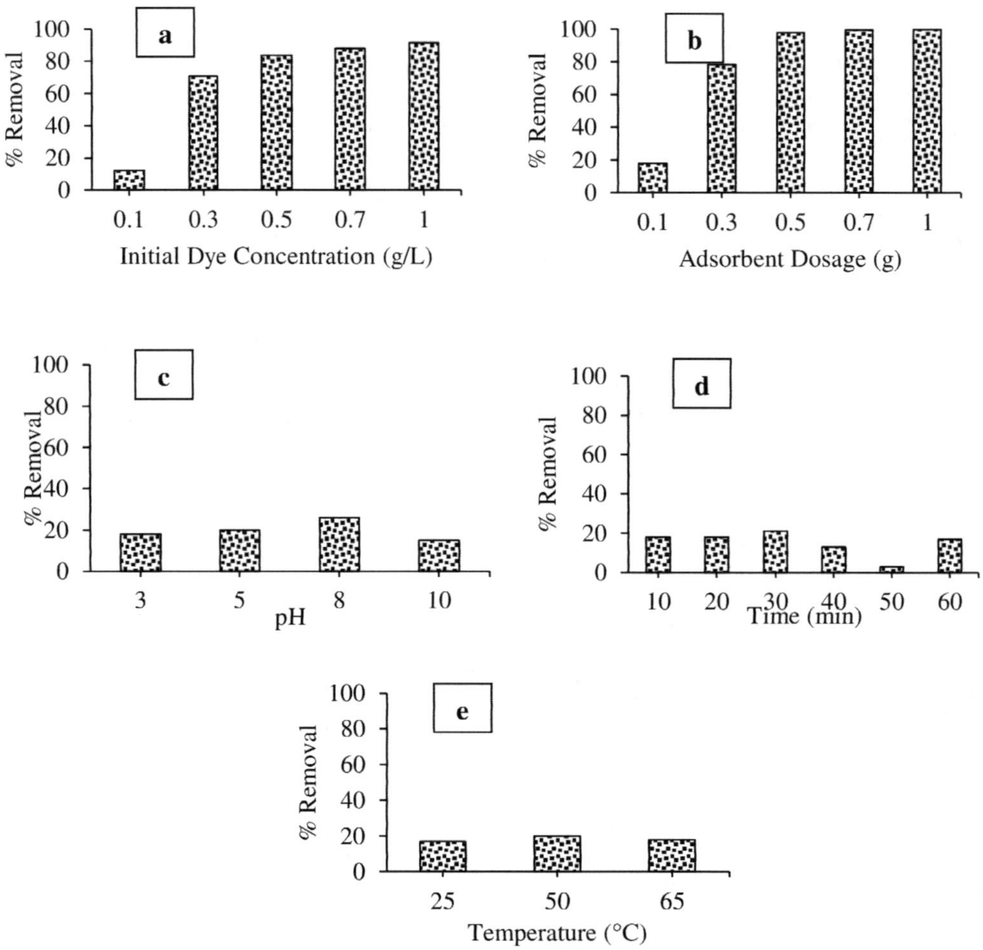

Figure 1. Effect of parameters on percentage removal of MB. a) Initial dye concentration b) Adsorbent dosage c) pH d) Time e) Temperature

3.2. Characterization of banana peels waste adsorbent (BPWA)

3.2.1. Morphology analysis

Scanning electron microscopy (SEM) is one of the important tool for characterizing the surface morphology of adsorbent surface. The SEM of BPWA was recorded and showed in Figure 2. The analysis of BPWA showed an uneven surface with many small pores sized 2 μm to 3 μm. The existence of the pores sizes can be due to the carbonization processes [17]. Generally, the percentage removal of MB highly related with the pores size of the adsorbent. The appropriate pores size can make MB particles entrance into the internal parts of the adsorbent easier and helpful in the sorption process [18].

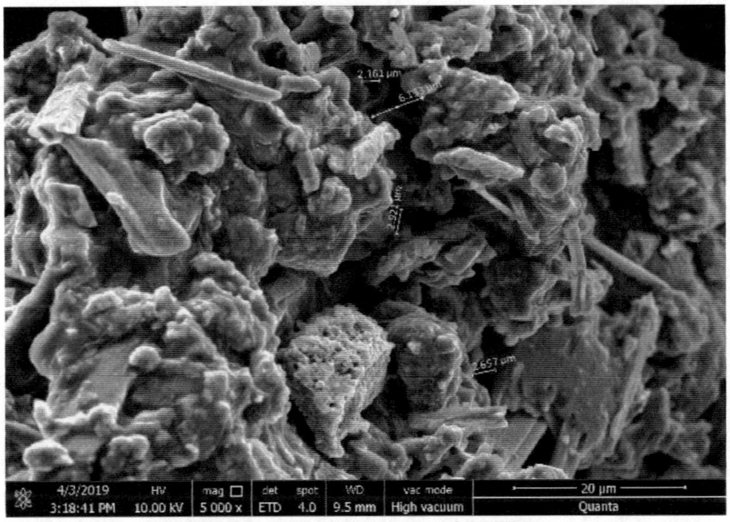

Figure 2. SEM image of BPWA.

3.2.2. N_2 adsorption-desorption analysis

The N_2 adsorption-desorption curve (Figure 3), provides qualitative information on the BPWA. From the Figure 3, it can be summaries that BPWA adsorbed large amount of N_2 molecules at relatively high pressure. From the figure, the best-fit for N_2 adsorption-desorption curve for the adsorbent is type II isotherm, which refers to unrestricted mono-multilayer adsorption and adsorbent is macro pores. The surface physical parameters obtained from the N_2 adsorption-desorption isotherms are summarized in Table 1. From the data, it is marked that the BET surface area of BPWA was 9.6363 $m^2 \cdot g^{-1}$. Therefore, it can be the reason for the higher percentage removal of MB during the adsorption reaction.

Table 1. Characteristic of BPWA samples.

Parameter	Method	Characteristic
Total surface area, , SBET/$m^2 \cdot g^{-1}$	BET	9.6363
Micropore surface area/$m^2 \cdot g^{-1}$	t-plot	NA
Microporous volume/$cm^3 \cdot g^{-1}$	t-plot	NA
Mesoporous volume/$cm^3 \cdot g^{-1}$	BJH adsorption	0.014834

In this study, the BJH method has been used to determine the pore distribution of the adsorbent (Figure 3). From the plotted curves, the pore size distribution curves showed the highest peaks at pore widths of 1884.7 nm. According to the classification of IUPAC-pore dimensions, the pores of adsorbents were grouped into micropore ($d < 2$ nm), mesopore ($d = 2\text{-}50$ nm) and macropore ($d > 50$ nm). Thus, the adsorbent was identified as a macro pores material. From the table above, it shows that the surface area of the adsorbent is considers as large.

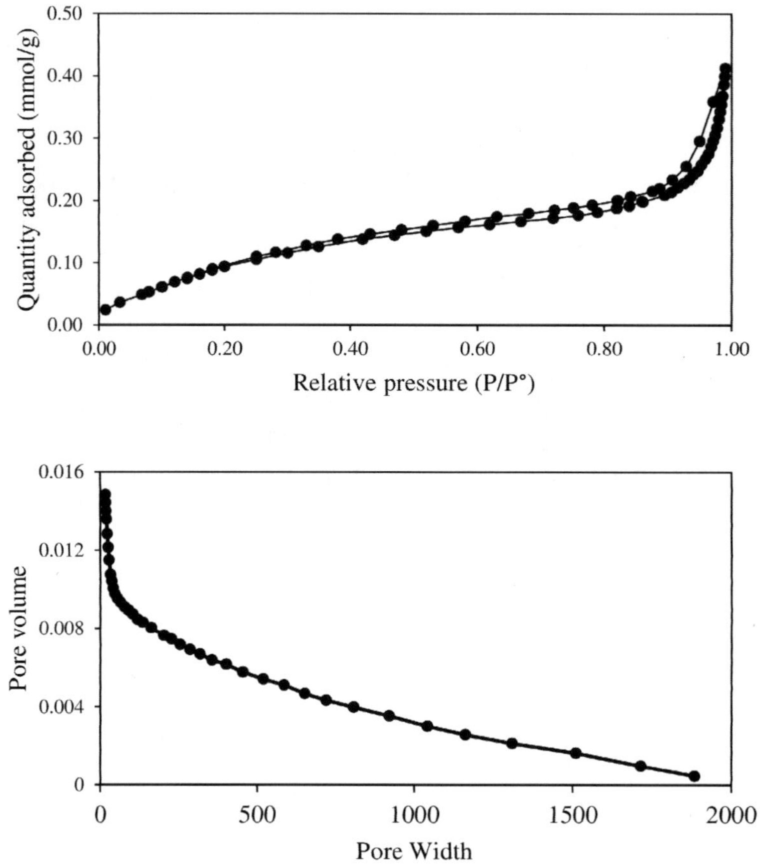

Figure 3. N_2 adsorption-desorption isotherms for BPWA, as well as pore size distribution of BPWA.

3.2.3. Spectroscopic study

The purpose of spectroscopic study is to examine the functional groups presence in the BPWA surface. In BPWA, it is observed that various peaks present which indicates the of multiple functional groups in the adsorbent. A wide peak at 3447.05 cm^{-1} for BPWA is due to the free hydroxyl group of polymeric compounds [19,20]. This is likely caused by inter/intra-hydrogen bonding (O-H) stretching which is associated to hydroxyl groups that can be found in alcohols, phenols and carboxylic structures. The O-H group has the ability to adsorb MB. The adsorbent also showed stretching adsorption bands at 1636.79 cm^{-1}. This can be correlate to carboxyl group C=O present in acetyl derivatives, ketone groups, aldehydes and esters. Weak bands observed at around 600 cm^{-1} to 700 cm^{-1} for the adsorbent were ascribed as amines groups. It can be found that BPWA has hydroxyl and carbonyl groups. These groups have negative charge where can be good sites for adsorption of MB.

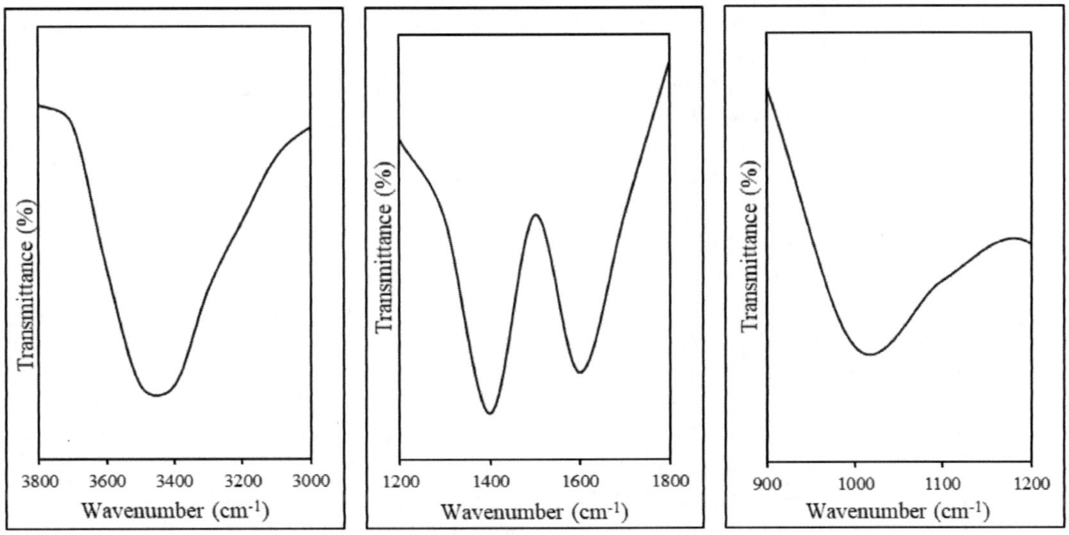

Figure 4. FTIR-spectra for BPWA.

4. Conclusions

As the conclusion, the BPW adsorbent show good percentage removal of MB. From the BET test, it can be confirmed that total surface area and mesoporous volume of BPW adsorbent was higher. Presences of hydroxyl, carbonyl and amines in BPW adsorbent were confirmed through FTIR test. These groups have negative charge where can be good sites for adsorption of MB. The optimum condition for adsorption study by using BPW adsorbent are 1 g L^{-1} MB, 1 g of adsorbent for 1 hr at pH 8 in 65 °C.

References

[1] Rehman R, Farooq S and Mahmud T 2019 Use of Agro-waste Musa acuminata and Solanum tuberosum peels for economical sorptive removal of Emerald green dye in ecofriendly way *J. Clean. Prod.* **206** 819–26

[2] Chaukura N, Edna C, Murimba E C and Gwenzi W 2017 Sorptive removal of methylene blue from simulated wastewater using ciochars derived from pulp and paper sludge *Environ. Technol. Innov.* **8** 132–40

[3] Budnyak T M, Aminzadeh S, Pylypchuk I V., Sternik D, Tertykh V A, Lindström M E and Sevastyanova O 2018 Methylene Blue dye sorption by hybrid materials from technical lignins *J. Environ. Chem. Eng.* **6** 4997–5007

[4] Wong S, Yac'cob N A N, Ngadi N, Hassan O and Inuwa I M 2018 From pollutant to solution of wastewater pollution: Synthesis of activated carbon from textile sludge for dye adsorption *Chinese J. Chem. Eng.* **26** 870–8

[5] Mahindrakar K V and Rathod V K 2018 Utilization of banana peels for removal of strontium (II) from water *Environ. Technol. Innov.* **11** 371–83

[6] Sohaimi K S A, Ngadi N, Mat H, Inuwa I M and Wong S 2017 Synthesis, characterization and application of textile sludge biochars for oil removal *J. Environ. Chem. Eng.* **5** 1415–22

[7] Ibupoto A S, Qureshi U A, Ahmed F, Khatri Z, Khatri M, Maqsood M, Brohi R Z and Kim I S 2018 Reusable carbon nanofibers for efficient removal of methylene blue from aqueous solution *Chem. Eng. Res. Des.* **136** 744–52

[8] Zou H, Ning X an, Wang Y and Zhou F 2019 The agricultural use potential of the detoxified textile dyeing sludge by integrated Ultrasound/Fenton-like process: A comparative study *Ecotoxicol. Environ. Saf.* **172** 26–32

[9] Hao Y, Wang Z, Wang Z and He J 2019 Preparation of hierarchically porous carbon from

cellulose as highly efficient adsorbent for the removal of organic dyes from aqueous solutions *Ecotoxicol. Environ. Saf.* **168** 298–303

[10] Oyewo O A, Onyango M S and Wolkersdorfer C 2016 Application of banana peels nanosorbent for the removal of radioactive minerals from real mine water *J. Environ. Radioact.* **164** 369–76

[11] Wang Y, Zhang Y, Li S, Zhong W and Wei W 2018 Enhanced methylene blue adsorption onto activated reed-derived biochar by tannic acid *J. Mol. Liq.* **268** 658–66

[12] Salman M, Waheed-uz-Zaman, Anwar J, Anwar S, Shafique U and Dar A 2009 Removal of Pb(II) and Cd(II) from water by adsorption on peels of banana *Bioresour. Technol.* **101** 1752–5

[13] Islam M A, Ahmed M J, Khanday W A, Asif M and Hameed B H 2017 Mesoporous activated carbon prepared from NaOH activation of rattan (Lacosperma secundiflorum) hydrochar for methylene blue removal *Ecotoxicol. Environ. Saf.* **138** 279–85

[14] Mouni L, Belkhiri L, Bollinger J-C, Bouzaza A, Assadi A, Tirri A, Dahmoune F, Madani K and Remini H 2018 Removal of Methylene Blue from aqueous solutions by adsorption on Kaolin: Kinetic and equilibrium studies *Appl. Clay Sci.* **153** 38–45

[15] Mohammed R R and Chong M F 2014 Treatment and decolorization of biologically treated Palm Oil Mill Effluent (POME) using banana peel as novel biosorbent *J. Environ. Manage.* **132** 237–49

[16] Amel K, Hassen M A and Kerroum D 2012 Isotherm and kinetics study of biosorption of cationic dye onto banana peel *Energy Procedia* **19** 286–95

[17] Yang B, Liu Y, Liang Q, Chen M, Ma L, Li L, Liu Q, Tu W, Lan D and Chen Y 2019 Evaluation of activated carbon synthesized by one-stage and two-stage co-pyrolysis from sludge and coconut shell *Ecotoxicol. Environ. Saf.* **170** 722–31

[18] Alaa El-Din G, Amer A A, Malsh G and Hussein M 2018 Study on the use of banana peels for oil spill removal *Alexandria Eng. J.* **57** 2061–8

[19] Kaushal I, Saharan P, Kumar V, Sharma A K and Umar A 2019 Superb sono-adsorption and energy storage potential of multifunctional Ag-Biochar composite *J. Alloys Compd.* **785** 240–9

[20] Sartova K, Omurzak E, Kambarova G, Dzhumaev I, Borkoev B and Abdullaeva Z 2019 Activated carbon obtained from the cotton processing wastes *Diam. Relat. Mater.* **91** 90–7

Acknowledgments

The authors extend their heartiest gratitude to Faculty of Civil Engineering Technology, Universiti Malaysia Pahang for laboratory facilities and funding the Senior Design Project 2 and Universiti Malaysia Pahang for the financial assistance under University Research Grant (RDU1803167).

IOP Conf. Series: Materials Science and Engineering **697** (2019) 012034 doi:10.1088/1757-899X/697/1/012034

Development of Robotic Rovers: A Review

Nurul Aqilah Herman[1], Roshahliza M Ramli[1], Mohd Azrul Hisham Mohd Adib[2] and Mirta Widia[3]
[1]Faculty of Electrical & Electronics Engineering Technology, University Malaysia Pahang,
[2]Faculty of Mechanical & Automotive Engineering Technology, University Malaysia Pahang,
[3]Faculty of Industrial Sciences & Technology, University Malaysia Pahang
26300 Pekan, Pahang, Malaysia.
*Email: roshahliza@ump.edu.my

Abstract. The robotic rover is defined as a small vehicle that can move over rough ground and it is initially used by NASA for space exploration. However, this type of rover is expensive, using complicated and complex materials and software as well as large in size. Robotic rover fills the gap without endangering personnel involves in life-threatening condition or scenario that human sometimes needs to deal with. It can be considered to be deployed in assisting safety authorities to collect information and insights as well as work lift in search and rescue operations. In this paper, the progress that has been made towards the development of robotic rover which has been done in previous years are reviewed, including the design, functions, specifications, and applications, to meet requirements in performing a variety of tasks including hazardous identification, site surveillance and monitoring. Subsequent to this, a few analyses on the design and development of the proposed robotic rover are also discussed. The ultimate aim is to come out with such rover that is portable, small in size and low cost, together with user-friendly functions and specifications for future potential applications.

1. Introduction

Robotic systems have been developed to assist humans in various tasks and applications, either in industrial or domestic purposes. These systems are categorized into Unmanned Aerial Vehicles (UAV), Unmanned Ground Vehicle (UGV), Under Sea Vehicles (USV), space robots and medical robots [1, 2]. By having these systems, certain dangerous and incapable tasks could be done to replace humans mainly in monitoring, inspections, repairs, and maintenance at risks places such as nuclear plants, oil rigs, construction sites, and others. This paper discusses and reviews types of UGV particularly in its designs, features, and applications, which initially it was used for the planetary surface on a lander-style spacecraft [3]. Remote OVerhead Extendable Robot (ROVER) is actually a term for a small-scale robotic UGV that had been built by NASA for space exploration designed to move across the surface of a planet. The popular rovers developed by NASA are Spirit and Opportunity that have been commissioned in Mars explorations [4]. Other than that, agile prototypes named The Light-weight Rover Unit (LRU) had been developed for autonomous planetary exploration purposes with relatively small in size [5]. As the results of successful robotic missions on Moon and Mars, future plans have been made for carrying out various activities and experiments. There are various tasks need to be done by the rover especially for atmospheric aspects like data supervision about air pressure, temperature climate and temperature [6]. Besides this, a rover has other potential wide applications in the real world such as in scientific exploration, safety and weapon system endeavor as well as in military applications [7]. Developments of robotic rovers are usually designed based on specific requirements needed by the authorities. Most of the developed rovers are large, bulky and take a lot of spaces to store, which are not suitable to be used in certain situations.

For example, the rovers would be deployed in many dangerous situations where humans fear to tread. In specific, the rovers could be used by the Hazardous Materials Management (HAZMAT) authorities in conducting their responsibilities to save lives. For HAZMAT personnel, the most threatening risk can be occurred on first entry by the team because of the type of materials that they are dealing with might be unknown. Therefore, a robotic rover can be sent to assess, reduce or eliminate the risks in dangerous situations and prevent human life-threatening risks of death that may lurk with wrong moves [8]. Therefore, this paper discusses and reviews the developments that have been made in unmanned ground vehicles or rovers, especially on its designs and applications, as well as potential implementations in the future.

2. Development of designs in the existing rover

In 1985, a robot has been developed by researchers in the University of Tokyo called AMOOTY, which named by using the acronyms of six researchers of the robot [9]. The design of this robot is shown in Figure 1, with a look like an elephant trunk and able to climb staircases. The researchers of this robot conducted research on the concept of maintenance in a nuclear reactor and came up with a locomotion system and nine degrees of freedom (DOF) for the robot arm. This work was an outstanding achievement whereby that time a robotic arm has not existed in nature yet. However, this robot still has many problems to be solved such as the speed is too slow because it is heavy and in bulky size which makes it hard to move around [10]. Another rover was developed under the Emergency Response Robotics project by Jet Propulsion Laboratory in 1991. As shown in Figure 2, this rover is a remote mobile robot named Andros for the deployment by the HAZMAT team [11]. By using this robot, the team had conducted several testing and trainings to determine other requirements needed by the HAZMAT team in order to use this robot in their missions efficiently.

Figure 1. AMOOTY [9]

The Andros robot was redesigned by adding other requirements such as track modifications and specialized key tools. Modification on the track was done by articulating front and rear sections of the robot for climbing stairs. Then, specialized key tools were developed to enable the robot in unlocking doors. Besides that, a winch system was added as an aid for the robot in opening doors. Based on these modifications highlighted by the HAZMAT team, the robot was renamed as HAZBOT II [12], as shown in Figure 3 (a). HAZBOT is also specifically known as Ground Emergency Response Vehicle, a type of UGV that is used only during emergency situations.

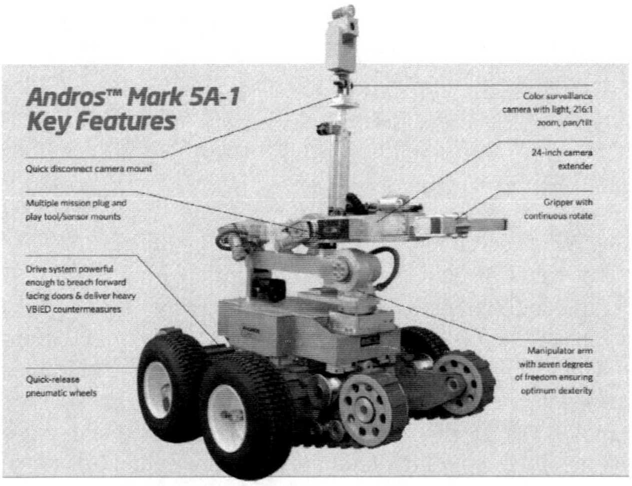

Figure 2. ANDROS robot [11]

(a) (b)

Figure 3. (a) HAZBOT II and (b) HAZBOT III [12]

However, the researchers found drawbacks in the design of HAZBOT II such as its dependent on the current configuration of manipulator at the operator control panel. Therefore, they decided to alter and redesign the robot in order to be operated in potential combustible atmospheres [13]. It needed to be redesigned with a smooth profile to ease the decontamination process if it is considered to be deployed in handling chemical gases [14]. In addition, the researchers were also considering to improve the speed of the robot and dexterity of the manipulator. The researchers have focused on implementing these modifications that led to the development of HAZBOT III, as shown in Figure 3(b).

Although HAZBOT III is the latest modification from the Jet Propulsion Laboratory (JPL) team members, they emphasized that there are four main factors need to be considered by designers in designing a rover for the future recommendations; robot speed, robot size, operator interface, and depth perception. As we knew, these emergency response robots are frequently exposed to the chemical gases and conduct operations such as unlocking doors, opening doors, and climbing stairs. It is, therefore, dexterity is very essential because these robots play the role to access the site at least less than an hour to collect information before the entry team is ready to go in [15]. Next, the size of the HAZBOT systems are too big and bulky, making them difficult to pass through doors and the robot arm is not required in most mission tasks. The development of the robot interface should be made very informative to provide the information required by the operator of the robot. By having all important information while operating the robot, it will help the operator of the robot to make decisions under critical circumstances. The improvement in depth perception is also needed to overcome the issue that the viewers faced difficulties when they have to view with a single video camera. Therefore, adding a simple depth cueing devices is the desirable solution.

There were few other robots were designed but for commercial value purposes such as LT2/F "Bulldog", as shown in Figure 4(a). This robot has many similar functions with previous HAZBOT II which it can open doors and maneuvers inside the house and its also equipped with various manipulator attachments to open doors. The specialty for this type of rover is that the arm can be equipped with a mount for a Carbon Fire 10 Disrupter. The supplier claimed that they can supply the disrupter mount with or without the disrupter to Explosive Ordinance Disposal (EOD) Departments. Hazardous device technicians used Explosive Ordinance Disposal (EOD) disrupters to neutralize explosive devices from a safe place [16]. The company which sells LT2 also designed a heavy-duty robot called HD2-s, as shown in Figure 4(b).

(a)

(b)

Figure 4. (a) LT2/F "Bulldog" and (b) HD2-s [16]

The development of UGV design intended for the use by authorities such as anti-terrorism and police. It commonly consists of mobility platform equipped with sensors and computer software, power system and additional features which are depending on the scope of the tasks to be conducted by the rover [17]. As an example, the SR-10 Inspector is equipped with crawler drive which makes the robot moves under various terrain conditions and also able to climb stairs with maximum speed beyond 16 km/h. Besides that, the sensors play a very important role to assist the robot to avoid collision and define the distance from the wall [18].

The ground robotic system [19] such as Teodor Robot as shown in Figure 5, is equipped with an array of metal detectors. It helps the robot to detect smaller anti-personal mines. Besides that, the chassis of Teodor is equipped with an extensive sensor arsenal [20], [21]. Teodor is known as search and rescue robot, so it really needs navigation assistance such as a real-time 3D laser mapping system [22], a real-time time-of-flight camera [23] and a real-time stereo vision system [24]. By implementing all these systems, it will ease search and rescue operations [24, 25] because of the robot itself capable to conduct autonomous navigation [26].

Figure 5. Teodor [22]

Based on all the models that had been reviewed, new inventions of the smart robotic rover was invented to meet the authorities' requirements. Every component and part of the product is designed and drawn with the precise dimension as the real product. Figure 6 shows the isometric view for the final design of robotic rover. The proposed rover consists of a robotic arm, gripper, track, chassis, DC motor, vision system, and controller.

The specification of the developed rover is explained in Table 1 [27]. The rover is considered lightweight and small in size even equipped with a robot arm, gripper, and camera system. This rover has a portable ground system to receive images from the camera installed on the rover. An experiment had been conducted on its ability to lift object where the robot arm with gripper can be lifted between 0.483 meters to 0.013 meters, as shown in Figure 7. This rover could lift small items such as a pen marker and a small aerosol spray that has weight around 750 gram, as shown in Figure 8 [28]. The robotic arm can also be moved in transverse motion but it needs to be modified by adding an additional degree of freedom (DOF) to make the motion of the arm more stable. The gripper attached at the end of the robot arm can be rotated in 360 degrees either in a clockwise or counter clockwise direction. This rover is connected to a joystick to control the movement of the rover as well as the robotic arm and gripper.

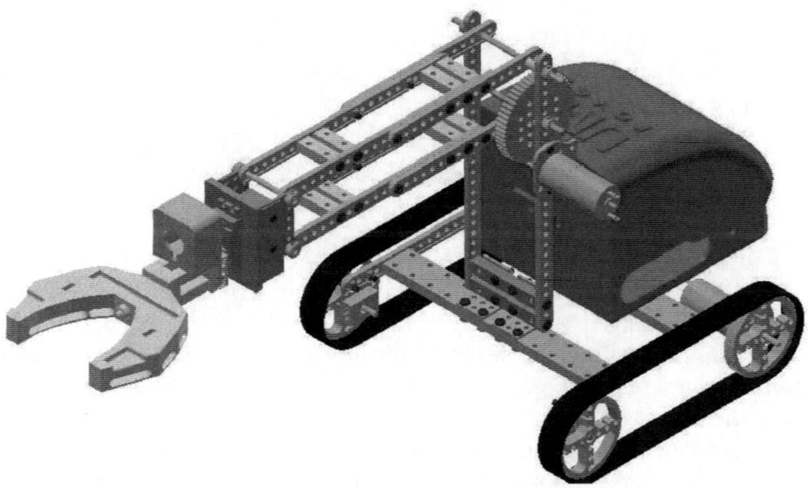

Figure 6. Robotic Rover

Table 1. Specifications of robotic rover

Specification	Description
Speed	Up to 5.8 mph (9.3 kph)
Dimension (Height x Width x Length)	216.06 mm × 132.99 mm × 264.67 mm
Weight	2.2 kg (rover), 3.0 kg (ground station)
Height of robotic arm	0.483 m (max), 0.013 m (min)

3. Applications of rover

This section reviews the applications of the rovers in the nuclear industry as well as search and rescue operations. Nuclear energy industries created dangerous environments that risk human if any casualties happen. As an alternative, robots are sent to access toxic and combustible atmospheres in nuclear plants. The robots or rovers that are deployed to this application are designed to detect nuclear radiation and air samples, as well as to conduct mechanical tasks in such circumstances. Robotic systems could be considered as the best solution to represent human in obtaining data from dangerous conditions in nuclear plants. Based on the literature [1], the rovers have been deployed in several nuclear plants such as at Three Mile Island Unit-2 plant in the USA, the Chernobyl plant in Ukraine, Tokaimura and Fukushima Daiichi plants in Japan. This literature also listed the details of the robots with its specifications related to previous references.

Besides that, the rovers were also used for Urban Search and Rescue (USAR) activities [29], [30]. The developed mobile robots should be robust in design and equipped with various functionality for survival during disasters. However, current rovers are still not ready to be deployed in different environments and they are deterred by narrow passages and wreckage. Even though there are many difficulties to conduct research related to this study, but the researchers still trying to find solutions especially urban search rescue and hazardous prevention in large-scale earthquakes and other types of disasters [31],[32]. The development of rover in search and rescue operation can be review in Table 2.

Table 2. Application of rovers in search and rescue operations

Location	Robot)	Operating area	Locomotion	Terrain	Developed yaer)	Ref
Defense Advanced Research Projects Agency (DARPA)	Inuktun micro-Tracs VGTV	Pile	Tracked	steep inclines/ Declines in	1989	[33]
	Micro-Tracs	Pile	Tracked	piping system	1989	[34]
	Foster-Miller					
	Solem	Pile	Tracked	Rough/slight	2001	
	Talon	Pile	Tracked	slopes/obstacles	2000	

4. Conclusions

Initially, rovers were used to replace man in space exploration. However, based on works of literature that have been reviewed in this paper, it can be also considered to be deployed in commercial applications such as site monitoring for several industries, search and rescue operations as well as conducting mechanical tasks using robotic arms and gripper attached to the rover. The goal of this paper is to analyze the specifications and features from existing rovers in order to invent simpler rover in terms of its system and manufacturing. In future work, we are planning to design more robust robotic rover for work lift, monitoring and safety purposes with robotic arm and gripper to collect and remove loads. Thorough analysis is required in achieving these aims, starting from designing, fabrication, material analysis as well as electrical aspects in order to manufacture this type of UGV in realization. Moreover, it could be considered to be used in monitoring situations during hazardous conditions or surveillance. By having this rover, it can reduce or eliminate risk to human life because it can be remotely controlled by using camera vision system and controller as what had been done by Teodor robot and robotic rover shown in Figure 6. Taking suggestions from recommendations of the HAZMAT teams, it is desirable to include additional features to the robotic rover such as gas detector, thermal camera, and rocker boogie track. The gas detector is suggested to be used for detection of toxic gases which are dangerous to humans in long exposure such as Carbon Monoxide (CO), Oxygen (O_2), Hydrogen Sulfide (H_2S), and other combustible gases which its concentration in air is defined as the Lower Explosive Limit (LEL). For example, if a gas is detected, electrodes on the LEL detector will send signals to alarms and can give indications for such gas levels between 0-100%. Thermal imaging on the rover would help to improve visibility in a dark environment during certain operations by detecting objects using infrared and creating an image based on that information. Movement of the rover also has to give attention and it is preferable if the rover could climb stairs and move on any type of land surfaces.

References

[1] Tsitsimpelis, I., et al., A review of ground-based robotic systems for the characterization of environments. Progress in Nuclear Energy, 2019. 111: p. 109-124.
[2] Habib, M.K. and Y.J.I.J.o.A.R.S. Baudoin, Robot-assisted risky intervention, search, rescue and environmental surveillance. 2010. 7(1): p. 10.
[3] Singh Sangeeta, H. and H. Kaur, Design and Analysis of Robotic Rover with Gripper Arm using Embedded C. Vol. 4. 2015. 149-152.
[4] Crisp, J.A., et al., Mars exploration rover mission. 2003. 108(E12).
[5] Schuster, M.J., et al. The LRU rover for autonomous planetary exploration and its success in the SpaceBotCamp challenge. in 2016 International Conference on Autonomous Robot Systems and Competitions (ICARSC). 2016. IEEE.
[6] Collins, C.L. and M.L. Robinson, Accuracy analysis and validation of the Mars Science Laboratory (MSL) robotic arm. 2013.
[7] Shao, C., et al., A tactile sensor network system using a multiple sensor platform with a dedicated CMOS-LSI for robot applications. 2017. 17(9): p. 1974.
[8] Welch, R. and G. Edmonds. Requirements for robots utilized at HAZMAT incident sites. in Space Programs and Technologies Conference and Exhibit. 1994.
[9] Iqbal, J., A.M. Tahir, and R. ul Islam. Robotics for nuclear power plants—challenges and future perspectives. in 2012 2nd international conference on applied robotics for the power industry (CARPI). 2012. IEEE.
[10] Wehe, D.K., et al., Intelligent robotics and remote systems for the nuclear industry. 1989.
[11] Welch, R.V., A Mobile Robot for Remote Response to Incidents Involving Hazardous Materials. 1994.
[12] Stone, H.W. and G. Edmonds. Hazbot: a hazardous materials emergency response mobile robot. in Proceedings 1992 IEEE International Conference on Robotics and Automation. 1992. IEEE.
[13] Matthies, L., et al., A portable, autonomous, urban reconnaissance robot. 2002. 40(2-3): p. 163-172.
[14] Meicran, H. Mobile Robots Responding to Toxic Chemical Spills. in Proceedings of the Forth American Nuclear Society Topical Meeting on Robotics and Remote Systems.
[15] Standard, E.R. and E.R. Plan, Hazardous Materials Regulations for New Jersey Emergency Responders.
[16] Russell, K., A Modular Explosive Ordnance Disposal Disrupter System. Vol. 6. 2013. 148-151.
[17] Habib, M.K., Y. Baudoin, and F. Nagata. Robotics for rescue and risky intervention. in IECON 2011-37th Annual Conference of the IEEE Industrial Electronics Society. 2011. IEEE.
[18] Maslowski, A.J.I.W.R., Intervention robotics-experience in Poland. 2006.
[19] Balta, H., et al., Integrated mobile robot system for landmine detection. 2014.
[20] De Cubber, G., et al. The EU-ICARUS project: developing assistive robotic tools for search and rescue operations. in 2013 IEEE international symposium on safety, security, and rescue robotics (SSRR). 2013. IEEE.
[21] De Cubber, G., et al. Search and rescue robots developed by the European Icarus project. in 7th

Int. Workshop on Robotics for Risky Environments. 2013. Citeseer.
[22] Będkowski, J., A. Masłowski, and G.J.I.R.A.I.J. De Cubber, Real time 3D localization and mapping for USAR robotic application. 2012. 39(5): p. 464-474
[23] Bertozzi, M., et al., Viac expedition toward autonomous mobility [from the field]. 2011. 18(3): p. 120-124.
[24] Balta, H., et al. Terrain traversability analysis for off-road robots using time-of-flight 3d sensing. in 7th IARP International Workshop on Robotics for Risky Environment-Extreme Robotics, Saint-Petersburg, Russia. 2013..
[25] De Cubber, G., H. Balta, and C. Lietart. Teodor: A semi-autonomous search and rescue and demining robot. in Applied Mechanics and Materials. 2014. Trans Tech Publ.
[26] Doroftei, D., et al., A Behaviour-BASED CONTROL AND SOFTWARE ARCHITECTURE FOR THE VISUALLY GUIDED ROBUDEM OUTDOOR MOBILE ROBOT. 2008. 2(4).
[27] Herman, N.A.B., Designing Chassis and electrical part casing, in Manufacturing in Production Technology. 2017, Universiti Malaysia Pahang. p. 17-72.
[28] Nazaruddin, A.S., Robot Arm and Gripper, in Manufacturing in Production Technology. 2017, Universiti Malaysia Pahang. p. 15-76.
[29] Huff, J., S. Conyers, and R. Voyles. MOTHERSHIP — A serpentine tread/limb hybrid marsupial robot for USAR. in 2012 IEEE International Symposium on Safety, Security, and Rescue Robotics (SSRR). 2012.
[30] White, T., et al. The design and operational performance of a climbing robot used for weld inspection in hazardous environments. in Proceedings of the 1998 IEEE International Conference on Control Applications (Cat. No.98CH36104). 1998.
[31] Ko, A., H.Y.J.I.J.o.A.S. Lau, and Technology, Robot assisted emergency search and rescue system with a wireless sensor network. 2009. 3: p. 69-78.
[32] Tadokoro, S. Rescue robotics challenge. in 2010 IEEE Workshop on Advanced Robotics and its Social Impacts. 2010. IEEE.
[33] Liu, Y., G.J.J.o.I. Nejat, and R. Systems, Robotic Urban Search and Rescue: A Survey from the Control Perspective. 2013. 72(2): p. 147-165.
[34] Murphy, R.R., Trial by fire [rescue robots]. IEEE Robotics & Automation Magazine, 2004. 11(3): p. 50-61.

AUTHOR INDEX

Institute of Physics
Dirac House, Temple Back
Bristol BS1 6BE UK

ISSN: 1757-8981
ISBN 978-1-7138-0989-0

9 781713 809890

CURRAN ASSOCIATES INC.
proceedings
.com